PRAISE FOR STATISTICS FOR PEOPLE WHO (THINK THEY) HATE STATISTICS

"*Statistics for People Who (Think They) Hate Statistics* really makes students learn and enjoy statistics and research in general. Students especially like the Ten Commandments and Internet sites."

—Professor Valarie Janesick
Professor of Educational Leadership
University of South Florida

"I just wanted to say that as a SUNY Delhi online RN-to-BSN student one day into Statistics 101—your book has saved my career! I put off my BSN due to statistics, even enrolling and then backing out a couple of times. I have read the first two chapters and already "get it." I know it will get harder, but I am so thankful for your easy-to-understand method. I told my husband last night I actually might like statistics and enjoy it. I was also thankful for the basic math review. No one ever broke it down like that for me, to the point where I was in remedial math in high school and still never got it. I no longer fear math or statistics."

—Meghan Wheeler, RN

"I truly appreciate your accessibility and help. I am learning to use SPSS in preparation for a doctoral program that will began in the fall and it has been twenty years since I have taken a statistics course. I am thankful for this straightforward book to help me catch up with current trends."

—Sylvia Miller-Martin

"I have loved statistics ever since my second undergraduate course. Your book *Statistics for People Who (Think They) Hate Statistics* has cleared up confusion and partial understandings that I have had for years. It is a must for anyone beginning or continuing their journey in this science. I love it, and will use it for all of the foreseeable future."

—Ronald A. Straube
Mission Texas Regional Medical Center

"Dr. Salkind, I just felt compelled to send you a note thanking you for such a great book—*Statistics for People Who (Think They) Hate Statistics*.

"I bought a house two years ago. The people who lived there previously left the book behind. I didn't throw it out because I am a book nut.

"Anyway, I have started work on a graduate degree in psychology and decided to pull your book out. This book has been a godsend. It is absolutely the best statistics book I have ever encountered when it comes to explaining things in understandable terms.

"It was well worth the 100K for the house, LOL!"

Bless you!!
Brian Wright

"The project team of Denise, Renee, Shawn, and Trish stated for their research hypothesis that brownies made with regular flour would be preferred to those made with gluten-free flour. The brownie recipe chosen was "The Reward" in Appendix F. Denise made the gluten-free brownies, Renee made the regular brownies, and our sample was our fellow students at Tusculum College. We used an ordinal survey process for rating the brownies on a scale from 1 to 5, with 1 being the worst and 5 being the best brownies you ever had. The gluten-free brownies won, disproving the research hypothesis. The mean and mode were the chosen method of comparison. The gluten-free brownies had a mean/mode of 4, and the regular brownies had a 3. The range for the gluten-free brownies was wider than the range for the regular-flour brownies. All who participated in the survey LOVED the brownies.

"This came about because I asked our instructor if we were going to use the information in Appendix F. Neither my instructor nor my classmates had checked out this particular appendix. The instructor told me I could make the brownies and bring them to class. That is when I told my instructor that I had celiac disease and only had gluten-free flour in my home. Usually gluten-free items are not preferred because of their texture. The instructor had always wanted to try something that was gluten-free, and that was how our in-class experiment was born."

—Denise Proske
Tesculum College

"I just wanted to take a moment of your time to inform you that I have selected your book, *Statistics for People Who (Think They) Hate Statistics*, to use in my course. I truly agree with the direction you have taken with your book and I know that our students will appreciate it just the same."

—Karl R. Krawitz, EdD
Baker University, Overland Park, KS

"I am a 'nontraditional' (that's how the nice folks at the University of Dayton refer to 'older') grad student enjoying your *Statistics for People Who (Think They) Hate Statistics*. Although I publicize research in my job, being involved in research and statistics myself is an entirely new challenge. So please count me as one of the countless who appreciate your approach to statistics with a sense of humor—it definitely helps alleviate the intimidation factor of the subject.

"Thanks again for taking on this (and other topics) in such a 'human' way :-)."

Best regards,
Pamela Gregg
Communication Administrator
University of Dayton Research Institute

"I just thought I would send a little positive reinforcement your way! As an undergraduate psychology student, I was urged by a friend to purchase your book but not, as you may think, for a stats class. I had taken the required stats class two years prior and had learned NOTHING! As I embarked on my senior honors thesis, I began to feel slightly—maybe more like extremely—overwhelmed by all of the data analysis I was about to undertake. That was when a friend of mine suggested I buy your book. My first reaction? 'I'm not buying another statistics book just for the fun of it!' Well after much prodding, I eventually bought it (the second edition at the time). Now as I take on statistics (multivariate, yikes!) yet again, only this time as a third-year graduate student, I find myself keeping your book alongside as an anxiety-reducing companion!

"Thanks for making statistics bearable for all these years!"

—Ashley Shier, MEd
University of Cincinnati
School of Psychology Doctoral Student

"Hello Dr. Salkind! Just wanted to thank you for putting together a great resource in your *Statistics for People* . . . . I use it to teach my grad course in Quantitative Research Methods in anthropology here at Northern Arizona University. We fondly refer to your book as 'Haters.'

"Thanks again."

Britton L. Shepardson, PhD
Lecturer & Assistant Chair
Department of Anthropology
Northern Arizona University

"Salkind's examples assist with the application of key concepts and tests. The book is easy to read due to the way information is presented, such as the Tech Talk, Things to Remember, the Key to Difficulty Index, the various 10 lists, the icons, and the illustrations—including the cartoons. Even the title brings laughter to students—and humor can be a great antidote to stress!"

—Mary Beth Zeni
Florida State School of Nursing

"Hi, Mr. Salkind,

"I am a full-time registered nurse of 19 years and have recently begun my journey of obtaining my Bachelor of Science in Nursing. Tomorrow is my first statistics class. I have just read your 'note to students' and wanted to write to you and inform you that you have described my symptoms to a T. My classmates and I are extremely anxious about our course and what we are in for over the next 3 months. After reading these two pages, I wanted to tell you that you have alleviated some of my anxiety and allowed me to stop fretting over the unknown and begin to read on. Thank you for that. I am working a night shift tonight; hopefully I will find time to read my required chapters with less anxiety and actually absorb some of the material I am reading. Thank you again; I will try and look forward to learning from your book, my excellent instructor, and my classmates."

Sincerely,
Lori Vajda, RN

"Vast quantities of statistical information consumed."

—Beldar from Remulak

"Dear Prof Salkind,

"I just want to thank you for the amazing book, *Statistics for People Who (Think They) Hate Statistics*. I definitely used to be one among them who hated statistics and used to ignore it so far. Now, as I am almost in the finishing stages of my PhD, I was thinking it would be a shame if I don't have a minimum knowledge of statistics. The book has not just helped my understanding in the subject, but it inspires me to do further reading in statistics. I have even recommended the book to a few within 2 days! Thank you so much for such a wonderful work!"

—A. J. Padman

"I just wanted to send a little 'thank you' your way for writing an extremely user-friendly book, *Statistics for People (Who Think) They Hate Statistics*. I'm a psychology major doing an independent study over break (at Alverno, a statistics course is a prerequisite for a class I'm taking this spring, experimental psychology). In other words, I'm pretty much learning this on my own (with a little guidance from my mentor), so I appreciate having a book that presents the material in a simple, sometimes humorous manner. I only suggest writing another textbook at a higher level of statistics so I can read that one too!"

Sincerely,
Jenny Saucerman

"I liked its humorous approach, which indeed helps to reduce statistical anxiety. The design of the book is inviting and relaxing, which is a plus. The writing style is great, and the presentation is appropriate for my students. A fun and well-written book, it is easy to read and use and presents statistics in a user-friendly way. . . . I would recommend it for sure."

—Minjuan Wang
San Diego State University

"Let me thank you for a wonderful textbook. Of all the texts I have used over the years, I would have to rate yours #1 for presenting material that can be followed and understood."

—Carolyn Letsche
MA Student in School Counseling

"Salkind's book is in a class by itself. It is easily the best book of its kind that I have come across. I enthusiastically recommend it for anyone interested in the subject and even (and especially) for those who aren't!"

—Russ Shafer-Landau
University of Wisconsin

"*Statistics for People Who (Think They) Hate Statistics* is definitely the right book for people who have to overcome that familiar anxious feeling when opening a standard statistics book and who having finally managed to do so are still not able to make much sense of it all. The book by Salkind is easy and pleasant to read and one that hardly needs any pre-knowledge of the field to be able to follow the author's train of thoughts. Salkind has managed to bring statistics home to people who hate statistics or thought they did."

From a review in *Statistical Methods in Medical Research*
(Arnold Publications)
—Dr. Andrea Winkler
Maudsley and Bethlem Hospital
London, UK

SAGE was founded in 1965 by Sara Miller McCune to support the dissemination of usable knowledge by publishing innovative and high-quality research and teaching content. Today, we publish over 900 journals, including those of more than 400 learned societies, more than 800 new books per year, and a growing range of library products including archives, data, case studies, reports, and video. SAGE remains majority-owned by our founder, and after Sara's lifetime will become owned by a charitable trust that secures our continued independence.

Los Angeles | London | New Delhi | Singapore | Washington DC | Melbourne

6
EDITION

Statistics for
People Who *(Think They)*
Hate Statistics

In honor and memory of Shane J. Lopez and welcome to Bella.

EDITION

Statistics for People Who *(Think They)* Hate Statistics

Neil J. Salkind

University of Kansas

Los Angeles | London | New Delhi
Singapore | Washington DC | Melbourne

FOR INFORMATION:

SAGE Publications, Inc.
2455 Teller Road
Thousand Oaks, California 91320
E-mail: order@sagepub.com

SAGE Publications Ltd.
1 Oliver's Yard
55 City Road
London, EC1Y 1SP
United Kingdom

SAGE Publications India Pvt. Ltd.
B 1/I 1 Mohan Cooperative Industrial Area
Mathura Road, New Delhi 110 044
India

SAGE Publications Asia-Pacific Pte. Ltd.
3 Church Street
#10-04 Samsung Hub
Singapore 049483

Copyright © 2017 by SAGE Publications, Inc.

All rights reserved. No part of this book may be reproduced or utilized in any form or by any means, electronic or mechanical, including photocopying, recording, or by any information storage and retrieval system, without permission in writing from the publisher.

This book includes screenshots of Microsoft Excel 2010 to illustrate the methods and procedures described in the book. Microsoft Excel is a product of the Microsoft Corporation.

SPSS is a registered trademark of International Business Machines Corporation.

Portions of information contained in this publication/book are printed with permission of Minitab Inc. All such material remains the exclusive property and copyright of Minitab Inc. All rights reserved.

Printed in the United States of America

Library of Congress Cataloging-in-Publication Data

Salkind, Neil J.
Statistics for people who (think they) hate statistics / Neil J. Salkind. — Sixth edition.
pages cm
Includes bibliographical references and index.

ISBN 978-1-5063-3383-0
ISBN 978-1-5063-3382-3 web pdf

This book is printed on acid-free paper.

SUSTAINABLE FORESTRY INITIATIVE
Certified Chain of Custody
Promoting Sustainable Forestry
www.sfiprogram.org
SFI-01268

SFI label applies to text stock

Acquisitions Editor: Helen Salmon
Editorial Assistant: Chelsea Pearson
eLearning Editor: Katie Ancheta
Production Editor: Libby Larson
Copy Editor: Paula L. Fleming
Typesetter: C&M Digitals (P) Ltd.
Proofreader: Scott Oney
Indexer: Will Ragsdale
Cover Designer: Candice Harman
Marketing Manager: Susannah Goldes

16 17 18 19 20 10 9 8 7 6 5 4 3 2 1

Brief Contents

Detailed Contents

PART V

Archeologists dig up the first edition of *Statistics for People Who (Think They) Hate Statistics.*

A Note to the Student:
Why I Wrote This Book

With another new edition (now the sixth), I welcome you to what I hope will be, in all ways, a good learning experience. I am overwhelmed by the opportunity to continue to revise this book and the pleasure it brings me and, I hope, brings you.

What many students of introductory statistics (be they new to the subject or just reviewing the material) have in common (at least at the beginning of their studies) is a relatively high level of anxiety, the origin of which is, more often than not, what they've heard from their fellow students. Often, a small part of what they have heard is true—learning statistics takes an investment of time and effort (and there's the occasional monster for a teacher).

But most of what they've heard (and where most of the anxiety comes from)—that statistics is unbearably difficult and confusing—is just not true. Thousands of fear-struck students have succeeded where they thought they would fail. They did it by taking one thing at a time, pacing themselves, seeing illustrations of basic principles as they are applied to real-life settings, and even having some fun along the way. That's what I tried to do in writing the first five editions of *Statistics for People Who (Think They) Hate Statistics*, and I tried even harder in completing this revision.

After a great deal of trial and error, and some successful and many unsuccessful attempts, and a ton of feedback from students and teachers at all levels of education, I have attempted to teach statistics in a way that I (and many of my students) think is unintimidating and informative. I have tried my absolute best to incorporate all of that experience into this book.

What you will learn from this book is the information you need to understand what the field and study of basic statistics is all about. You'll learn about the fundamental ideas and the most commonly used techniques to organize and make sense out of data. There's very little theory (but some), and there are few mathematical proofs or discussions of the rationale for certain mathematical routines.

Why isn't this theory stuff and more in *Statistics for People Who (Think They) Hate Statistics?* Simple. Right now, you don't need it. It's not that I don't think it's important. Rather, at this point and time in your studies, I want to offer you material at a level I think you can understand and learn with some reasonable amount of effort, while at the same time not be scared off from taking additional courses in the future. I (and your professor) want you to succeed.

So, if you are looking for a detailed unraveling of the derivation of the analysis of variance F ratio, go find another good book from SAGE (I'll be glad to refer you to one). But if you want to learn why and how statistics can work for you, you're in the right place. This book will help you understand the material you read in journal articles, explain what the results of many statistical analyses mean, and teach you how to perform basic statistical tasks.

And, if you want to talk about any aspect of teaching or learning statistics, feel free to contact me. You can do this through my email address at school (njs@ku.edu). Good luck, and let me know how I can improve this book to even better meet the needs of the beginning statistics student. And, if you want the data files that will help you succeed, either go to the Sage website at **edge.sagepub.com/salkind6e**, or contact me via email and let me know the edition you are using.

AND A (LITTLE) NOTE TO THE INSTRUCTOR

I would like to share two things.

First, I applaud your efforts at teaching basic statistics. Although this topic may be easier for some students, most find the material very challenging. Your patience and hard work are appreciated by all, and if there is anything I can do to help, please send me a note.

Second, *Statistics for People Who (Think They) Hate Statistics* is not meant to be a dumbed-down book similar to others you may have seen. Nor is the title meant to convey anything other than the fact that many students new to the subject are very anxious about what's to come. This is not an academic or textbook version of a "book for dummies" or anything of the kind. I have made every effort to address students with the respect they deserve, not to patronize them, and to ensure that the material is approachable. How well I did in these regards is up to you, but I want to convey my very clear intent that this book contain the information needed in an introductory course, and even though my approach involves some humor, nothing about my intent is anything other than serious. Thank you.

Acknowledgments

Everybody at SAGE deserves a great deal of thanks for providing me with the support, guidance, and professionalism that takes a mere idea (way back before the first edition) and makes it into a book like the one you are now reading—and then makes it successful. From Johnny Garcia, who heads up the distribution center, to Vanessa Vondressa, who handles financials—this book would not be possible without their hard work.

However, some people have to be thanked individually for their special care and hard work. Helen Salmon, Senior Acquisitions Editor—Research Methods and Statistics, has shepherded this edition, being always available to discuss new ideas and seeing to it that everything got done on time and done well. She is the editor whom every author wants. C. Deborah Laughton, Lisa Cuevas Shaw, and Vicki Knight—all previous editors—helped this book along the way, and to them, I am forever grateful. Others who deserve a special note are Katie Ancheta, associate editor; Chelsea Pearson, editorial assistant; and Libby Larson, production editor supreme. Special, special thanks goes to Paula Fleming for her sharp eye and sound copyediting, which make this material read as well as it does. Libby and Paula are the best in the galactic empire. And special thanks to Dr. Patrick Ament, University of Central Missouri, who, along with his very capable students, took the time to send me detailed feedback about typos, suggestions for changes, and more, making this edition much more accurate and complete than the previous one. Thanks to Patrick and his students.

SAGE would like to thank the following peer reviewers for their editorial insight and guidance:

Charles E. Baukal Jr., Oklahoma State University

Michael E. Cox, Online Instructor, multiple universities

Jonathan Allen Kringen, University of New Haven

Tim W. Ficklin, Chaminade University of Honolulu

Qingwen Dong, University of the Pacific

Charles J. Fountaine, University of Minnesota–Duluth

De'Arno De'Armond, West Texas A&M University

Laura Anderson, School of Nursing, University of Buffalo

Jordan K. Aquino, California State University–Fullerton

And Now, About the Sixth Edition . . .

What you read above about this book reflects my thoughts about why I wrote this book in the first place. But it tells you little about this sixth edition.

Any book is always a work in progress, and this latest edition of *Statistics for People Who (Think They) Hate Statistics* is no exception. Over the past 17 years or so, many people have told me how helpful this book is, and others have told me how they would like it to change and why. In revising this book, I am trying to meet the needs of all audiences. Some things remain the same, and some have indeed changed.

There are always new things worth consideration and different ways to present old themes and ideas. Here's a list of what you'll find that's new in the sixth edition of *Statistics for People Who (Think They) Hate Statistics*.

- Information on levels of measurement has been moved to Chapter 2 from Chapter 6 because it seemed to fit better and many students and faculty thought moving it would be a good idea.
- "Understanding the SPSS Output" is a new section that I've added to Chapters 10 through 17 that I hope provides useful information about IBM® SPSS® Statistics* output and what it means.
- When appropriate, a section on effect size has been added to the coverage of each inferential statistic (starting in Chapter 10). That seems to be an increasingly important topic even at the introductory level.
- A discussion of partial correlation has been included in Chapter 5.
- A new chapter (Chapter 19) has been added as an introduction to data mining using SPSS. It is a mini treatment of this vast topic but at least lets introductory students know that the topic exists and is important to consider in their studies.
- Finally, there are about 20% more additional exercises at the end of each chapter.

This sixth edition features SPSS 23, the latest version that SPSS offers. For the most part, you can use a version of SPSS that is as early

---

*SPSS is a registered trademark of International Business Machines Corporation.

as version 11 to do most of the work, and these earlier versions can read the data files created with the later versions. If the reader needs help with SPSS, we offer a mini course in Appendix A, which is also available online.

- Maybe the most interesting (and perhaps coolest) thing about this edition is that it is available as an interactive ebook. The interactive ebook version includes a variety of teaching and substantive features, such as links to everything from animated chapter summaries to book reviews to step-by-step "Getting Started" videos about how to carry out many of the procedures covered in the book as well as practice problems at the end of the chapter. Not just a flat PDF of these existing hard-copy pages of the edition, this is a "rich" media version, with many interactive features to help you get the most out of this textbook. We all hope you find it as exciting as we do and hope it helps you in understanding the material.

You can find the interactive eBook icons in the print and digital versions of the text. In the interactive eBook, all of these icons are hyperlinked to multimedia content. Many of these media resources are also available on the open-access SAGE edge site at edge.sagepub.com/salkind6e.

Below is a guide to the icons:

 Video

 Web site

 SAGE journal article

Any typos and such that appear in this edition of the book are entirely my fault, and I apologize to the professors and students who are inconvenienced by their appearance. And I so appreciate any letters, calls, and emails pointing out these errors. You can see corrections of all these errors at http://www.statisticsforpeople.com, and I welcome any additional corrections. We have all made every effort in this edition to correct previous errors and hope we did a reasonably good job. Let me hear from you with suggestions, criticisms, nice notes, and so on. Good luck.

Neil J. Salkind
University of Kansas
njs@ku.edu

About the Author

Neil J. Salkind received his PhD in human development from the University of Maryland, and after teaching for 35 years at the University of Kansas, he remains as Professor Emeritus in the Department of Psychology and Research in Education, where he continues to collaborate with colleagues and work with students. His early interests were in the area of children's cognitive development, and after research in the areas of cognitive style and (what was then known as) hyperactivity, he was a postdoctoral fellow at the University of North Carolina's Bush Center for Child and Family Policy. His work then changed direction to focus on child and family policy, specifically the impact of alternative forms of public support on various child and family outcomes. He has delivered more than 150 professional papers and presentations; written more than 100 trade and textbooks; and is the author of *Statistics for People Who (Think They) Hate Statistics* (SAGE), *Theories of Human Development* (SAGE), and *Exploring Research* (Prentice Hall). He has edited several encyclopedias, including the *Encyclopedia of Human Development*, the *Encyclopedia of Measurement and Statistics,* and the *Encyclopedia of Research Design.* He was editor of *Child Development Abstracts and Bibliography* for 13 years. He now lives in Lawrence, Kansas, where he likes to read, swim with the River City Sharks, work as the proprietor and sole employee of big boy press, bake brownies (see www.statisticsforpeople.com for the recipe), and poke around old Volvos and old houses.

PART I

Yippee! I'm in Statistics

"Well, if it isn't a significant subset
of our study group!"

N ot much to shout about, you might say? Let me take a minute and show you how some very accomplished scientists use this widely used set of tools we call *statistics*.

- Michelle Lampl is a physician, the Samuel Candler Dobbs Professor of Anthropology and the co-director of the Emory-Georgia Tech Predictive Health Initiative. She was having coffee with a friend, who commented on how quickly her young infant was growing. In fact, the new mother spoke as if her son was "growing like a weed." Being a curious scientist (as all scientists should be), Dr. Lampl thought she might actually examine how rapid this child's growth, and that of others, was during infancy. She proceeded to measure a group of children's growth on a daily basis and found, much to her surprise, that some infants grew as much as 1 inch overnight! Some growth spurt.

  Want to know more? Why not read the original work? You can find more about this in Lampl, M., Veldhuis, J. D., & Johnson, M. L. (1992). Saltation and stasis: A model of human growth. *Science, 258,* 801–803.

- Sue Kemper is the Roberts Distinguished Professor of Psychology at the University of Kansas and has worked on the most interesting of projects. She and several other researchers studied a group of nuns and examined how their early experiences, activities, personality characteristics, and other information related to their health during their late adult years. Most notably, this diverse group of scientists (including psychologists, linguists, neurologists, and others) wanted to know how well all this information predicts the occurrence of Alzheimer's disease. Kemper and her colleagues found that the complexity of the nuns' writing during their early 20s was related to the nuns' risk for Alzheimer's 50, 60, and 70 years later.

  Want to know more? Why not read the original work? You can find more about this in Snowdon, D. A., Kemper, S. J., Mortimer, J. A., Greiner, L. H., Wekstein, D. R., & Markesbery, W. R. (1996). Linguistic ability in early life and cognitive function and Alzheimer's disease in late life: Findings from the nun study. *Journal of the American Medical Association, 275,* 528–532.

- Aletha Huston (a very distinguished professor emerita) from the University of Texas in Austin devoted a good deal of her professional work to understanding what effects television watching has on young children's psychological development. Among other things, she and her late husband, John C. Wright, investigated the impact that the amount of educational television programs watched during the early preschool years might have on outcomes in the later school years. They found convincing evidence that children who watch educational programs such as *Mr. Rogers* and *Sesame Street* do better in school than those who do not.

Want to know more? Why not read the original work? You can find more about this in Collins, P. A., Wright, J. C., Anderson, R., Huston, A. C., Schmitt, K., & McElroy, E. (1997, April). *Effects of early childhood media use on adolescent achievement.* Paper presented at the biennial meeting of the Society for Research in Child Development, Washington, DC.

All of these researchers had a specific question they found interesting and used their intuition, curiosity, and excellent training to answer it. As part of their investigations, they used this set of tools we call *statistics* to make sense out of all the information they collected. Without these tools, all this information would have been just a collection of unrelated outcomes. The outcomes would be nothing that Lampl could have used to reach a conclusion about children's growth, or Kemper could have used to better understand aging and cognition (and perhaps Alzheimer's disease), or Huston and Wright could have used to better understand the impact of watching television on young children's achievement and social development.

Statistics—the science of organizing and analyzing information to make it more easily understood—made these tasks doable. The reason that any of the results from such studies are useful is that we can use statistics to make sense out of them. And that's exactly the goal of this book—to provide you with an understanding of these basic tools and how researchers use them and, of course, how to use them yourself.

In this first part of *Statistics for People Who (Think They) Hate Statistics,* you will be introduced to what the study of statistics is about and why it's well worth your efforts to master the basics—the important terminology and ideas that are central to the field. This part gives you a solid preparation for the rest of the book.

1 Statistics or Sadistics?

It's Up to You

Difficulty Scale ☺ ☺ ☺ ☺ ☺ (really easy)

WHAT YOU'LL LEARN ABOUT IN THIS CHAPTER

✦ What statistics is all about

✦ Why you should take statistics

✦ How to succeed in this course

WHY STATISTICS?

You've heard it all before, right?

Introduction to
Chapter 1

"Statistics is difficult."

"The math involved is impossible."

"I don't know how to use a computer."

"What do I need this stuff for?" "What do I do next?"

And the famous cry of the introductory statistics student: "I don't get it!"

Well, relax. Students who study introductory statistics find themselves, at one time or another, thinking at least one of the above and quite possibly sharing the thought with another student, their spouse, a colleague, or a friend.

And all kidding aside, some statistics courses can easily be described as *sadistics*. That's because the books are repetitiously boring and the authors have no imagination.

That's not the case for you. The fact that you or your instructor has selected *Statistics for People Who (Think They) Hate Statistics* shows that you're ready to take the right approach—one that is unintimidating, informative, and applied (and even a little fun) and that tries to teach you what you need to know about using statistics as the valuable tool that it is.

If you're using this book in a class, it also means that your instructor is clearly on your side. He or she knows that statistics can be intimidating but has taken steps to see that it is not intimidating for you. As a matter of fact, we'll bet there's a good chance (as hard as it may be to believe) that you'll be enjoying this class in just a few short weeks.

And Why SPSS?

Throughout this book, you'll be shown how to use SPSS, a statistical analysis tool, for the analysis of data. No worries; you'll also be shown how to do the same analysis by hand to assure you of an understanding of both.

Why SPSS? Simple. It's one of the most popular, most powerful analytic tools available today, and it can be an exceedingly important and valuable tool for learning how to use basic and some advanced statistics. In fact, many stats courses taught at the introductory level use SPSS as their primary computational tool, and you can look to Appendix A for a refresher on some basic SPSS tasks. Also, the way that technology is advancing, few opportunities to use statistics in research, administration, and everyday work will not require some knowledge of how and when to use tools such as SPSS. That's why we're including it in this book! We will show you how to use it to make your statistics learning experience a better one.

A 5-MINUTE HISTORY OF STATISTICS

Why Statistics
Is Important

Before you read any further, it would be useful to have some historical perspective about this topic called statistics. After all, almost every undergraduate in the social, behavioral, and biological sciences and every graduate student in education, nursing, psychology, social welfare and social services, anthropology, and . . . (you get the picture) is required to take this course. Wouldn't it be nice to have some idea from whence the topic it covers came? Of course it would.

Way, way back, as soon as humans realized that counting was a good idea (as in "How many of these do you need to trade for one of those?"), collecting information became a useful skill.

If counting counted, then one would know how many times the sun would rise in one season, how much food was needed to last the winter, and what amount of resources belonged to whom.

That was just the beginning. Once numbers became part of language, it seemed like the next step was to attach these numbers to outcomes. That started in earnest during the 17th century, when the first set of data pertaining to populations was collected. From that point on, scientists (mostly mathematicians, but then physical and biological scientists) needed to develop specific tools to answer specific questions. For example, Francis Galton (a half-cousin of Charles Darwin, by the way), who lived from 1822 to 1911, was very interested in the nature of human intelligence. He also speculated that hair loss was due to the intense energy that went into thinking. No, really. But back to statistics.

To explore one of his primary questions regarding the similarity of intelligence among family members, he used a specific statistical tool called the correlation coefficient (first developed by mathematicians), and then he popularized its use in the behavioral and social sciences.

You'll learn all about this tool in Chapter 5. In fact, most of the basic statistical procedures that you will learn about were first developed and used in the fields of agriculture, astronomy, and even politics. Their application to human behavior came much later.

The past 100 years have seen great strides in the invention of new ways to use old ideas. The simplest test for examining the differences between the averages of two groups was first advanced during the early 20th century. Techniques that build on this idea were offered decades later and have been greatly refined. And the introduction of personal computers and such programs as Excel has opened up the use of sophisticated techniques to anyone who wants to explore these fascinating topics.

The introduction of these powerful personal computers has been both good and bad. It's good because most statistical analyses no longer require access to a huge and expensive mainframe computer. Instead, a simple personal computer costing less than $250 or a cloud account can do 95% of what 95% of the people need. On the other hand, less than adequately educated students (such as your fellow students who passed on taking this course!) will take any old data they have and think that by running them through some sophisticated analysis, they will have reliable, trustworthy, and meaningful outcomes—not true. What your professor would say is "Garbage in, garbage out"; if you don't start with reliable and trustworthy data, what you'll have after your data are analyzed is unreliable and untrustworthy results.

Today, statisticians in all different areas, from criminal justice to geophysics to psychology to determining whether the "hot" hand really exists in the NBA (no kidding—see the *Wall Street Journal* article at http://www.wsj.com/articles/SB10001424052702304071004579409074

Cool Ideas and
Where They
Come From

1015745370), find themselves using basically the same techniques to answer different questions. There are, of course, important differences in how data are collected, but for the most part, the analyses (the plural of *analysis*) that are done following the collection of data (the plural of *datum*) tend to be very similar, even if called something different. The moral here? This class will provide you with the tools to understand how statistics are used in almost any discipline. Pretty neat, and all for just three or four credits.

If you want to learn more about the history of statistics and see a historical time line, great places to start are Saint Anselm's College at www.anselm.edu/homepage/jpitocch/biostatshist.html and the University of California–Los Angeles at www.stat.ucla.edu/history.

Okay. Five minutes is up, and you know as much as you need to know about the history of statistics. Let's move on to what it is (and isn't).

STATISTICS: WHAT IT IS (AND ISN'T)

Statistics Is
Useful

Statistics for People Who (Think They) Hate Statistics is a book about basic statistics and how to apply them to a variety of different situations, including the analysis and understanding of information.

In the most general sense, **statistics** describes a set of tools and techniques that are used for describing, organizing, and interpreting information or data. Those data might be the scores on a test taken by students participating in a special math curriculum, the speed with which problems are solved, the number of side effects when patients use one type of drug rather than another, the number of errors in each inning of a World Series game, or the average price of a dinner in an upscale restaurant in Santa Fe, New Mexico (not cheap).

In all of these examples, and the million more we could think of, data are collected, organized, summarized, and then interpreted. In this book, you'll learn about collecting, organizing, and summarizing data as part of descriptive statistics. And then you'll learn about interpreting data when you learn about the usefulness of inferential statistics.

What Are Descriptive Statistics?

Descriptive statistics are used to organize and describe the characteristics of a collection of data. The collection is sometimes called a **data set** or just **data**.

For example, the following list shows you the names of 22 college students, their major areas of study, and their ages. If you needed to describe what the most popular college major is, you could use a descriptive statistic that summarizes their most frequent choice (called the mode). In this case, the most common major is psychology. And if you wanted to know the average age, you could easily compute another descriptive statistic that identifies this variable (that one's called the mean). Both of these simple descriptive statistics are used to describe data. They do a fine job allowing us to represent the characteristics of a large collection of data such as the 22 cases in our example.

Statistics All Around Us

Name	Major	Age	Name	Major	Age
Richard	Education	19	Elizabeth	English	21
Sara	Psychology	18	Bill	Psychology	22
Andrea	Education	19	Hadley	Psychology	23
Steven	Psychology	21	Buffy	Education	21
Jordan	Education	20	Chip	Education	19
Pam	Education	24	Homer	Psychology	18
Michael	Psychology	21	Margaret	English	22
Liz	Psychology	19	Courtney	Psychology	24
Nicole	Chemistry	19	Leonard	Psychology	21
Mike	Nursing	20	Jeffrey	Chemistry	18
Kent	History	18	Emily	Spanish	19

So watch how simple this is. To find the most frequently selected major, just find the one that occurs most often. And to find the average age, just add up all the age values and divide by 22. You're right—the most often occurring major is psychology (9 times), and the average age is 20.3 (actually 20.27). Look, Ma! No hands—you're a statistician.

What Are Inferential Statistics?

Inferential statistics are often (but not always) the next step after you have collected and summarized data. Inferential statistics are used to make inferences based on a smaller group of data (such as our group of 22 students) about a possibly larger one (such as all the undergraduate students in the College of Arts and Sciences).

A smaller group of data is often called a sample, which is a portion, or a subset, of a population. For example, all the fifth graders in Newark (your author's fair city of origin), New Jersey, would be a population (the population is all the occurrences with certain characteristics, in this case, being in fifth grade and attending school in Newark), whereas a selection of 150 of these students would be a sample.

Let's look at another example. Your marketing agency asks you (a newly hired researcher) to determine which of several names is most appealing for a new brand of potato chip. Will it be Chipsters? FunChips? Crunchies? As a statistics pro (we know we're moving a bit ahead of ourselves, but keep the faith), you need to find a small group of potato chip eaters who are representative of all potato chip fans and ask these people to tell you which one of the three names they like the most. Then, if you do things right, you can easily extrapolate the findings to the huge group of potato chip eaters.

Or let's say you're interested in the best treatment for a particular type of disease. Perhaps you'll try a new drug as one alternative, a placebo (a substance that is known not to have any effect) as another alternative, and nothing as the third alternative to see what happens. Well, you find out that more patients get better when no action is taken and nature (and we assume that's the only factor or set of factors that differentiate the groups) just takes its course! The drug does not have any effect. Then, with that information, you can extrapolate to the larger group of patients who suffer from the disease, given the results of your experiment.

In Other Words . . .

Statistics is a tool that helps us understand the world around us. It does so by organizing information we've collected and then letting us make certain statements about how characteristics of those data are applicable to new settings. Descriptive and inferential statistics work hand in hand, and which statistic you use and when depends on the question you want answered.

And today, a knowledge of statistics is more important than ever because it provides us with the tools to make decisions that are based on empirical (observed) evidence and not our own biases or beliefs. Want to know whether early intervention programs work? Then test whether they work and provide that evidence to the court that will make a ruling on the viability of a new school bond issue that could pay for those programs.

WHAT AM I DOING IN A STATISTICS CLASS?

Job
Opportunities
in Statistics

You might find yourself using this book for many reasons. You might be enrolled in an introductory statistics class. Or you might be reviewing for your comprehensive exams. Or you might even be reading this on summer vacation (horrors!) in preparation for a more advanced class.

In any case, you are a statistics student, whether you have to take a final exam at the end of a formal course or you're just in it of your own accord. But there are plenty of good reasons to be studying this material—some fun, some serious, and some both.

Here's the list of some of the things that my students hear at the beginning of our introductory statistics course:

1. Statistics 101 or Statistics 1 or whatever it's called at your school looks great listed on your transcript. Kidding aside, this may be a required course for you to complete your major. But even if it is not, having these skills is definitely a big plus when it comes time to apply for a job or for further schooling. And with more advanced courses, your résumé will be even more impressive.

2. If this is not a required course, taking basic statistics sets you apart from those who do not. It shows that you are willing to undertake a course that is above average with regard to difficulty and commitment. And, as the political and economic (and sports!) worlds become more "accountable," more emphasis is being placed on analytic skills. Who knows, this course may be your ticket to a job!

3. Basic statistics is an intellectual challenge of a kind that you might not be used to. There's a good deal of thinking that's required, a bit of math, and some integration of ideas and application. The bottom line is that all this activity adds up to what can be an invigorating intellectual experience because you learn about a whole new area or discipline.

4. There's no question that having some background in statistics makes you a better student in the social or behavioral sciences, because you will have a better understanding not only of what you read in journals but also of what your professors and colleagues may be discussing and doing in and out of class. You will be amazed the first time you say to yourself, "Wow, I actually understand what they're talking about." And it will happen over and over again, because you will have the basic tools necessary to understand exactly how scientists reach the conclusions they do.

5. If you plan to pursue a graduate degree in education, anthropology, economics, nursing, sociology, or any one of many other social, behavioral, and biological pursuits, this course will give you the foundation you need to move further.

6. There are many different ways of thinking about, and approaching, different types of problems. The set of tools you learn about in this book (and this course) will help you look at interesting

Some Statistics
Humor

problems from a new perspective. And, while the possibilities may not be apparent now, this new way of thinking can be brought to new situations.

7. Finally, you can brag that you completed a course that everyone thinks is the equivalent of building and running a nuclear reactor.

TEN WAYS TO USE THIS BOOK (AND LEARN STATISTICS AT THE SAME TIME!)

Success in Your
Statistics Class

Yep. Just what the world needs—another statistics book. But this one is different. It is directed at the student, is not condescending, is informative, and is as basic as possible in its presentation. It makes no presumptions about what you should know before you start and proceeds in slow, small steps, which lets you pace yourself.

However, there has always been a general aura surrounding the study of statistics that it's a difficult subject to master. And we don't say otherwise, because parts of it are challenging. On the other hand, millions and millions of students have mastered this topic, and you can, too. Here are 10 hints to close this introductory chapter before we move on to our first topic.

1. **You're not dumb.** That's true. If you were, you would not have gotten this far in school. So, treat statistics as you would any other new course. Attend the lectures, study the material, do the exercises in the book and from class, and you'll do fine. Rocket scientists know statistics, but you don't have to be a rocket scientist to succeed in statistics.

2. **How do you know statistics is hard?** Is statistics difficult? Yes and no. If you listen to friends who have taken the course and didn't work hard and didn't do well, they'll surely volunteer to tell you how hard it was and how much of a disaster it made of their entire semester, if not their lives. And let's not forget—we always tend to hear from complainers. So, we'd suggest that you start this course with the attitude that you'll wait and see how it is and judge the experience for yourself. Better yet, talk to several people who have had the class and get a good idea of what they think. Don't base your expectations on just one spoilsport's experience.

3. **Don't skip lessons—work through the chapters in sequence.** *Statistics for People Who (Think They) Hate Statistics* is written so that each chapter provides a foundation for the next one in the book. When you are all done with the course, you will (I hope) continue to use this book as a reference. So if you need a particular value from a table, you might consult Appendix B. Or if you need to remember how to compute the standard deviation, you might turn to Chapter 3. But for now,

read each chapter in the sequence that it appears. It's okay to skip around and see what's offered down the road. Just don't study later chapters before you master earlier ones.

4. **Form a study group.** This is a big hint and one of the most basic ways to ensure some success in this course. Early in the semester, arrange to study with friends or classmates. If you don't have any friends who are in the same class as you, then make some new ones or offer to study with someone who looks as happy to be there as you are. Studying with others allows you to help them if you know the material better, or to benefit from those who know some material better than you. Set a specific time each week to get together for an hour and go over the exercises at the end of the chapter or ask questions of one another. Take as much time as you need. Studying with others is an invaluable way to help you understand and master the material in this course.

5. **Ask your teacher questions, and then ask a friend.** If you do not understand what you are being taught in class, ask your professor to clarify it. Have no doubt—if you don't understand the material, then you can be sure that others do not as well. More often than not, instructors welcome questions. And especially because you've read the material before class, your questions should be well informed and help everyone in class to better understand the material.

6. **Do the exercises at the end of a chapter.** The exercises are based on the material and the examples in the chapter they follow. They are there to help you apply the concepts that were taught in the chapter and build your confidence at the same time. If you can answer these end-of-chapter exercises, then you are well on your way to mastering the content of the chapter. Correct answers to each exercise are provided in Appendix D.

7. **Practice, practice, practice.** Yes, it's a very old joke:

Q. How do you get to Carnegie Hall?

A. Practice, practice, practice.

Well, it's no different with basic statistics. You have to use what you learn and use it frequently to master the different ideas and techniques. This means doing the exercises at the end of Chapters 1 through 17 and Chapter 19, as well as taking advantage of any other opportunities you have to understand what you have learned.

8. **Look for applications to make it more real.** In your other classes, you probably have occasion to read journal articles, talk about the results of research, and generally discuss the importance of the scientific method in your own area of study. These are all opportunities to see how your study of statistics can help you better understand the topics

under class discussion as well as the area of beginning statistics. The more you apply these new ideas, the fuller your understanding will be.

9. **Browse.** Read over the assigned chapter first; then go back and read it with more intention. Take a nice leisurely tour of *Statistics for People Who (Think They) Hate Statistics* to see what's contained in the various chapters. Don't rush yourself. It's always good to know what topics lie ahead as well as to familiarize yourself with the content that will be covered in your current statistics class.

10. **Have fun.** This might seem like a strange thing to say, but it all boils down to you mastering this topic rather than letting the course and its demands master you. Set up a study schedule and follow it, ask questions in class, and consider this intellectual exercise to be one of growth. Mastering new material is always exciting and satisfying—it's part of the human spirit. You can experience the same satisfaction here—just keep your eye on the ball and make the necessary commitment to stay current with the assignments and work hard.

ABOUT THOSE ICONS

An icon is a symbol. Throughout *Statistics for People . . .* , you'll see a variety of icons. Here's what each one is and what each represents:

 This icon represents information that goes beyond the regular text. At times we may want to elaborate on a particular point and find we can do so more easily outside of the flow of the usual material.

 Here, we discuss some more technical ideas and tips to give you a sampling of topics beyond the scope of this course. You might find these interesting and useful.

 Throughout *Statistics for People . . .* , you'll find a small-steps icon like the one you see here. This indicates that a set of steps is coming up that will direct you through a particular process. Sometimes you will use SPSS to do these steps. These steps have been tested and approved by whatever federal agency approves these things.

 That finger with the bow is a cute icon, but its primary purpose is to help reinforce important points about the topic that you just read about. Try to emphasize these points in your studying, because they are usually central to the topic.

Appendix A contains an introduction to SPSS. Working through this appendix is all you really need to do to be ready to use SPSS. If you have an earlier version of SPSS (or the Mac version), you will still find this material to be very helpful. In fact, the latest Windows and Mac versions of SPSS are almost identical in appearance and functionality.

Appendix B contains important tables you will learn about and need throughout the book.

And, in working through the exercises in this book, you will use the data sets in Appendix C. In the exercises, you'll find references to data sets with names like "Chapter 2 Data Set 1," and each of these sets is shown in Appendix C. You can either enter the data manually or download them from the publisher's site at **edge.sagepub.com/salkind6e** or get them directly from the author. Just send a note to njs@ku.edu and don't forget to mention the edition for which you need the data sets.

Appendix D contains answers to end-of-chapter questions.

Appendix E contains a primer on math for those who could use a refresher.

And Appendix F offers the long-sought-after brownie recipe (yes, you finally found it).

KEY TO DIFFICULTY ICONS

To help you along a bit, we placed a difficulty index at the beginning of each chapter. This adds some fun to the start of each chapter, but it's also a useful tip to let you know what's coming and how difficult chapters are in relation to one another.

☺ (very hard)

☺ ☺ (hard)

☺ ☺ ☺ (not too hard, but not easy either)

☺ ☺ ☺ ☺ (easy)

☺ ☺ ☺ ☺ ☺ (very easy)

GLOSSARY

Bolded terms in the text are included in the glossary at the back of the book.

REAL-WORLD STATS

Real-World Stats will appear at the end of every chapter as appropriate, and, hopefully, will provide you with a demonstration of how a particular method, test, idea, or some aspect of statistics is used in the everyday workplace of scientists, physicians, policy makers, government folks, and others. In this first such exploration, we look at a very short paper where the author recalls and shares the argument that the National Academy of Sciences (first chartered in 1863, by the way!) "shall, whenever called upon by any department of the Government, investigate, examine, experiment, and report upon any subject of science or art." This charter, some 50 years later in 1916, led to the formation of the National Research Council, another federal body that helped provide information that policy makers need to make informed decisions. And often this "information" takes the from of quantitative data—also referred to as statistics—that assist people in evaluating alternative approaches to problems that have a wide-ranging impact on the public. So, this article, as does your book and the class you are taking, points out how important it is to think clearly and use data to support your arguments.

Real-World
Stats

Want to know more? Go online or go to the library, and find . . .

Cicerone, R. (2010). The importance of federal statistics for advancing science and identifying policy options. *The Annals of the American Academy of Political and Social Science, 631,* 25–27.

SUMMARY

Chapter
Summary

That couldn't have been that bad, right? We want to encourage you to continue reading and not worry about what's difficult or time-consuming or too complex for you to understand and apply. Just take one chapter at a time, as you did this one.

TIME TO PRACTICE

Because there's no substitute for the real thing, Chapters 1 through 17 and Chapter 19 will end with a set of exercises that will help you review the material that was covered in the chapter. As noted above, the answers to these exercises can be found near the end of the book in Appendix D.

For example, here is the first set of exercises (but don't look for any answers for these because these are kind of "on your own" answers—each question and answer is highly tied to your own experiences and interest).

1. Interview someone who uses statistics in his or her everyday work. It might be your adviser, an instructor, a researcher who lives on your block, a health care analyst, a marketer for a company, a city planner, or . . . Ask the person what his or her first statistics course was like. Find out what the person liked and didn't like. See if this individual has any suggestions to help you succeed. And most important, ask the person about how he or she uses these new-to-you tools at work.

2. We hope that you are part of a study group or, if that is not possible, that you have a telephone, email, instant messaging, or webcam study buddy (or even more than one). And, of course, plenty of texting and Facebook friends. Talk to your group or a fellow student in your class about similar likes, dislikes, fears, etc., about the statistics course. What do you have in common? Not in common? Discuss with your fellow student strategies to overcome your fears.

3. Search through your local newspaper (or any other publication) and find the results of a survey or interview about any topic. Summarize the results and do the best job you can describing how the researchers who were involved, or the authors of the survey, came to the conclusions they did. Their methods and reasoning may or may not be apparent. Once you have some idea of what they did, try to speculate as to what other ways the same information might be collected, organized, and summarized.

4. Go to the library (either in person or online) and find a copy of a journal article in your own discipline. Then, go through the article and highlight the section (usually the "Results" section) where statistical procedures were used to organize and analyze the data. You don't know much about the specifics of this yet, but how many different statistical procedures (such as t-test, mean, and calculation of the standard deviation) can you identify? Can you take the next step and tell your instructor how the results relate to the research question or the primary topic of the research study?

5. Find five websites that contain data on any topic and write a brief description of what type of information is offered and how it is organized. For example, if you go to the mother of all data sites, the US Census (http://www.census.gov), you'll find links to hundreds of databases, tables, and other informative tools. Try to find data and information that fit in your own discipline.

6. And the big extra-credit assignment is to find someone who actually uses SPSS for daily data analysis needs. Ask if there is anything specific about SPSS that makes it stand out as a tool for their type of data analysis. You may very well find these good folks in everything from political science to nursing, so search widely!

7. Finally, as your last in this first set of exercises, come up with five of the most interesting questions you can about your own area of study or interest. Do your best to come up with questions for which you would want real, existing information or data to answer. Be a scientist!

STUDENT STUDY SITE

⑤SAGE edge™

Get the tools you need to sharpen your study skills! Visit **edge.sagepub.com/ salkind6e** to access practice quizzes, eFlashcards, original and curated videos, data sets, journal articles, and more!

PART II

Σigma Freud and Descriptive Statistics

"Yes, it's all becoming clear."

One of the things that Sigmund Freud, the founder of psycho-analysis, did quite well was to observe and describe the nature of his patients' conditions. An astute observer, he used his skills to develop the first systematic and comprehensive theory of personality. Regardless of what you may think about the validity of his ideas, he was a good scientist.

Back in the early 20th century, courses in statistics (like the one you are taking) were not offered as part of undergraduate or graduate curricula. The field was relatively new, and the nature of scientific explorations did not demand the precision that this set of tools brings to the scientific arena.

But things have changed. Now, in almost any endeavor, numbers count (as Francis Galton, the inventor of correlation and a first cousin to Charles Darwin, said as well). This section of *Statistics for People Who (Think They) Hate Statistics* is devoted to understanding how we can use statistics to describe an outcome and better understand it, once the information about the outcome is organized.

Chapter 2 discusses measures of central tendency, how computing one of several different types of averages gives you the one best data point that represents a set of scores and when to use what. Chapter 3 completes the coverage of tools we need to fully describe a set of data points in its discussion of variability, including the standard deviation and variance. When you get to Chapter 4, you will be ready to learn how distributions, or sets of scores, differ from one another and what this difference means. Chapter 5 deals with the nature of relationships between variables, namely, correlations. And finally, Chapter 6 introduces you to the importance of reliability and validity when we are describing some of the qualities of effective measurement tools.

When you finish Part II, you'll be in excellent shape to start understanding the role that probability and inference play in the social, behavioral, and other sciences.

2 Means to an End

Computing and Understanding Averages

Difficulty Scale ☺ ☺ ☺ ☺ (moderately easy)

WHAT YOU'LL LEARN ABOUT IN THIS CHAPTER

✦ Understanding measures of central tendency

✦ Computing the mean for a set of scores

✦ Computing the median for a set of scores

✦ Computing the mode for a set of scores

✦ Understanding and applying scales or levels of measurement

✦ Selecting a measure of central tendency

You've been very patient, and now it's finally time to get started working with some real, live data. That's exactly what you'll do in this chapter. Once data are collected, a usual first step is to organize the information using simple indexes to describe the data. The easiest way to do this is through computing an average, of which there are several different types.

An **average** is the one value that best represents an entire group of scores. It doesn't matter whether the group of scores is the number correct on a spelling test for 30 fifth graders or the batting percentage of each of the New York Yankees (which, by the way, was not very good during the 2015 season) or the number of people who registered as Democrats or Republicans for the upcoming elections. In all of these examples, groups of data can be summarized using an average. You can also think of an average as the "middle" space or as a fulcrum on a seesaw. It's the point where all the values in a set of values are balanced.

Introduction to
Chapter 2

21

Averages, also called **measures of central tendency**, come in three flavors: the mean, the median, and the mode. Each provides you with a different type of information about a distribution of scores and is simple to compute and interpret.

COMPUTING THE MEAN

Computing the Mean

The **mean** is the most common type of average that is computed. It is simply the sum of all the values in a group, divided by the number of values in that group. So, if you had the spelling scores for 30 fifth graders, you would simply add up all the scores to get a total, and then divide by the number of students, which is 30.

The formula for computing the mean is shown in Formula 2.1.

$$\overline{X} = \frac{\Sigma X}{n},\qquad(2.1)$$

where

- the letter X with a line above it (also sometimes called "X bar") is the mean value of the group of scores or the mean;
- the Σ, or the Greek letter sigma, is the summation sign, which tells you to add together whatever follows it;
- the X is each individual score in the group of scores; and
- the n is the size of the sample from which you are computing the mean.

To compute the mean, follow these steps:

1. List the entire set of values in one or more columns. These are all the Xs.
2. Compute the sum or total of all the values.
3. Divide the total or sum by the number of values.

For example, if you needed to compute the average number of shoppers at three different locations, you would compute a mean for that value.

Location	Number of Annual Customers
Lanham Park Store	2,150
Williamsburg Store	1,534
Downtown Store	3,564

The mean or average number of shoppers in each store is 2,416. Formula 2.2 shows how this average was computed using the formula you saw in Formula 2.1:

$$\overline{X} = \frac{\Sigma X}{n} = \frac{2,150 + 1,534 + 3,564}{3} = \frac{7,248}{3} = 2,416 \qquad (2.2)$$

Or, if you needed to compute the average number of students in grades kindergarten through 6, you would follow the same procedure.

Grade	Number of Students
Kindergarten	18
1	21
2	24
3	23
4	22
5	24
6	25

The mean or average number of students in each class is 22.43. Formula 2.3 shows how this average was computed using the formula you saw in Formula 2.1:

$$\overline{X} = \frac{\Sigma X}{n} = \frac{18 + 21 + 24 + 23 + 22 + 24 + 25}{7} = \frac{157}{7} = 22.43 \qquad (2.3)$$

See, we told you it was easy. No big deal.

- The mean is sometimes represented by the letter M and is also called the typical, average, or most central score. If you are reading another statistics book or a research report and you see something like $M = 45.87$, it probably means that the mean is equal to 45.87.
- In the formula, a small n represents the sample size for which the mean is being computed. A large N (← like this) would represent the population size. In some books and in some journal articles, no distinction is made between the two.

- The sample mean is the measure of central tendency that most accurately reflects the population mean.
- The mean is like the fulcrum on a seesaw. It's the centermost point where all the values on one side of the mean are equal in weight to all the values on the other side of the mean.
- Finally, for better or worse, the mean is very sensitive to extreme scores. An extreme score can pull the mean in one or the other direction and make it less representative of the set of scores and less useful as a measure of central tendency. This, of course, all depends on the values for which the mean is being computed. And, if you have extreme scores and the mean won't work as well as you want, we have a solution! More about that later.

The mean is also referred to as the **arithmetic mean**, and there are other types of means that you may read about, such as the harmonic mean. Those are used in special circumstances and need not concern you here. And if you want to be technical about it, the arithmetic mean (which is the one that we have discussed up to now) is also defined as the point about which the sum of the deviations is equal to zero (whew!). So, if you have scores like 3, 4, and 5 (whose mean is 4), the sum of the deviations about the mean (−1, 0, and +1) is 0.

Remember that the word *average* means only the one measure that best represents a set of scores and that there are many different types of averages. Which type of average you use depends on the question that you are asking and the type of data that you are trying to summarize. This is a levels of measurement issue which we will cover later in this chapter when we talk about when to use which measure.

Computing a Weighted Mean

You've just seen an example of how to compute a simple mean. But there may be situations where you have the occurrence of more than one value and you want to compute a weighted mean. A weighted mean can be easily computed by multiplying the value by the frequency of its occurrence, adding the total of all the products, and then dividing by the total number of occurrences. It beats adding up every individual data point.

To compute a weighted mean, follow these steps:

1. List all the values in the sample for which the mean is being computed, such as those shown in the column labeled Value (the value of X) in the following table.

2. List the frequency with which each value occurs.

3. Multiply the value by the frequency, as shown in the third column.

4. Sum all the values in the Value × Frequency column.

5. Divide by the total frequency.

For example, here's a table that organizes the values and frequencies in a flying proficiency test for 100 airline pilots.

Value	Frequency	Value × Frequency
97	4	388
94	11	1,034
92	12	1,104
91	21	1,911
90	30	2,700
89	12	1,068
78	9	702
60 (Don't fly with this guy.)	1	60
Total	100	8,967

The weighted mean is 8,967/100, or 89.67. Computing the mean this way is much easier than entering 100 different scores into your calculator or computer program.

In basic statistics, an important distinction is made between those values associated with samples (a part of a population) and those associated with populations. To do this, statisticians use the following conventions. For a sample statistic (such as the mean of a sample), Roman letters are used. For a population parameter (such as the mean of a population), Greek letters are used. So, for example, the mean for the spelling score for a sample of 100 fifth graders is represented as $\bar{X}_5$, whereas the mean for the spelling score for the entire population of fifth graders is represented, using the Greek letter mu, as μ_5.

COMPUTING THE MEDIAN

Computing the
Median

The median is also an average but of a very different kind. The **median** is defined as the midpoint in a set of scores. It's the point at which one half, or 50%, of the scores fall above and one half, or 50%, fall below. It's got some special qualities that we will talk about later in this section, but for now, let's concentrate on how it is computed. There's no standard formula for computing the median.

To compute the median, follow these steps:

1. List the values in order, from either highest to lowest or lowest to highest.
2. Find the middle-most score. That's the median.

For example, here are the incomes from five different households:

$135,456

$25,500

$32,456

$54,365

$37,668

Here is the list ordered from highest to lowest:

$135,456

$54,365

$37,668

$32,456

$25,500

There are five values. The middle-most value is $37,668, and that's the median.

Now, what if the number of values is even? Let's add a value ($34,500) to the list so there are six income levels. Here they are sorted with the largest value first:

$135,456

$54,365

$37,668

$34,500

$32,456

$25,500

When there is an even number of values, the median is simply the mean of the two middle values. In this case, the middle two cases are $34,500 and $37,668. The mean of those two values is $36,084. That's the median for that set of six values.

What if the two middle-most values are the same, such as in the following set of data?

$45,678

$25,567

$25,567

$13,234

Then the median is same as both of those middle-most values. In this example, it's $25,567.

If we had a series of values that was the number of days spent in rehabilitation for a sports-related injury for seven different patients, the numbers might look like this:

43

34

32

12

51

6

27

As we did before, we can order the values (51, 43, 34, 32, 27, 12, 6) and then select the middle value as the median, which in this case is 32. So, the median number of days spent in rehab is 32.

If you know about medians, you should also know about **percentile points**. Percentile points are used to define the percentage of cases equal to and below a certain point in a distribution or set of scores. For example, if a score is "at the 75th percentile," it means that the score is at or above 75% of the other scores in the distribution. The median is also known as the 50th percentile, because it's the point below which 50% of the cases in the distribution fall. Other percentiles are useful as well, such as the 25th percentile, often called Q_1, and the 75th percentile, referred to as Q_3. So what's Q_2? The median, of course.

Here comes the answer to the question you've probably had in the back of your mind since we started talking about the median. Why use the median instead of the mean? For one very good reason. The median is insensitive to extreme scores, whereas the mean is not.

When you have a set of scores in which one or more scores are extreme, the median better represents the centermost value of that set of scores than any other measure of central tendency. Yes, even better than the mean.

What do we mean by *extreme*? It's probably easiest to think of an extreme score as one that is very different from the group to which it belongs. For example, consider the list of five incomes that we worked with earlier (shown again here):

$135,456

$54,365

$37,668

$32,456

$25,500

The value $135,456 is more different from the other four than is any other value in the set. We would consider that an extreme score.

The best way to illustrate how useful the median is as a measure of central tendency is to compute both the mean and the median for a set of data that contains one or more extreme scores and then compare them to see which one best represents the group. Here goes.

The mean of the set of five scores you see above is the sum of the set of five divided by 5, which turns out to be $57,089. On the other hand, the median for this set of five scores is $37,668. Which is more representative of the group? The value $37,668, because it clearly lies more in the middle of the group, and we like to think about "the average" (in this case, we are using the median as a measure of average) as being representative or assuming a central position. In fact, the mean value of $57,089 falls above the fourth highest value ($54,365) and is not very central or representative of the distribution.

It's for this reason that certain social and economic indicators (often involving income) are reported using a median as a measure of central tendency—"The median income of the average American family is . . ."—rather than using the mean to summarize the values. There are just too many extreme scores that would **skew**, or significantly distort, what is actually a central point in the set or distribution of scores.

You learned earlier that sometimes the mean is represented by the capital letter *M* instead of $\bar{X}$. Well, other symbols are used for the median as well. We like the letter *M*, but some people confuse it with the mean, so they use *Med* or *Mdn* for median. Don't let that throw you—just remember what the median is and what it represents, and you'll have no trouble adapting to different symbols.

Here are some interesting and important things to remember about the median.

- While the mean is the middle point of a set of values, the median is the middle point of a set of cases.
- Because the median cares about how many cases there are, and not the values of those cases, extreme scores (sometimes called **outliers**) don't count.

COMPUTING THE MODE

The third and last measure of central tendency that we'll cover, the mode, is the most general and least precise measure of central tendency, but it plays a very important part in understanding the characteristics of a set of scores. The **mode** is the value that occurs most frequently. There is no formula for computing the mode.

To compute the mode, follow these steps:

1. List all the values in a distribution but list each value only once.
2. Tally the number of times that each value occurs.
3. The value that occurs most often is the mode.

Using SPSS for Descriptive Statistics

For example, an examination of the political party affiliation of 300 people might result in the following distribution of scores.

Party Affiliation	Number or Frequency
Democrats	90
Republicans	70
Independents	140

The mode is the value that occurs most frequently, which in the preceding example is Independents. That's the mode for this distribution.

If we were looking at the modal response on a 100-item multiple-choice test, we might find that alternative A was chosen more frequently than any other. The data might look like this.

Item Alternative Selected	A	B	C	D
Number of Times	57	20	12	11

On this 100-item multiple-choice test where each item has four choices (A, B, C, and D), A was the answer selected 57 times. It's the modal response.

Want to know the easiest and most commonly made mistake made when computing the mode? It's selecting the number of *times* a category occurs, rather than the *label* of the category itself. Instead of the mode being Independents (in our first example above), it's easy for someone to conclude the mode is 140. Why? Because he or she is looking at the number of times the value occurred, not the value that occurred most often! This is a simple mistake to make, so be on your toes when you are asked about these things.

Apple Pie à la Bimodal

If every value in a distribution contains the same number of occurrences, then there really isn't a single mode. But if more than one value appears with equal frequency, the distribution is multimodal. The set of scores can be bimodal (with two modes), as the following set of data using hair color illustrates.

Hair Color	Number or Frequency
Red	7
Blond	12
Black	45
Brown	45

In the above example, the distribution is bimodal because the frequency of the values of black and brown hair occurs equally. You can even have a bimodal distribution when the modes are relatively close together but not exactly the same, such as 45 people with black hair and 44 with brown hair. The question becomes, How much does one class of occurrences stand apart from another?

Can you have a trimodal distribution? Sure—where three values have the same frequency. It's unlikely, especially when you are dealing with a large set of **data points**, or observations, but certainly possible. The real answer to the above stand-apart question is that categories have to be mutually exclusive—you simply cannot have both black and red hair (although if you look around the classroom, you may think differently). Of course, you can have those two colors, but each person's hair color is forced into only one category.

WHEN TO USE WHAT MEASURE OF CENTRAL TENDENCY (AND ALL YOU NEED TO KNOW ABOUT SCALES OF MEASUREMENT FOR NOW)

Which measure of central tendency you use depends upon certain characteristics of the data you are working with—specifically the **scale of measurement** at which that data occurs. And that scale or level dictates the specific measure of central tendency you will use.

But let's step back for just a minute and make sure that we have some vocabulary straight, beginning with the idea of what measurement is.

Measurement is the assignment of values to outcomes following a set of rules—simple. The results are the different scales we'll define in a moment, and an outcome is anything we are interested in measuring, such as hair color, gender, test score, or height.

These scales of measurement, or rules, are the particular levels at which outcomes are observed. Each level has a particular set of characteristics, and scales of measurement come in four flavors (there are four types): nominal, ordinal, interval, and ratio.

Let's move on to a brief discussion and examples of the four scales of measurement and then discuss how these levels of scales fit with the different measures of central tendency discussed earlier.

A Rose by Any Other Name: The Nominal Level of Measurement

The **nominal level of measurement** is defined by the characteristics of an outcome that fit into one and only one class or category. For example, gender can be a nominal variable (female and male), as can ethnicity (Caucasian or African American), as can political affiliation (Republican, Democrat, or Independent). Nominal-level variables are "names" (*nominal* in Latin), and the nominal level can be the least precise level of measurement. Nominal levels of measurement have categories that are mutually exclusive; for example, political affiliation cannot be both Republican and Democrat.

Any Order Is Fine With Me: The Ordinal Level of Measurement

The *ord* in **ordinal level of measurement** stands for *order*, and the characteristic of things being measured here is that they are ordered. The perfect example is a rank of candidates for a job. If we know that Russ is ranked #1, Sheldon is ranked #2, and Hannah is ranked #3, then this is an ordinal arrangement. We have no idea how much higher on this scale Russ is relative to Sheldon than Sheldon is relative to Hannah. We just know that it's "better" to be #1 than #2 than #3, but not by how much.

$1 + 1 = 2$: The Interval Level of Measurement

Now we're getting somewhere. When we talk about the **interval level of measurement**, a test or an assessment tool is based on some underlying continuum such that we can talk about how much more a higher performance is than a lesser one. For example, if you get 10 words correct on a vocabulary test, that is twice as many as getting 5 words correct. A distinguishing characteristic of interval-level scales is that the intervals or spaces or points along the scale are equal to one another. Ten words correct is 2 more than 8 correct, which is 3 more than 5 correct.

Can Anyone Have Nothing of Anything? The Ratio Level of Measurement

Well, here's a little conundrum for you. An assessment tool at the **ratio level of measurement** is characterized by the presence of an absolute zero on the scale. What that zero means is the absence of any of the trait that is being measured. The conundrum? Are there outcomes we measure where it is possible to have nothing of what is being measured? In some disciplines, that can be the case. For example, in the physical and biological sciences, you can have the absence of a characteristic, such as absolute zero (no molecular movement) or zero light. In the social and behavioral sciences, it's a bit harder. Even if you score zero on that spelling test or miss every item of an IQ test (in Russian), that does not mean that you have no spelling ability or no intelligence, right?

In Sum . . .

These scales of measurement, or rules, represent particular levels at which outcomes are measured. And, in sum, we can say the following:

- Any outcome can be assigned to one of four scales of measurement.
- Scales of measurement have an order, from the least precise being nominal to the most precise being ratio.
- The "higher up" the scale of measurement, the more precise the data being collected, and the more detailed and informative the data are. It may be enough to know that some people are rich and some poor (and that's a nominal or categorical distinction), but it's much better to know exactly how much money they make (ratio). We can always make the "rich" versus "poor" distinction if we want to once we have all the information.
- Finally, the more precise scales contain all the qualities of the scales below them; for example, the interval scale includes the characteristics of the ordinal and nominal scales. If you know that the Cubs' batting average is .350, you know it is better than that of the Tigers (who hit .250) by 100 points, but you also know that the Cubs are better than the Tigers (but not by how much) and that the Cubs are different from the Tigers (but there's no direction to the difference).

Okay, we've defined levels of measurement and discussed three different measures of central tendency and given you fairly clear examples of each. But the most important question remains unanswered. That is, "When do you use which measure?"

In general, which measure of central tendency you use depends on the type of data that you are describing, which in turn means at what level of measurement the data occur. Unquestionably, a measure of central tendency for qualitative, categorical, or nominal data (such as racial group, eye color, income bracket, voting preference, and neighborhood location) can be described using only the mode.

For example, you can't be interested in the most central measure that describes which political affiliation is most predominant in a group and use the mean—what in the world could you conclude, that everyone is half Republican? Rather, saying that out of 300 people, almost half (140) are Independent seems to be the best way to describe the value of this variable. In general, the median and mean are best used with quantitative data, such as height, income level in dollars (not categories), age, test score, reaction time, and number of hours completed toward a degree.

It's also fair to say that the mean is a more precise measure than the median and that the median is a more precise measure than the mode. This means that all other things being equal, use the mean, and indeed, the mean is the most often used measure of central tendency. However, we do have occasions when the mean would not be appropriate as a measure of central tendency—for example, when we have categorical or nominal data, such as hospitalized versus non-hospitalized people. Then we use the mode.

So, here is a set of three guidelines that may be of some help. And remember, there can always be exceptions.

1. Use the mode when the data are categorical in nature and values can fit into only one class, such as hair color, political affiliation, neighborhood location, and religion. When this is the case, these categories are called mutually exclusive.

2. Use the median when you have extreme scores and you don't want to distort the average (computed as the mean), such as when the variable of interest is income expressed in dollars.

3. Finally, use the mean when you have data that do not include extreme scores and are not categorical, such as the numerical score on a test or the number of seconds it takes to swim 50 yards.

USING THE COMPUTER TO COMPUTE DESCRIPTIVE STATISTICS

 If you haven't already, now would be a good time to check out Appendix A starting on page 377 so you can become familiar with the basics of using SPSS. Then come back here.

Let's use SPSS to compute some descriptive statistics. The data set we are using is named Chapter 2 Data Set 1, and it is a set of 20 scores on a test of prejudice. All of the data sets are available in Appendix C and from the SAGE website **edge.sagepub.com/salkind6e** or from the author at njs@ku.edu. There is one variable in this data set:

Variable	Definition
Prejudice	The value on a test of prejudice as measured on a scale from 1 to 100

Here are the steps to compute the measures of central tendency that we discussed in this chapter. Follow along and do it yourself. With this and all exercises, including data that you enter or download, we'll assume that the data set is already open in SPSS.

1. Click Analyze → Descriptive Statistics → Frequencies.

2. Double-click on the variable named Prejudice to move it to the Variable(s) box.

3. Click Statistics, and you will see the Frequencies: Statistics dialog box shown in Figure 2.1.

4. Under Central Tendency, click the Mean, Median, and Mode boxes.

5. Click Continue.

6. Click OK.

The SPSS Output

Figure 2.2 shows you selected output from the SPSS procedure for the variable named Prejudice.

In the Statistics part of the output, you can see how the mean, median, and mode are all computed along with the sample size and the fact that there were no missing data. SPSS does not use symbols such as X in its output. Also listed in the output are the frequency of each value and the percentage of times it occurs, all useful descriptive information.

| Figure 2.1 | The Frequencies: Statistics Dialog Box From SPSS |

| Figure 2.2 | Descriptive Statistics From SPSS |

⟶ **Frequencies**

[DataSet1] C:\Textbook Stuff\Stat for People 6e\Chapter

Statistics

Predjudice

N	Valid	20
	Missing	0
Mean		84.70
Median		87.00
Mode		87

Predjudice

		Frequency	Percent	Valid Percent	Cumulative Percent
Valid	55	1	5.0	5.0	5.0
	64	1	5.0	5.0	10.0
	67	1	5.0	5.0	15.0
	76	1	5.0	5.0	20.0
	77	1	5.0	5.0	25.0
	81	2	10.0	10.0	35.0
	82	1	5.0	5.0	40.0
	87	4	20.0	20.0	60.0
	89	1	5.0	5.0	65.0
	93	1	5.0	5.0	70.0
	94	2	10.0	10.0	80.0
	96	1	5.0	5.0	85.0
	99	3	15.0	15.0	100.0
	Total	20	100.0	100.0	

It's a bit strange, but if you select Analyze → Descriptive Statistics → Descriptives (instead of clicking Frequencies last) in SPSS and then click Options, there's no median or mode option, which you might expect because they are basic descriptive statistics. The lesson here? Statistical analysis programs are usually quite different from one another, use different names for the same things, and make different assumptions about what's where. If you can't find what you want, it's probably there. Just keep hunting. Also, be sure to use the Help feature to get help navigating through all this new information until you find what you need.

Understanding the SPSS Output

This SPSS output is pretty straightforward and easy to interpret.

The average mean score for the 20 scores is 84.70 (and remember that the span of possible scores is from 0 to 100). The median, or the point at which 50% of the scores fall above and 50% fall below, is 87 (which is pretty close to the mean), and the most frequently occurring score, or the mode, is 87.

SPSS output can be full of information or just give you the basics. It all depends upon the type of analysis that you are conducting. In the above example, we have just the basics and frankly, just what we need. Throughout *Statistics for People . . .* , you will be seeing output and then learning about what it means, but in some cases, discussing the entire collection of output information is far beyond the scope of the book. We'll focus on output that is directly related to what you learned in the chapter.

REAL-WORLD STATS

Few applications of descriptive statistics would make more sense than using them in a survey or poll, and there have been literally millions of such surveys (as in every US presidential election). In an article, Roger Morrell and his colleagues examined Internet use patterns in 550 adults in several age-groups, including middle-aged (ages 40–59), young-old (ages 60–74), and old-old (ages 75–92). With a response rate of 71%, which is pretty good, they found a few very interesting (but not entirely unexpected) outcomes:

- There are distinct age and demographic differences among individuals who use the Internet.
- Middle-aged and older Web users are similar in their use patterns.
- The two primary predictors for not using the Internet are lack of access to a computer and lack of knowledge about the Internet.
- Old-old adults have the least interest in using the Internet compared with middle-aged and young adults.
- The primary content areas in learning how to use the Internet are using electronic mail and accessing health information and information about traveling for pleasure.

This survey, which primarily used descriptive statistics to reach its conclusions, was done almost 20 years ago, and many, many things have changed since then. But the reason we are including it in this example of real-world statistics is that it is a good illustration of a historical anchor point that can be used for comparative purposes in future studies—a purpose for which descriptive statistics and such studies are often used.

Real-World
Stats

Want to know more? Go online or go to the library and read . . .

Morrell, R. W., Mayhorn, C. B., & Bennett, J. (2000). A survey of World Wide Web Use in middle-aged and older adults. *Human Factors, 42,* 175–182.

SUMMARY

Chapter
Summary

No matter how fancy schmancy your statistical techniques are, you will almost always start by simply describing what's there—hence the importance of understanding the simple notion of central tendency. From here, we go to another important descriptive construct: variability, or how different scores are from one another. That's what we'll explore in Chapter 3!

TIME TO PRACTICE

1. By hand, compute the mean, median, and mode for the following set of 40 chemistry final scores.

93	85	99	77
94	99	86	76
95	99	97	84
91	89	77	87

97	83	80	98
75	94	81	85
78	92	89	94
76	94	96	94
90	79	80	92
77	86	83	81

2. Compute the mean, median, and mode for the following three sets of scores saved as Chapter 2 Data Set 2. Do it by hand or use a computer program such as SPSS. Show your work, and if you use SPSS, print out a copy of the output.

Problem 2

Score 1	Score 2	Score 3
3	34	154
7	54	167
5	17	132
4	26	145
5	34	154
6	25	145
7	14	113
8	24	156
6	25	154
5	23	123

3. Compute the means for the following set of scores saved as Chapter 2 Data Set 3 using SPSS. Print out a copy of the output.

Number of Beds (Infection Rate)	Infection Rate (per 1,000 Admissions)
234	1.7
214	2.4
165	3.1
436	5.6
432	4.9
342	5.3
276	5.6
187	1.2
512	3.3
553	4.1

4. You are the manager of a fast-food restaurant. Part of your job is to report to the boss at the end of each day which special is selling best. Use your vast knowledge of descriptive statistics and write one paragraph to let the boss know what happened today. Here are the data. Don't use SPSS to compute important values; rather, do it by hand. Be sure to include a copy of your work.

Special	Number Sold	Cost
Huge Burger	20	$2.95
Baby Burger	18	$1.49
Chicken Littles	25	$3.50
Porker Burger	19	$2.95
Yummy Burger	17	$1.99
Coney Dog	20	$1.99
Total specials sold	119	

5. Imagine you are the CEO of a huge corporation and you are planning an expansion. You'd like your new store to post similar numbers as the other three that are in your empire. By hand, provide some idea of what you want the store's financial performance to look like. And remember that you have to select whether to use the mean, the median, or the mode as an average. Good luck, young Jedi.

Average	Store 1	Store 2	Store 3	New Store
Sales (in thousands of dollars)	323.6	234.6	308.3	
Number of items purchased	3,454	5,645	4,565	
Number of visitors	4,534	6,765	6,654	

Problem 6

6. Here are ratings (on a scale from 1 through 5) for various Super Bowl party foods. You have to decide which food is rated highest (5 is a winner and 1 a loser). Decide what type of average you will use and why. Do this by hand or use SPSS.

Snack Food	North Fans	East Fans	South Fans	West Fans
Loaded Nachos	4	4	5	4
Fruit Cup	2	1	2	1
Spicy Wings	4	3	3	3
Gargantuan Overstuffed Pizza	3	4	4	5
Beer Chicken	5	5	5	4

7. Under what conditions would you use the median rather than the mean as a measure of central tendency? Why? Provide an example of two situations in which the median might be more useful than the mean as a measure of central tendency.

8. Suppose you are working with a data set that has some very "different" (much larger or much smaller than the rest of the data) scores. What measure of central tendency would you use and why?

9. For this exercise, use the following set of 16 scores (ranked) that consists of income levels ranging from about $50,000 to about $200,000. What is the best measure of central tendency and why?

$199,999

$98,789

$90,878

$87,678

$87,245

$83,675

$77,876

$77,743

$76,564

$76,465

$75,643

$66,768

$65,654

$58,768

$54,678

$51,354

10. Use the data in Chapter 2 Data Set 4 and, manually, compute the average attitude scores (with a score of 10 being positive and 1 being negative) for three groups of individuals' attitudes reflecting their experience with urban transportation.

Problem 11

11. Take a look at the following number of pie orders from the Lady Bird diner and determine the average number of orders for each week.

Week	Chocolate Silk	Apple	Douglas County Pie
1	12	21	7
2	14	15	12
3	18	14	21
4	27	12	15

STUDENT STUDY SITE

ⓢSAGE edge™

Get the tools you need to sharpen your study skills! Visit **edge.sagepub.com/ salkind6e** to access practice quizzes, eFlashcards, original and curated videos, data sets, journal articles, and more!

3

Vive la Différence

Understanding Variability

Difficulty Scale ☺☺☺☺ (moderately easy, but not a cinch)

WHAT YOU'LL LEARN ABOUT IN THIS CHAPTER

✦ Understanding the value of variability as a descriptive tool

✦ Computing the range

✦ Computing the standard deviation

✦ Computing the variance

✦ Understanding what the standard deviation and variance have in common—and how they are different

WHY UNDERSTANDING VARIABILITY IS IMPORTANT

In Chapter 2, you learned about different types of averages, what they mean, how they are computed, and when to use them. But when it comes to descriptive statistics and describing the characteristics of a distribution, averages are only half the story. The other half is measures of variability.

Introduction to Chapter 3

In the simplest of terms, **variability** reflects how scores differ from one another. For example, the following set of scores shows some variability:

$$7, 6, 3, 3, 1$$

The following set of scores has the same mean (4) but has less variability than the previous set:

$$3, 4, 4, 5, 4$$

The next set has no variability at all—the scores do not differ from one another—but it also has the same mean as the other two sets we just showed you.

$$4, 4, 4, 4, 4$$

Variability (also called spread or dispersion) can be thought of as a measure of how different scores are from one another. It's even more accurate (and maybe even easier) to think of variability as how different scores are from one particular score. And what "score" do you think that might be? Well, instead of comparing each score to every other score in a distribution, the one score that could be used as a comparison is—that's right—the mean. So, variability becomes a measure of how much each score in a group of scores differs from the mean. More about this in a moment.

Remember what you already know about computing averages—that an average (whether it is the mean, the median, or the mode) is a representative score of a set of scores. Now, add your new knowledge about variability—that it reflects how different scores are from one another. Each is an important descriptive statistic. Together, these two (average and variability) can be used to describe the characteristics of a distribution and show how distributions differ from one another.

Three measures of variability are commonly used to reflect the degree of variability, spread, or dispersion in a group of scores. These are the range, the standard deviation, and the variance. Let's take a closer look at each one and how each one is used.

Actually, how data points differ from one another is a central part of understanding and using basic statistics. But when it comes to differences between individuals and groups (a mainstay of most social and behavioral sciences), the whole concept of variability becomes really important. Sometimes it's called fluctuation, or lability, or error, or one of many other terms, but the fact is, variety is the spice of life, and what makes people different from one another also makes understanding them and their behavior all the more challenging (and interesting). Without variability either in a set of data or between individuals and groups, things are just boring.

COMPUTING THE RANGE

The **range** is the most general measure of variability. It gives you an idea of how far apart scores are from one another. The range is computed simply by subtracting the lowest score in a distribution from the highest score in the distribution.

In general, the formula for the range is

$$r = h - l, \tag{3.1}$$

where

- r is the range;
- h is the highest score in the data set; and
- l is the lowest score in the data set.

Take the following set of scores, for example (shown here in descending order):

$$98, 86, 77, 56, 48$$

In this example, $98 - 48 = 50$. The range is 50. In a set of 500 numbers, where the largest is 98 and the smallest is 37, the range is 61.

There really are two kinds of ranges. One is the *exclusive range,* which is the highest score minus the lowest score (or $h - l$) and the one we just defined. The second kind of range is the *inclusive range,* which is the highest score minus the lowest score plus 1 (or $h - l + 1$). You most commonly see the exclusive range in research articles, but the inclusive range is also used on occasion if the researcher prefers it.

The range tells you how different the highest and lowest values in a data set are from one another—that is, the range shows how much spread there is from the lowest to the highest point in a distribution. So, although the range is fine as a general indicator of variability, it should not be used to reach any conclusions regarding how individual scores differ from one another. And you will usually never see it reported as the only measure of variability but only as one of several—which brings us to . . .

COMPUTING THE STANDARD DEVIATION

Now we get to the most frequently used measure of variability, the standard deviation. Just think about what the term implies—it's a deviation from something (guess what?) that is standard. Actually, the **standard deviation** (abbreviated as *s* or *SD*) represents the average amount of variability in a set of scores. In practical terms, it's the average distance from the mean. The larger the standard deviation, the larger the average distance each data point is from the mean of the distribution and the more variable the set of scores is.

So, what's the logic behind computing the standard deviation? Your initial thoughts may be to compute the mean of a set of scores and then subtract each individual score from the mean. Then, compute the average of that distance.

That's a good idea—you'll end up with the average distance of each score from the mean. But it won't work (see if you know why, even though we'll show you why in a moment).

First, here's the formula for computing the standard deviation:

$$s = \sqrt{\frac{\Sigma\left(X - \overline{X}\right)^2}{n-1}}, \tag{3.2}$$

where

- *s* is the standard deviation;
- Σ is sigma, which tells you to find the sum of what follows;
- *X* is each individual score;
- $\overline{X}$ is the mean of all the scores; and
- *n* is the sample size.

This formula finds the difference between each individual score and the mean ($X - \overline{X}$), squares each difference, and sums them all together.

1. List each score. It doesn't matter whether the scores are in any particular order.

2. Compute the mean of the group.

3. Subtract the mean from each score.

4. Square each individual difference. The result is the column marked $(X - \overline{X})^2$.

5. Sum all the squared deviations about the mean. As you can see, the total is 28.

6. Divide the sum by $n - 1$, or $10 - 1 = 9$, so then $28/9 = 3.11$.

7. Compute the square root of 3.11, which is 1.76 (after rounding). That is the standard deviation for this set of 10 scores.

Then, it divides the sum by the size of the sample (minus 1) and takes the square root of the result. As you can see, and as we mentioned earlier, the standard deviation is an average deviation from the mean.

Unbiased Versus Biased

Here are the data we'll use in the following step-by-step explanation of how to compute the standard deviation:

$$5, 8, 5, 4, 6, 7, 8, 8, 3, 6$$

Here's what we've done so far, where $X - \bar{X}$ represents the difference between the actual score and the mean of all the scores, which is 6.

X	$\bar{X}$	$X - \bar{X}$
8	6	$8 - 6 = +2$
8	6	$8 - 6 = +2$
8	6	$8 - 6 = +2$
7	6	$7 - 6 = +1$
6	6	$6 - 6 = 0$
6	6	$6 - 6 = 0$
5	6	$5 - 6 = -1$
5	6	$5 - 6 = -1$
4	6	$4 - 6 = -2$
3	6	$3 - 6 = -3$

X	$(X - \bar{X})$	$(X - \bar{X})^2$
8	+2	4
8	+2	4
8	+2	4
7	+1	1
6	0	0
6	0	0
5	-1	1
5	-1	1
4	-2	4
3	-3	9
Sum	0	28

What we now know from these results is that each score in this distribution differs from the mean by an average of 1.76 points.

They're important to review and will increase your understanding of what the standard deviation is.

First, why didn't we just add up the deviations from the mean? Because the sum of the deviations from the mean is always equal to zero. Try it by summing the deviations $(2 + 2 + 2 + 1 + 0 + 0 - 1 - 1 - 2 - 3)$. In fact, that's the best way to check whether you computed the mean correctly.

There's another type of deviation that you may read about, and you should know what it means. The **mean deviation** (also called the mean absolute deviation) is the sum of the absolute value of the deviations from the mean divided by the number of scores. You already know that the sum of the deviations from the mean must equal zero (otherwise, the mean is computed incorrectly). Instead, let's take the *absolute* value of each deviation (which is the value regardless of the sign). Sum the absolute values and divide by the number of data points, and you have the mean deviation. So, if you have a set of scores such as 3, 4, 5, 5, 8 and the arithmetic mean is 5, the mean deviation is the sum of 2 (the absolute value of $5 - 3$), 1, 0, 0, and 3, for a total of 6. Then divide this by 5 to get the result of 1.2. (Note: The absolute value of a number is usually represented as that number with a vertical line on each side of it, such as $|5|$. For example, the absolute value of -6, or $|-6|$, is 6.)

Second, why do we square the deviations? Because we want to get rid of the negative sign so that when we do eventually sum them, they don't add up to 0.

And finally, why do we eventually end up taking the square root of the entire value in Step 7? Because we want to return to the same units with which we originally started. We squared the deviations from the mean in Step 4 (to get rid of negative values) and then took the square root of their total in Step 7. Pretty tidy.

Why n − 1? What's Wrong With Just n?

You might have guessed why we square the deviations about the mean and why we go back and take the square root of their sum. But how about subtracting the value of 1 from the denominator of the formula? Why do we divide by $n - 1$ rather than just plain ol' n? Good question.

The answer is that s (the standard deviation) is an estimate of the population standard deviation, and it is an **unbiased estimate** at that, but only when we subtract 1 from n. By subtracting 1 from the denominator, we artificially force the standard deviation to be larger than it would be otherwise. Why would we want to do that? Because, as good scientists, we are conservative. Being conservative means that if we have to err (and there is always a pretty good chance that we will), we will do so on the side of overestimating what the standard deviation of the population is. Dividing by a smaller denominator lets us do so. Thus, instead of dividing by 10, we divide by 9. Or instead of dividing by 100, we divide by 99.

 Biased estimates are appropriate if your intent is only to describe the characteristics of the sample. But if you intend to use the sample as an estimate of a population parameter, then it's best to calculate the unbiased statistic.

Take a look in the following table and see what happens as the size of the sample gets larger (and moves closer to the population in size). The $n - 1$ adjustment has far less impact on the difference between the biased and the unbiased estimates of the standard deviation (the bold column in the table) as the sample size increases.

All other things being equal, then, the larger the size of the sample, the less difference there is between the biased and the unbiased estimates of the standard deviation.

Sample Size	Value of Numerator in Standard Deviation Formula	Biased Estimate of the Population Standard Deviation (dividing by n)	Unbiased Estimate of the Population Standard Deviation (dividing by $n - 1$)	Difference Between Biased and Unbiased Estimates
10	500	7.07	7.45	0.38
100	500	2.24	2.25	0.01
1,000	500	0.7071	0.7075	0.0004

The moral of the story? When you compute the standard deviation for a sample, which is an estimate of the population, the closer to the size of the population the sample is, the more accurate the estimate will be.

What's the Big Deal?

The computation of the standard deviation is very straightforward. But what does it mean? As a measure of variability, all it tells us is how much each score in a set of scores, on the average, varies from the mean. But it has some very practical applications, as you will find out in Chapter 4. Just to whet your appetite, consider this: The standard deviation can be used to help us compare scores from different distributions, *even when the means and standard deviations are different.* Amazing! This, as you will see, can be very cool.

- The standard deviation is computed as the average distance from the mean. So, you will need to first compute the mean as a measure of central tendency. Don't fool around with the median or the mode in trying to compute the standard deviation.
- The larger the standard deviation, the more spread out the values are, and the more different they are from one another.
- Just like the mean, the standard deviation is sensitive to extreme scores. When you are computing the standard deviation of a sample and you have extreme scores, note that fact somewhere in your written report and in your interpretation of what the data mean.
- If $s = 0$, there is absolutely no variability in the set of scores, and the scores are essentially identical in value. This will rarely happen.

So, what's the big deal about variability and its importance? As a statistical concept, you learned above that it is a measure of dispersion or how scores differ from one another. But, "shakin' it up" really is the spice of life, and as you may have learned in your biology or psychology classes or from your own personal reading, variability is *the* key component of evolution, for without change (which variability always accompanies), organisms cannot adapt. All very cool.

COMPUTING THE VARIANCE

Here comes another measure of variability and a nice surprise. If you know the standard deviation of a set of scores and you can square a number, you can easily compute the **variance** of that same set of scores. This third measure of variability, the variance, is simply the standard deviation squared.

In other words, it's the same formula you saw earlier but without the square root bracket, like the one shown in Formula 3.3:

$$s^2 = \frac{\Sigma(X - \overline{X})^2}{n-1} \tag{3.3}$$

If you take the standard deviation and never complete the last step (taking the square root), you have the variance. In other words, $s^2 = s \times s$, or the variance equals the standard deviation times itself (or squared). In our earlier example, where the standard deviation was equal to 1.76 (before rounding), the variance is equal to 1.76^2, or 3.11. As another example, let's say that the standard deviation of a set of 150 scores is 2.34. Then, the variance would be 2.34^2, or 5.48.

You are not likely to see the variance mentioned by itself in a journal article or see it used as a descriptive statistic. This is because the variance is a difficult number to interpret and apply to a set of data. After all, it is based on squared deviation scores.

But the variance is important because it is used both as a concept and as a practical measure of variability in many statistical formulas and techniques. You will learn about these later in *Statistics for People Who (Think They) Hate Statistics*.

The Standard Deviation Versus the Variance

How are standard deviation and the variance the same, and how are they different?

Well, they are both measures of variability, dispersion, or spread. The formulas used to compute them are very similar. You see them (but mostly the standard deviation) reported all over the place in the "Results" sections of journals.

Computing Standard Deviation

They are also quite different.

First, and most important, the standard deviation (because we take the square root of the average summed squared deviation) is stated in the original units from which it was derived. The variance is stated in units that are squared (the square root of the final value is never taken).

What does this mean? Let's say that we need to know the variability of a group of production workers assembling circuit boards. Let's say that they average 8.6 boards per hour and the standard deviation is 1.59. The value 1.59 means that the difference in the average number of boards assembled per hour is about 1.59 circuit boards from the mean. This kind of information is quite valuable when trying to understand the overall performance of groups.

Let's look at an interpretation of the variance, which is 1.59^2, or 2.53. This would be interpreted as meaning that the average difference between the workers is about 2.53 circuit boards *squared* from the mean. Which of these two makes more sense?

USING THE COMPUTER TO COMPUTE MEASURES OF VARIABILITY

Let's use SPSS to compute some measures of variability. We are using the file named Chapter 3 Data Set 1.

There is one variable in this data set:

Computing
Measures of
Variability

Variable	Definition
ReactionTime	Reaction time on a tapping task in seconds

Here are the steps to compute the measures of variability that we discussed in this chapter.

1. Open the file named Chapter 3 Data Set 1.
2. Click Analyze → Descriptive Statistics → Frequencies.
3. Double-click on the ReactionTime variable to move it to the Variable(s) box.
4. Click on the Statistics button, and you will see the Frequencies: Statistics dialog box. This dialog box is used to select the variables and procedures you want to perform.
5. Under Dispersion, click Std. deviation.
6. Under Dispersion, click Variance.
7. Under Dispersion, click Range.
8. Click Continue.
9. Click OK.

The SPSS Output

Figure 3.1 shows selected output from the SPSS procedure for ReactionTime.

Figure 3.1 SPSS Output for the Variable Reaction Time

Statistics

ReactionTime

N	Valid	30
	Missing	0
Std. Deviation		.70255
Variance		.494
Range		2.60

Source: IBM.

Understanding the SPSS Output

There are 30 valid cases with no missing cases, and the standard deviation is 0.70255. The variance equals 0.494 (or s^2), and the range is 2.60. As you already know, the standard deviation, variance, and range are measures of dispersion or variability. In the case of this set of 30 observations, the standard deviation (the most commonly used of all these measures) is equal to 0.703.

More on Measures of Dispersion

We will go into much greater detail as to how this value is used in Chapter 8, but for now, the value of 0.703 represents the average amount each score is from the centermost point in the set of scores, or the mean. For this reaction time data, it means that the average reaction time varies from the mean by about 0.703 seconds.

Let's try another one, using the data titled Chapter 3 Data Set 2. There are two variables in this data set:

Variable	Definition
Math_Score	Score on a mathematics test
Reading_Score	Score on a reading test

Follow the same set of instructions as given previously, only in Step 3, select both variables (either by double-clicking on each one or by selecting it and clicking the "move" arrow).

More SPSS Output

The SPSS output is shown in Figure 3.2, where you can see selected output from the SPSS procedure for these two variables. There are 30 valid cases with no missing cases, and the standard deviation for math scores is 12.36 with a variance of 152.70 and a range of 43. For reading scores, the standard deviation is 18.70, the variance is a whopping 349.69 (that's pretty big), and the range is 76 (which is large as well, reflecting the similarly large variance).

Figure 3.2   Output for the Variables Math_Score and Reading_Score

Statistics

		Math_Score	Reading_Score
N	Valid	30	30
	Missing	0	0
Std. Deviation		12.357	18.700
Variance		152.700	349.689
Range		43	76

Source: IBM.

Understanding the SPSS Output

As with the example shown in Figure 3.1, the interpretation of the SPSS output is relatively straightforward.

On average, the amount of deviation from the mean for math sores is 12.3, and on average, the amount of deviation for reading scores is 18.70. Whether these measures of dispersion are considered "large" or "small" is a question that can only be answered relative to a variety of factors, including what's being measured, the number of observations, and the range of possible scores. Context is everything.

REAL-WORLD STATS

If you were like a mega stats person, then you might be interested in the properties of measures of variability for the sake of those properties. That's what mainline statisticians spend lots of their time doing—looking at the characteristics and performance and assumptions (and the violation thereof) of certain statistics.

But we're more interested in how these tools are used, so let's take a look at one such study that actually focused on variability as an outcome. And, as you read earlier, variability among scores is interesting for sure, but when it comes to understanding the reasons for variability among substantive performances and people, then the topic becomes really interesting.

This is exactly what Nicholas Stapelberg and his colleagues in Australia did when they looked at variability in heart rate as it related to coronary heart disease. Now, they did not look at this phenomenon directly, but they entered the search terms "heart rate variability," "depression," and "heart disease" into multiple electronic databases and found that decreased heart rate variability is found in conjunction with both major depressive disorders and coronary heart disease.

Why might this be the case? The researchers think that both diseases disrupt control feedback loops that help the heart function efficiently. This is a terrific example of how looking at variability can be the focal point of a study, rather than an accompanying descriptive statistic.

Want to know more? Go online or to the library and read . . .

Real-World
Stats

Stapelberg, N. J., Hamilton-Craig, I., Neumann, D. L., Shum, D. H., & McConnell, H. (2012). Mind and heart: Heart rate variability in major depressive disorder and coronary heart disease—a review and recommendations. *The Australian and New Zealand Journal of Psychiatry, 46,* 946–957.

SUMMARY

Measures of variability help us even more fully understand what a distribution of data points looks like. Along with a measure of central tendency, we can use these values to distinguish distributions from one another and effectively describe what a collection of test scores, heights, or measures of personality looks like and what those individual scores represent. Now that we can think and talk about distributions, let's explore ways we can look at them.

Chapter Summary

TIME TO PRACTICE

1. Why is the range the most convenient measure of dispersion, yet the most imprecise measure of variability? When would you use the range?

2. Fill in the exclusive and inclusive ranges for the following items.

High Score	Low Score	Inclusive Range	Exclusive Range
12.1	3		
92	51		
42	42		
7.5	6		
27	26		

3. Why would you expect more variability on a measure of personality in college freshmen than you would on a measure of height?

4. Why does the standard deviation get smaller as the individuals in a group score more similarly on a test? And why would you expect the amount of variability on a measure to be relatively less with a larger number of observations than with a smaller one?

5. For the following set of scores, compute the range, the unbiased and the biased standard deviations, and the variance. Do the exercise by hand.

94, 86, 72, 69, 93, 79, 55, 88, 70, 93

6. In #5 just above, why is the unbiased estimate greater than the biased estimate?

Problem 7

7. Use SPSS to compute all the descriptive statistics for the following set of three test scores over the course of a semester. Which test had the highest average score? Which test had the smallest amount of variability?

Test 1	Test 2	Test 3
50	50	49
48	49	47
51	51	51
46	46	55
49	48	55
48	53	45
49	49	47
49	52	45
50	48	46
50	55	53

8. For the following set of scores, compute by hand the unbiased estimates of the standard deviation and variance.

58

56

48

76

69

76

78

45

66

9. The variance for a set of scores is 36. What is the standard deviation, and what is the range?

10. Find the inclusive range, the sample standard deviation, and the sample variance of each of the following sets of scores:

a. 5, 7, 9, 11

b. 0.3, 0.5, 0.6, 0.9

c. 6.1, 7.3, 4.5, 3.8

d. 435, 456, 423, 546, 465

11. This practice problem uses the data contained in the file named Chapter 3 Data Set 3. There are two variables in this data set:

Problem 11

Variable	Definition
Height	Height in inches
Weight	Weight in pounds

Using SPSS, compute all of the measures of variability you can for height and weight.

12. How can you tell whether SPSS produces a biased or an unbiased estimate of the standard deviation?

13. Compute the biased and unbiased values of the standard deviation and variance for the set of accuracy scores shown in Chapter 3 Data Set 4. Use SPSS if you can; otherwise, do it by hand. Which one is smaller and why?

14. On a spelling test, the standard deviation is equal to 0.94. What does this mean?

STUDENT STUDY SITE

Ⓢ**SAGE** edge™

Get the tools you need to sharpen your study skills! Visit **edge.sagepub.com/ salkind6e** to access practice quizzes, eFlashcards, original and curated videos, data sets, journal articles, and more!

4 A Picture Really Is Worth a Thousand Words

Difficulty Scale ☺ ☺ ☺ ☺
(moderately easy, but not a cinch)

WHAT YOU'LL LEARN ABOUT IN THIS CHAPTER

✦ Understanding why a picture is really worth a thousand words

✦ Creating a histogram and a polygon

✦ Understanding the different shapes of different distributions

✦ Using SPSS to create incredibly cool charts

✦ Creating different types of charts and understanding their application and uses

WHY ILLUSTRATE DATA?

In the previous two chapters, you learned about two important types of descriptive statistics—measures of central tendency and measures of variability. Both of these provide you with the one best measure for describing a group of data (central tendency) and a measure of how diverse, or different, scores are from one another (variability).

Introduction to Chapter 4

What we did not do, and what we will do here, is examine how differences in these two measures result in different-looking distributions. Numbers alone (such as $M = 3$ and $s = 3$) may be important, but a visual representation is a much more effective way of examining the characteristics of a distribution as well as the characteristics of any set of data.

Interactive Visualizations

So, in this chapter, we'll learn how to visually represent a distribution of scores as well as how to use different types of graphs to represent different types of data.

TEN WAYS TO A GREAT FIGURE
(EAT LESS AND EXERCISE MORE?)

Whether you create illustrations by hand or use a computer program, the same principles of decent design apply. Here are 10 to copy and put above your desk.

1. **Minimize chart or graph junk.** "Chart junk" (a close cousin to "word junk") happens when you use every function, every graph, and every feature a computer program has to make your charts busy, full, and uninformative. More is definitely less.

2. **Plan out your chart before you start creating the final copy.** Use graph paper even if you will be using a computer program to generate the graph. And, in fact, just use your computer to generate and print out graph paper (try www.printfreegraphpaper.com).

3. **Say what you mean and mean what you say—no more and no less.** There's nothing worse than a cluttered (with too much text and fancy features) graph to confuse the reader.

4. **Label everything so nothing is left to the misunderstanding of the audience.**

5. **A graph should communicate only one idea.**

6. **Keep things balanced.** When you construct a graph, center titles and axis labels.

7. **Maintain the scale in a graph.** The scale refers to the relationship between the horizontal and vertical axes. This ratio should be about 3 to 4, so a graph that is 3 inches wide will be about 4 inches tall.

8. **Simple is best and less is more.** Keep the chart simple, but not simplistic. Convey one idea as straightforwardly as possible, with distracting information saved for the accompanying text. Remember, a chart or graph should be able to stand alone, and the reader should be able to understand the message.

9. **Limit the number of words you use.** Too many words, or words that are too large (both in terms of physical size and idea-wise), can detract from the visual message your chart should convey.

10. **A chart alone should convey what you want to say.** If it doesn't, go back to your plan and try it again.

FIRST THINGS FIRST: CREATING
A FREQUENCY DISTRIBUTION

The most basic way to illustrate data is through the creation of a frequency distribution. A **frequency distribution** is a method of tallying and representing how often certain scores occur. In the creation of a frequency distribution, scores are usually grouped into class intervals, or ranges of numbers.

Here are 50 scores on a test of reading comprehension on which a frequency distribution is based:

47	10	31	25	20
2	11	31	25	21
44	14	15	26	21
41	14	16	26	21
7	30	17	27	24
6	30	16	29	24
35	32	15	29	23
38	33	19	28	20
35	34	18	29	21
36	32	16	27	20

And here's the frequency distribution. You can see that for each range of scores, there are associated frequency counts.

Class Interval	Frequency
45–49	1
40–44	2
35–39	4
30–34	8
25–29	10
20–24	10
15–19	8
10–14	4
5–9	2
0–4	1

The Classiest of Intervals

As you can see from the above table, a **class interval** is a range of numbers, and the first step in the creation of a frequency distribution is to define how large each interval will be. As you can see in the frequency distribution that we created, each interval spans five possible scores, such as 5–9 (which includes scores 5, 6, 7, 8, and 9) and 40–44 (which includes scores 40, 41, 42, 43, and 44). How did we decide to have an interval that includes only five scores? Why not five intervals, each consisting of 10 scores? Or two intervals, each consisting of 25 scores?

Here are some general rules to follow in the creation of a class interval, regardless of the size of values in the data set you are dealing with.

1. Select a class interval that has a range of 2, 5, 10, 15, or 20 data points. In our example, we chose 5.

2. Select a class interval so that 10 to 20 such intervals cover the entire range of data. A convenient way to do this is to compute the range and then divide by a number that represents the number of intervals you want to use (between 10 and 20). In our example, there are 50 scores and we wanted 10 intervals: 50/10 = 5, which is the size of each class interval. If you had a set of scores ranging from 100 to 400, you could start with an estimate of 20 intervals and see if the interval range makes sense for your data: 300/20 = 15, so 15 would be the class interval.

3. Begin listing the class interval with a multiple of that interval. In our frequency distribution of reading comprehension test scores, the class interval is 5, and we started the lowest class interval at 0.

4. Finally, the largest interval goes at the top of the frequency distribution.

Simply put, there are no hard-and-fast rules about creating class intervals on the way to creating a frequency distribution. Here are six general rules:

1. Determine the range.

2. Decide on the number of class intervals.

3. Decide on the size of the class interval.

4. Decide the starting point for the first class.

5. Create the class intervals.

6. Put the data into the class intervals.

Once class intervals are created, it's time to complete the frequency part of the frequency distribution. That's simply counting the number of times a score occurs in the raw data and entering that number in each of the class intervals represented by the count.

In the frequency distribution that we created for our reading comprehension data, the number of scores that occur between 30 and 34 and thus are in the 30–34 class interval is 8. So, an 8 goes in the column marked Frequency. There's your frequency distribution.

Sometimes it is a good idea to graph your data first and then do whatever calculations or analysis is called for. By first looking at the data, you may gain insight into the relationship between variables, what kind of descriptive statistic is the right one to use to describe the data, and so on. This extra step might increase your insights and the value of what you are doing.

THE PLOT THICKENS: CREATING A HISTOGRAM

Now that we've got a tally of how many scores fall in what class intervals, we'll go to the next step and create what is called a **histogram,** a visual representation of the frequency distribution where the frequencies are represented by bars.

Depending on the book or journal article or report you read and the software you use, visual representations of data are called graphs (such as in SPSS) or charts (such as in the Microsoft spreadsheet Excel). It really makes no difference. All you need to know is that a graph or a chart is the visual representation of data.

To create a histogram, do the following:

1. Using a piece of graph paper, place values at equal distances along the x-axis, as shown in Figure 4.1. Now, identify the **midpoint** of each class interval, which is the middle point in the interval. It's pretty easy to just eyeball, but you can also just add the top and bottom values of the class interval and divide by 2. For example, the midpoint of the class interval 0–4 is the average of 0 and 4, or 4/2 = 2.

2. Draw a bar or column centered on each midpoint that represents the entire class interval to the height representing the frequency of that class interval. For example, in Figure 4.2, you can see that in our first entry, the class interval of 0–4 is represented by the frequency of 1 (representing the one time a value between 0 and 4 occurs). Continue drawing bars or columns until each of the frequencies for each of the class intervals is represented. Here's a nice hand-drawn (really!) histogram for the frequency distribution of the 50 scores that we have been working with so far.

Notice that each class interval is represented by a range of scores along the x-axis.

Figure 4.1	Class Intervals Along the x-Axis

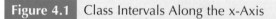
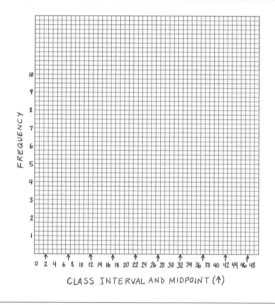

The Tallyho Method

You can see by the simple frequency distribution at the beginning of the chapter that you already know more about the distribution of scores than you'd learn from just a simple listing of them. You have a good idea of what values occur with what frequency. But another visual representation (besides a histogram) can be done by using tallies for each of the occurrences, as shown in Figure 4.3.

We used tallies that correspond with the frequency of scores that occur within a certain class. This gives you an even better visual representation of how often certain scores occur relative to other scores.

Figure 4.2 A Hand-Drawn Histogram

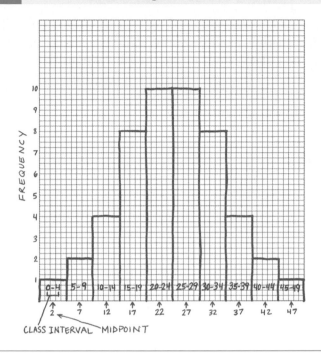

Figure 4.3 Tallying Scores

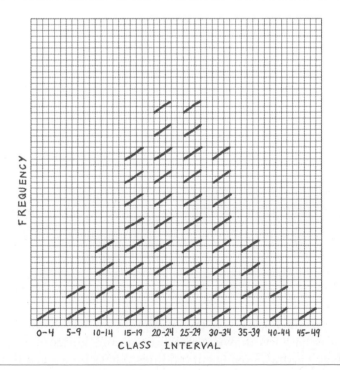

THE NEXT STEP: A FREQUENCY POLYGON

Creating a histogram or a tally of scores wasn't so difficult, and the next step (and the next way of illustrating data) is even easier. We're going to use the same data—and, in fact, the histogram that you just saw created—to create a frequency polygon. A **frequency polygon** is a continuous line that represents the frequencies of scores within a class interval, as shown in Figure 4.4.

| Figure 4.4 | A Hand-Drawn Frequency Polygon |

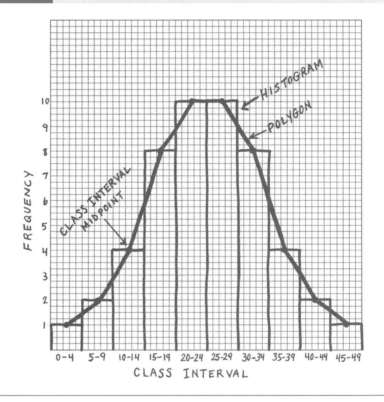

How did we draw this? Here's how.

1. Place a midpoint at the top of each bar or column in a histogram (see Figure 4.2).
2. Connect the lines and you've got it—a frequency polygon!

Note that in Figure 4.4, the histogram on which the frequency polygon is based is drawn using vertical and horizontal lines, and the polygon is drawn using curved lines. That's because, although we want you to see what a frequency polygon is based on, you usually don't see the underlying histogram.

Why use a frequency polygon rather than a histogram to represent data? It's more a matter of preference than anything else. A frequency polygon appears more dynamic than a histogram (a line that represents change in frequency always looks neat), but you are basically conveying the same information.

Cumulating Frequencies

Once you have created a frequency distribution and have visually represented those data using a histogram or a frequency polygon, another option is to create a visual representation of the cumulative frequency of occurrences by class intervals. This is called a **cumulative frequency distribution**.

Data, Data, Everywhere

A cumulative frequency distribution is based on the same data as a frequency distribution but with an added column (Cumulative Frequency), as shown below.

Class Interval	Frequency	Cumulative Frequency
45–49	1	50
40–44	2	49
35–39	4	47
30–34	8	43
25–29	10	35
20–24	10	25
15–19	8	15
10–14	4	7
5–9	2	3
0–4	1	1

The cumulative frequency distribution begins with the creation of a new column labeled Cumulative Frequency. Then, we add the frequency in a class interval to all the frequencies below it. For example, for the class interval of 0–4, there is 1 occurrence and none below it, so the cumulative frequency is 1. For the class interval of 5–9, there are 2 occurrences in that class interval and one below it for a total of 3 (2 + 1) occurrences. The last class interval (45–49) contains 1 occurrence, and there is a total of 50 occurrences at or below that class interval.

Once we create the cumulative frequency distribution, then the data can be plotted just as they were in a histogram or a frequency polygon. Only this time, we'll skip right ahead and plot the midpoint of each class interval as a function of the cumulative frequency of that class interval. You can see the cumulative frequency distribution in Figure 4.5 based on the 50 scores from the beginning of this chapter.

Figure 4.5 A Hand-Drawn Cumulative Frequency Distribution

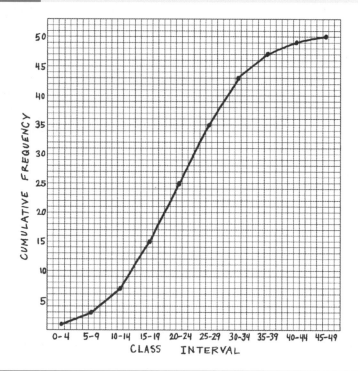

Another name for a cumulative frequency polygon is an **ogive**. And, if the distribution of the data is normal or bell shaped (see Chapter 8 for more on this), then the ogive represents what is popularly known as a bell curve or a normal distribution. SPSS creates a really nice ogive—it's called a P-P plot (for probability plot) and is easy to create. See Appendix A for an introduction to creating graphs using SPSS, as well as the material toward the end of this chapter.

OTHER COOL WAYS TO CHART DATA

What we did so far in this chapter is take some data and show how charts such as histograms and polygons can be used to communicate visually. But several other types of charts are also used in the behavioral and social sciences, and although it's not necessary for you to know exactly how to create them (manually), you should at least be familiar with their names and what they do. So here are some popular charts, what they do, and how they do it.

There are several very good personal computer applications for creating charts, among them the spreadsheet Excel (a Microsoft product)—the author's personal favorite—and, of course, SPSS. The charts in the "Using the Computer to Illustrate Data" section were created using SPSS as well.

Moments

Bar Charts

A bar or column chart should be used when you want to compare the frequencies of different categories with one another. Categories are organized horizontally on the *x*-axis, and values are shown vertically on the *y*-axis. Here are some examples of when you might want to use a column chart:

- Number of participants in different water exercise activities
- The sales of three different types of products
- Number of children in each of six different grades

Figure 4.6 shows a graph of number of participants in different water activities.

Figure 4.6   A Bar Chart That Compares Different Water Activities

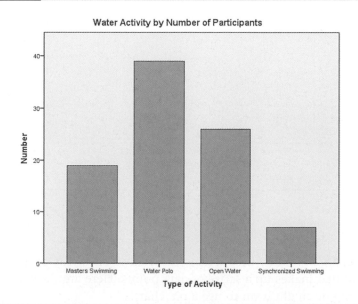

Column Charts

A column chart is identical to a bar chart, but in this chart, categories are organized on the y-axis (which is the vertical one), and values are shown on the x-axis (the horizontal one).

Line Charts

A line chart should be used when you want to show a trend in the data at equal intervals. Here are some examples of when you might want to use a line chart:

- Number of cases of mononucleosis (mono) per season among college students at three state universities
- Toy sales for the T&K company over four quarters
- Number of travelers on two different airlines for each quarter

In Figure 4.7, you can see a chart of sales in units over four quarters.

Figure 4.7   Using a Line Chart to Show a Trend Over Time

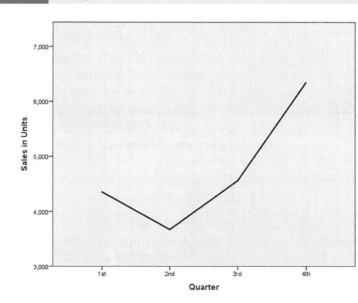

Pie Charts

A pie chart should be used when you want to show the proportion of an item that makes up a series of data points. Here are some examples of when you might want to use a pie chart:

- Percentage of children living in poverty by ethnicity
- Proportion of night and day students enrolled
- Age of participants by gender

Note that a pie chart only offers counts by the nominal classes (such as ethnicity, tie of enrollment and gender) the chart represents.

In Figure 4.8, you can see a pie chart of voter preference. And we did a few fancy-schmancy things, such as separating and labeling the slices.

Figure 4.8	A Pie Chart Illustrating the Relative Proportion of One Category to Others

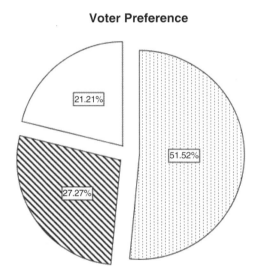

Voter Preference

⊡ Crazy Pile-on Party
⊠ Nothing Better to Do Party
☐ What, Me Worry? Party

21.21%
51.52%
27.27%

USING THE COMPUTER (SPSS, THAT IS) TO ILLUSTRATE DATA

Now let's use SPSS and go through the steps in creating some of the charts that we explored in this chapter. First, here are some general SPSS charting guidelines.

1. The newer versions of SPSS come with a Chart Builder option on the Graphs menu. This is the easiest way to get started, and that's what we will use.

2. In general, you click Graphs → Chart Builder, and you see a dialog box from which you will select the type of graph you want to create.

3. Click the type of graph you want to create and then select the specific design of that type of graph.

4. Drag the variable names to the axis where each belongs.

5. Click OK, and you'll see your graph.

Let's practice.

Creating a Histogram

1. Enter the data you want to use to create the graph.

2. Click Graphs → Chart Builder and you will see the Chart Builder dialog box, as shown in Figure 4.9. If you see any other screen, click OK.

3. Click the Histogram option in the Choose from: list and double-click the first image.

4. Drag the variable named Score to the x-axis? location in the preview window.

5. Click OK and you will see the histogram, as shown in Figure 4.10.

Creating a Histogram

The histogram in Figure 4.10 looks a bit different from the hand-drawn one representing the same data shown earlier in this chapter, in Figure 4.2. The difference is that SPSS defines class intervals using its own idiosyncratic method. SPSS took as the middle of a class interval the bottom number of the interval (such as 10), rather than the midpoint (such as 12.5). Consequently, scores are allocated to different groups. The lesson here? How you group data makes a big difference in the way they look in a histogram. And, once you get to know SPSS well, you can make all kinds of fine-tuned adjustments to make graphs appear exactly as you want them.

Figure 4.9 The Chart Builder Dialog Box

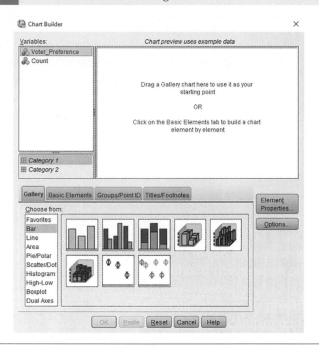

Figure 4.10 A Histogram Created Using the Chart Builder

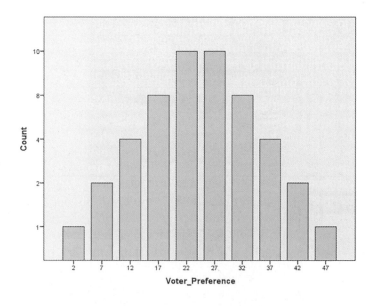

Creating a Bar Graph

To create a bar graph, follow these steps:

1. Enter the data you want to use to create the graph. We used Average Voter Attitude Score (by party), as you see here:

Republican	Democrat	Independent
54	63	19

2. Click Graphs → Chart Builder, and you will see the Chart Builder dialog box, as shown in Figure 4.11. If you see any other screen, click OK.
3. Click the Bar option in the Choose from: list and double-click the first image.
4. Drag the variable named Party to the x-axis? location in the preview window.
5. Drag the variable named Number to the Count axis.
6. Click OK and you will see the bar graph, as shown in Figure 4.12.

Understanding Average Value

| Figure 4.11 | The Chart Builder Dialog Box |

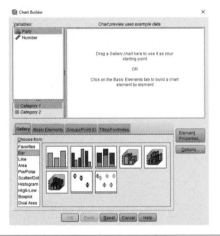

| Figure 4.12 | A Bar Graph Created Using the Chart Builder |

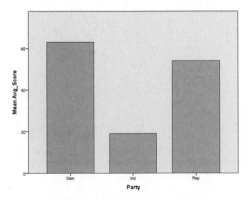

Creating a Line Graph

To create a line graph, follow these steps:

1. Enter the data you want to use to create the graph. In this example, we will be using the percentage in attendance of the total student body over the duration of a 10-year program. Here are the data:

Year	Percent Attending
1	87
2	88
3	89
4	76
5	80
6	96
7	91
8	97
9	89
10	79

Creating a Line Graph

2. Click Graphs → Chart Builder and you will see the Chart Builder dialog box, as shown in Figure 4.11. If you see any other screen, click OK.

3. Click the Line option in the Choose from: list and double-click the first image.

4. Drag the variable named Year to the x-axis? location in the preview window.

5. Drag the variable named Attendance to the y-axis? location.

6. Click OK, and you will see the line graph, as shown in Figure 4.13. We used the SPSS Chart Editor to change the minimum and maximum values on the y-axis.

Figure 4.13 A Line Graph Created Using the Chart Builder

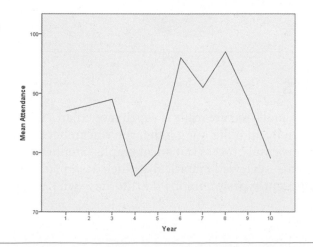

Creating a Pie Chart

To create a pie chart, follow these steps:

1. Enter the data you want to use to create the chart. In this example, the pie chart represents the percentage of people buying different brands of doughnuts. Here are the data:

2. Click Graphs → Chart Builder, and you will see the Chart Builder dialog box, as shown in Figure 4.11. If you see any other screen, click OK.

3. Click the Pie/Polar option in the Choose from: list and double-click the only image.

4. Drag the variable named Brand to the Slice by? axis label.

5. Drag the variable named Percentage to the Angle Variable? axis label.

6. Click OK, and you will see the pie chart, as shown in Figure 4.14.

Creating a
Pie Chart

Brand	Percentage
Krispies	55
Dunks	35
Other	10

Figure 4.14 A Pie Chart Created Using the Chart Builder

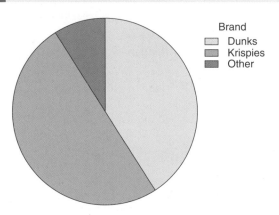

REAL-WORLD STATS

Graphs work, and a picture really is worth more than a thousand words.

In this article, an oldie but goodie, the researchers examined how people perceive and process statistical graphs. Stephen Lewandowsky and Ian Spence reviewed empirical studies designed to explore how suitable different types of graphs are and how what is known about human perception can have an impact on the design and utility of these charts.

They focused on some of the theoretical explanations for why certain elements work and don't, the use of pictorial symbols (like a happy face symbol, which could make up the bar in a bar chart), and multivariate displays where more than one set of data needs to be represented. And, as is very often the case with any paper, they concluded that not enough data were available yet. Given the increasingly visual world in which we live (emoticons, anyone? ☹ ☺), this is interesting and useful reading to gain a historical perspective on how information was (and still is) discussed as a scientific topic.

Want to know more? Go online or to the library and find . . .

Real-World Stats

Lewandowsky, S., & Spence, I. (1989). The perception of statistical graphs. *Sociological Methods Research, 18,* 200–242.

SUMMARY

There's no question that charts are fun to create and can add enormous understanding to what might otherwise appear to be disorganized data. Follow our suggestions in this chapter and use charts well, but only when they enhance, not just add to, what's already there.

Chapter Summary

TIME TO PRACTICE

1. A data set of 50 comprehension scores (named Comprehension Score) called Chapter 4 Data Set 1 is available in Appendix C and on the website. Answer the following questions and/or complete the following tasks:

    Problem 1

    a. Create a frequency distribution and a histogram for the set.

    b. Why did you select the class interval you used?

    c. Is this distribution skewed? How do you know?

2. Here is a frequency distribution. Create a histogram by hand or by using SPSS.

3. A third-grade teacher wants to improve her students' level of engagement during group discussions and instruction. She keeps track of each of the 15 third graders' number of responses every day for 1 week, and the data are available as Chapter 4 Data Set 2. Use SPSS to create a bar chart with one bar for each day (and warning—this may be a toughie).

    Problem 3

Class Interval	Frequency
261–280	140
241–260	320
221–240	3380
201–220	600
181–200	500
161–180	410
141–160	315
121–140	300
100–120	200

4. Identify whether these distributions are negatively skewed, positively skewed, or not skewed at all and explain why you describe them that way.

   a. This talented group of athletes scored very high on the vertical jump task.

   b. On this incredibly crummy test, everyone received the same score.

   c. On the most difficult spelling test of the year, the third graders wept as the scores were delivered and then their parents complained.

Problem 5

5. Use the data available as Chapter 4 Data Set 3 on pie preference to create a pie chart ☺ using SPSS.

6. For each of the following, indicate whether you would use a pie, line, or bar chart and why.

   a. The proportion of freshmen, sophomores, juniors, and seniors in a particular university

   b. Change in temperature over a 24-hour period

   c. Number of applicants for four different jobs

   d. Percentage of test takers who passed

   e. Number of scores in each of 10 categories

7. Provide an example of when you might use each of the following types of charts. For example, you would use a pie chart to show the proportion of children in Grades 1 through 6 who receive a reduced-price lunch. When you are done, draw the fictitious chart by hand.

   a. Line

   b. Bar

   c. Scatter/Dot (extra credit)

8. Go to the library and find a journal article in your area of interest that contains empirical data but does not contain any visual representation of them. Use the data to create a chart. Be sure to specify what type of chart you are creating and why you chose the one you did. You can create the chart manually or using SPSS or Excel.

9. Create the worst-looking chart that you can, crowded with chart and font junk. Nothing makes as lasting an impression as a *bad* example.

10. And, finally, what is the purpose of a chart or graph?

STUDENT STUDY SITE

⑤SAGE edge™

Get the tools you need to sharpen your study skills! Visit **edge.sagepub.com/salkind6e** to access practice quizzes, eFlashcards, original and curated videos, data sets, journal articles, and more!

Ice Cream and Crime

Computing Correlation Coefficients

Difficulty Scale ☺ ☺ (moderately hard)

WHAT YOU'LL LEARN ABOUT IN THIS CHAPTER

✦ Understanding what correlations are and how they work

✦ Computing a simple correlation coefficient

✦ Interpreting the value of the correlation coefficient

✦ Understanding what other types of correlations exist and when they should be used

---

WHAT ARE CORRELATIONS ALL ABOUT?

Measures of central tendency and measures of variability are not the only descriptive statistics that we are interested in using to get a picture of what a set of scores looks like. You have already learned that knowing the values of the one most representative score (central tendency) and a measure of spread or dispersion (variability) is critical for describing the characteristics of a distribution.

Introduction to
Chapter 5

However, sometimes we are as interested in the relationship between variables—or, to be more precise, how the value of one variable changes when the value of another variable changes. The way we express this interest is through the computation of a simple correlation coefficient. For example, what's the relationship between age and strength? Income and education? Memory skills and drug use? Voting preferences and attitude toward regulation?

A **correlation coefficient** is a numerical index that reflects the relationship between two variables. The value of this descriptive statistic ranges between −1.00 and +1.00 A correlation between two variables is sometimes referred to as a bivariate (for two variables) correlation. Even more specifically, the type of correlation that we will talk about in the majority of this chapter is called the **Pearson product-moment correlation**, named for its inventor, Karl Pearson.

The Pearson correlation coefficient examines the relationship between two variables, but both of those variables are continuous in nature. In other words, they are variables that can assume any value along some underlying continuum; examples include height (you really can be 5 feet and 6.1938574673 inches tall), age, test score, and income. But a host of other variables are not continuous. They're called discrete or categorical variables, and examples are race (such as black and white), social class (such as high and low), and political affiliation (such as Democrat and Republican). You need to use other correlational techniques, such as the point-biserial correlation, in these cases. These topics are for a more advanced course, but you should know they are acceptable and very useful techniques. We mention them briefly later on in this chapter.

There are other types of correlation coefficients that measure the relationship between more than two variables, and we'll leave those for the next statistics course (which you are looking forward to already, right?).

Types of Correlation Coefficients: Flavor 1 and Flavor 2

A correlation reflects the dynamic quality of the relationship between variables. In doing so, it allows us to understand whether variables tend to move in the same or opposite directions when they change. If variables change in the same direction, the correlation is called a **direct correlation** or a **positive correlation**. If variables change in opposite directions, the correlation is called an **indirect correlation** or a **negative correlation**. Table 5.1 shows a summary of these relationships.

Now, keep in mind that the examples in the table reflect generalities, for example, regarding time to complete a test and the number of items correct on that test. In general, the less time that is taken on a test, the lower the score. Such a conclusion is not rocket science, because the faster one goes, the more likely one is to make careless mistakes such as not reading instructions correctly. But of course, some people can go very fast and do very well. And other people go very slowly and don't

Table 5.1 Types of Correlations

What Happens to Variable X	What Happens to Variable Y	Type of Correlation	Value	Example
X increases in value.	Y increases in value.	Direct or positive	Positive, ranging from .00 to +1.00	The more time you spend studying, the higher your test score will be.
X decreases in value.	Y decreases in value.	Direct or positive	Positive, ranging from .00 to +1.00	The less money you put in the bank, the less interest you will earn.
X increases in value.	Y decreases in value.	Indirect or negative	Negative, ranging from −1.00 to .00	The more you exercise, the less you will weigh.
X decreases in value.	Y increases in value.	Indirect or negative	Negative, ranging from −1.00 to .00	The less time you take to complete a test, the more items you will get wrong.

do well at all. The point is that we are talking about the performance of a group of people on two different variables. We are computing the correlation between the two variables for the group, not for any one particular person.

There are several easy (but important) things to remember about the correlation coefficient:

- A correlation can range in value from −1.00 to +1.00.
- The absolute value of the coefficient reflects the strength of the correlation. So a correlation of −.70 is stronger than a correlation of +.50. One frequently made mistake regarding correlation coefficients occurs when students assume that a direct or positive correlation is always stronger (i.e., "better") than an indirect or negative correlation because of the sign and nothing else.
- A correlation always reflects a situation in which there are at least two data points (or variables) per case.
- Another easy mistake is to assign a value judgment to the sign of the correlation. Many students assume that a negative relationship is not good and a positive one is good. That's why, instead of using the terms *negative* and *positive*, we may prefer to use the terms *indirect* and *direct* to communicate meaning more clearly.

- The Pearson product-moment correlation coefficient is represented by the small letter *r* with a subscript representing the variables that are being correlated. For example,

r_{xy} is the correlation between variable *X* and variable *Y*.

$r_{weight\text{-}height}$ is the correlation between weight and height.

$r_{SAT\text{-}GPA}$ is the correlation between SAT score and grade point average (GPA).

The correlation coefficient reflects the amount of variability that is shared between two variables and what they have in common. For example, you can expect an individual's height to be correlated with an individual's weight because these two variables share many of the same characteristics, such as the individual's nutritional and medical history, general health, and genetics. However, if one variable does not change in value and therefore has nothing to share, then the correlation between it and another variable is zero. For example, if you computed the correlation between age and number of years of school completed, and everyone was 25 years old, there would be no correlation between the two variables because there's literally nothing (no variability) in age available to share.

Likewise, if you constrain or restrict the range of one variable, the correlation between that variable and another variable will be less than if the range is not constrained. For example, if you correlate reading comprehension and grades in school for very high-achieving children, you'll find the correlation to be lower than if you computed the same correlation for children in general. That's because the reading comprehension score of very high-achieving students is quite high and much less variable than it would be for all children. The moral? When you are interested in the relationship between two variables, try to collect sufficiently diverse data—that way, you'll get the truest representative result. And how do you do that? Measure a variable as precisely as possible.

COMPUTING A SIMPLE CORRELATION COEFFICIENT

The computational formula for the simple Pearson product-moment correlation coefficient between a variable labeled *X* and a variable labeled *Y* is shown in Formula 5.1.

$$r_{xy} = \frac{n\Sigma XY - \Sigma X \Sigma Y}{\sqrt{\left[n\Sigma X^2 - (\Sigma X)^2\right]\left[n\Sigma Y^2 - (\Sigma Y)^2\right]}},$$ (5.1)

Computing
a Simple
Correlation
Coefficient

where

- r_{xy} is the correlation coefficient between X and Y;
- n is the size of the sample;
- X is the individual's score on the X variable;
- Y is the individual's score on the Y variable;
- XY is the product of each X score times its corresponding Y score;
- X^2 is the individual's X score, squared; and
- Y^2 is the individual's Y score, squared.

Here are the data we will use in this example:

X	Y	X²	Y²	XY
2	3	4	9	6
4	2	16	4	8
5	6	25	36	30
6	5	36	25	30
4	3	16	9	12
7	6	49	36	42
8	5	64	25	40
5	4	25	16	20
6	4	36	16	24
7	5	49	25	35
Total, Sum, or Σ 54	43	320	201	247

Before we plug the numbers in, let's make sure you understand what each one represents:

- ΣX, or the sum of all the X values, is 54.
- ΣY, or the sum of all the Y values, is 43.
- ΣX^2, or the sum of each X value squared, is 320.
- ΣY^2, or the sum of each Y value squared, is 201.
- ΣXY, or the sum of the products of X and Y, is 247.

It's easy to confuse the sum of a set of values squared and the sum of the squared values. The sum of a set of values squared is taking values such as 2 and 3, summing them (to be 5), and then squaring that (which is 25). The sum of the squared values is taking values such as 2 and 3, squaring them (to get 4 and 9, respectively), and then adding those together (to get 13). Just look for the parentheses as you work.

Here are the steps in computing the correlation coefficient:

1. List the two values for each participant. You should do this in a column format so as not to get confused. Use graph paper if working manually or SPSS or some other data analysis tool if working digitally.
2. Compute the sum of all the X values and compute the sum of all the Y values.
3. Square each of the X values and square each of the Y values.
4. Find the sum of the XY products.

Computing the Correlation

These values are plugged into the equation you see in Formula 5.2:

$$r_{xy} = \frac{(10 \times 247) - (54 \times 43)}{\sqrt{[(10 \times 320) - 54^2][(10 \times 201) - 43^2]}} \quad (5.2)$$

Ta-da! And you can see the answer in Formula 5.3:

$$r_{xy} = \frac{148}{213.83} = .692 \quad (5.3)$$

What's really interesting about correlations is that they measure the amount of distance that one variable *covaries* in relation to another. So, if both variables are highly variable (have lots of wide-ranging values), the correlation between them is more likely to be high than if not. Now, that's not to say that lots of variability guarantees a higher correlation, because the scores have to vary in a systematic way. But if the variance is constrained in one variable, then no matter how much the other variable changes, the correlation will be lower. For example, let's say you are examining the correlation between academic achievement in high school and first-year grades

in college and you look at only the top 10% of the class.
Well, that top 10% is likely to have very similar grades,
introducing no variability and no room for the one variable to vary as a function of the other. Guess what you
get when you correlate one variable with another variable that does not change (that is, has no variability)?
$r_{xy} = 0$, that's what. The lesson here? Variability works,
and you should not artificially limit it.

A Visual Picture of a Correlation: The Scatterplot

There's a very simple way to visually represent a correlation: Create what
is called a **scatterplot**, or **scattergram** (in SPSS lingo it's a Scatter/Dot
graph). This is simply a plot of each set of scores on separate axes.

Here are the steps to complete a scattergram like the one you see in
Figure 5.1, which plots the 10 sets of scores for which we computed
the sample correlation earlier.

Figure 5.1   A Simple Scattergram

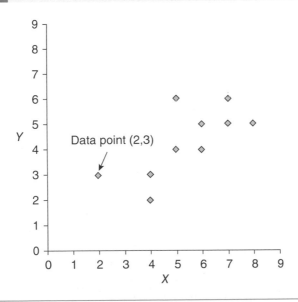

1. Draw the x-axis and the y-axis. Usually, the X variable goes on the horizontal axis and the Y variable goes on the vertical axis.
2. Mark both axes with the range of values that you know to be the case for
   the data. For example, the value of the X variable in our example ranges
   from 2 to 8, so we marked the x-axis from 0 to 9. There's no harm in
   marking the axes a bit low or high—just as long as you allow room for the

values to appear. The value of the *Y* variable ranges from 2 to 6, and we marked that axis from 0 to 9. Having similarly labeled (and scaled) axes can sometimes make the finished scatterplot easier to understand.

3. Finally, for each pair of scores (such as 2 and 3, as shown in Figure 5.1), we entered a dot on the chart by marking the place where 2 falls on the *x*-axis and 3 falls on the *y*-axis. The dot represents a data point, which is the intersection of the two values.

When all the data points are plotted, what does such an illustration tell us about the relationship between the variables? To begin with, the general shape of the collection of data points indicates whether the correlation is direct (positive) or indirect (negative).

A positive slope occurs when the data points group themselves in a cluster from the lower left-hand corner on the *x*- and *y*-axes through the upper right-hand corner. A negative slope occurs when the data points group themselves in a cluster from the upper left-hand corner on the *x*- and *y*-axes through the lower right-hand corner.

Here are some scatterplots showing very different correlations where you can see how the grouping of the data points reflects the sign and strength of the correlation coefficient.

Figure 5.2 shows a perfect direct correlation where $r_{xy} = 1.00$ and all the data points are aligned along a straight line with a positive slope.

Figure 5.2   A Perfect Direct, or Positive, Correlation

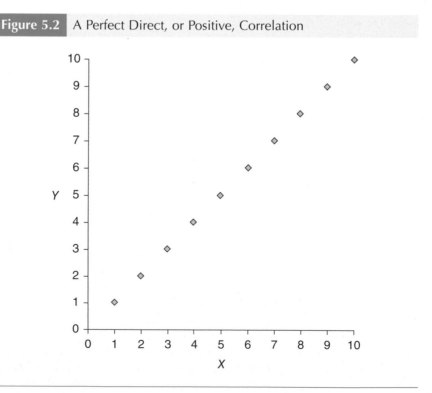

If the correlation were perfectly indirect, the value of the correlation coefficient would be −1.00, and the data points would align themselves in a straight line as well but from the upper left-hand corner of the chart to the lower right. In other words, the line that connects the data points would have a negative slope.

Don't ever expect to find a perfect correlation between any two variables in the behavioral or social sciences. Such a correlation would say that two variables are so perfectly related, they share everything in common. In other words, knowing one is like knowing the other. Just think about your classmates. Do you think they all share any one thing in common that is perfectly related to another of their characteristics across all those different people? Probably not. In fact, r values approaching .7 and .8 are just about the highest you'll see.

In Figure 5.3, you can see the scatterplot for a strong (but not perfect) direct relationship where $r_{xy} = .70$. Notice that the data points align themselves along a positive slope, although not perfectly.

Now, we'll show you a strong indirect, or negative, relationship in Figure 5.4, where $r_{xy} = -.82$. Notice that the data points align themselves on a negative slope from the upper left-hand corner of the chart to the lower right-hand corner.

Figure 5.3 A Strong, but Not Perfect, Direct Relationship

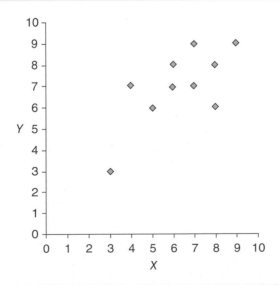

Figure 5.4   A Strong, but Not Perfect, Indirect Relationship

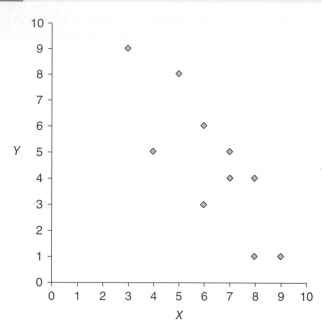

That's what different types of correlations look like, and you can really tell the general strength and direction by examining the way the points are grouped.

Not all correlations are reflected by a straight line showing the X and the Y values in a relationship called a **linear correlation** (see Chapter 16 for tons of fun stuff about this). The relationship may not be linear and may not be reflected by a straight line. Let's take the correlation between age and memory. For the early years, the correlation is probably highly positive—the older children get, the better their memories. Then, into young and middle adulthood, there isn't much of a change or much of a correlation, because most young and middle adults maintain a good (but not necessarily increasingly variable) memory. But with old age, memory begins to suffer, and there is an indirect relationship between memory and aging in the later years. If you take these together and look at the relationship over the life span, you find that the correlation between memory and age tends to look something like a curve where memory increases, levels off, and then decreases. It's a *curvilinear* relationship, and sometimes, the best description of a relationship is that it is curvilinear.

Bunches of Correlations: The Correlation Matrix

What happens if you have more than two variables? How are the correlations illustrated? Use a **correlation matrix** like the one shown in Table 5.2—a simple and elegant solution.

Table 5.2 Correlation Matrix

	Income	Education	Attitude	Vote
Income	1.00	.574	−.08	−.291
Education	.574	1.00	−.149	−.199
Attitude	−.08	−.149	1.00	−.169
Vote	−.291	−.199	−.169	1.00

As you can see, there are four variables in the matrix: level of income (Income), level of education (Education), attitude toward voting (Attitude), and whether the individual voted in the most recent election (Vote).

For each pair of variables, there is a correlation coefficient. For example, the correlation between income level and education is .574. Similarly, the correlation between income level and whether the person participated in the most recent election is −.291 (meaning that the higher the level of income, the less likely people are to vote).

In such a matrix with four variables, there are always 4!/(4 − 2)!2!, or four things taken two at a time for a total of six correlation coefficients. Because variables correlate perfectly with themselves (those are the 1.00s down the diagonal), and because the correlation between Income and Vote is the same as the correlation between Vote and Income, the matrix creates a mirror image of itself.

You can use SPSS—or almost any other statistical analysis package, such as Excel—to easily create a matrix like the one you saw earlier. In applications like Excel, you can use the Data Analysis ToolPak.

You will see such matrices (the plural of *matrix*) when you read journal articles that use correlations to describe the relationships among several variables.

UNDERSTANDING WHAT THE CORRELATION COEFFICIENT MEANS

Well, we have this numerical index of the relationship between two variables, and we know that the higher the value of the correlation (regardless of its sign), the stronger the relationship is. But because the correlation coefficient is a value that is not directly tied to the value of an outcome, just how can we interpret it and make it a more meaningful indicator of a relationship?

Here are different ways to look at the interpretation of that simple r_{xy}.

Using-Your-Thumb (or Eyeball) Method

Perhaps the easiest (but not the most informative) way to interpret the value of a correlation coefficient is by eyeballing it and using the information in Table 5.3.

So, if the correlation between two variables is .5, you could safely conclude that the relationship is a moderate one—not strong, but certainly not weak enough to say that the variables in question don't share anything in common.

This eyeball method is perfectly acceptable for a quick assessment of the strength of the relationship between variables, such as when you briefly evaluate data presented visually. But because this rule of thumb depends on a subjective judgment (of what's "strong" or "weak"), we would like a more precise method. That's what we'll look at now.

Table 5.3 Interpreting a Correlation Coefficient

Size of the Correlation	Coefficient General Interpretation
.8 to 1.0	Very strong relationship
.6 to .8	Strong relationship
.4 to .6	Moderate relationship
.2 to .4	Weak relationship
.0 to .2	Weak or no relationship

A DETERMINED EFFORT: SQUARING THE CORRELATION COEFFICIENT

Here's the much more precise way to interpret the correlation coefficient: computing the coefficient of determination. The **coefficient of determination** is the percentage of variance in one variable that is accounted for by the variance in the other variable. Quite a mouthful, huh?

Earlier in this chapter, we pointed out how variables that share something in common tend to be correlated with one another. If we correlated math and English grades for 100 fifth-grade students, we would find the correlation to be moderately strong, because many of the reasons why children do well (or poorly) in math tend to be the same reasons why they do well (or poorly) in English. The number of hours they study, how bright they are, how interested their parents are in their schoolwork, the number of books they have at home, and more are all related to both math and English performance and account for differences between children (and that's where the variability comes in).

The more these two variables share in common, the more they will be related. These two variables share variability—or the reason why children differ from one another. And on the whole, the brighter child who studies more will do better.

To determine exactly how much of the variance in one variable can be accounted for by the variance in another variable, the coefficient of determination is computed by squaring the correlation coefficient.

For example, if the correlation between GPA and number of hours of study time is .70 (or $r_{GPA \cdot time}$ = .70), then the coefficient of determination, represented by $r^2_{GPA \cdot time}$, is $.7^2$, or .49. This means that 49% of the variance in GPA can be explained by the variance in studying time. And the stronger the correlation, the more variance can be explained (which only makes good sense). The more two variables share in common (such as good study habits, knowledge of what's expected in class, and lack of fatigue), the more information about performance on one score can be explained by the other score.

However, if 49% of the variance can be explained, this means that 51% cannot—so even for a strong correlation of .70, many of the reasons why scores on these variables tend to be different from one another go unexplained. This amount of unexplained variance is called the **coefficient of alienation** (also called the **coefficient of nondetermination**). Don't worry. No aliens here. This isn't *X-Files* or *Walking Dead* stuff—it's just the amount of variance in Y not explained by X (and, of course, vice versa since the relationship goes both ways).

How about a visual presentation of this sharing variance idea? Okay. In Figure 5.5, you'll find a correlation coefficient, the corresponding coefficient of determination, and a diagram that represents how much variance is shared between the two variables. The larger the shaded area in each diagram (and the more variance the two variables share), the more highly the variables are correlated.

- The first diagram in Figure 5.5 shows two circles that do not touch. They don't touch because they do not share anything in common. The correlation is zero.

- The second diagram shows two circles that overlap. With a correlation of .5 (and $r^2_{xy} = .25$), they share about 25% of the variance between them.
- Finally, the third diagram shows two circles placed almost on top of each other. With an almost perfect correlation of $r_{xy} = .90$ ($r^2_{xy} = .81$), they share about 81% of the variance between them.

| Figure 5.5 | How Variables Share Variance and the Resulting Correlation |

Correlation	Coefficient of Determination	Variable *X*		Variable *Y*
$r_{xy} = 0$	$r^2_{xy} = 0$	◯	0% shared	◯
$r_{xy} = .5$	$r^2_{xy} = .25$ or 25%		25% shared	
$r_{xy} = .9$	$r^2_{xy} = .81$ or 81%		81% shared	

As More Ice Cream Is Eaten . . . the Crime Rate Goes Up (or Association vs. Causality)

Now here's the really important thing to be careful about when computing, reading about, or interpreting correlation coefficients.

Imagine this. In a small Midwestern town, a phenomenon occurred that defied any logic. The local police chief observed that as ice cream consumption increased, crime rates tended to increase as well. Quite simply, if you measured both, you would find the relationship was direct, meaning that as people eat more ice cream, the crime rate increases. And as you might expect, as they eat less ice cream, the crime rate goes down. The police chief was baffled until he recalled the Stats 1 class he took in college and still fondly remembered.

His wondering how this could be turned into an aha! "Very easily," he thought. The two variables must share something or have something in common with one another. Remember that it must be something that relates to both level of ice cream consumption and level of crime rate. Can you guess what that is?

The *outside temperature* is what they both have in common. When it gets warm outside, such as in the summertime, more crimes are committed (it stays light longer, people leave the windows open, bad guys and girls are out more, etc.). And because it is warmer, people enjoy the ancient treat and art of eating ice cream. Conversely, during the long and dark winter months, less ice cream is consumed and fewer crimes are committed as well.

Joe, recently elected as a city commissioner, learns about these findings and has a great idea, or at least one that he thinks his constituents will love. (Keep in mind, he skipped the statistics offering in college.) Why not just limit the consumption of ice cream in the summer months to reduce the crime rate? Sounds good, right? Well, on closer inspection, it really makes no sense at all.

That's because of the simple principle that correlations express the *association* that exists between two or more variables; they have nothing to do with *causality*. In other words, just because level of ice cream consumption and crime rate increase together (and decrease together as well) does not mean that a change in one results in a change in the other.

For example, if we took all the ice cream out of all the stores in town and no more was available, do you think the crime rate would decrease? Of course not, and it's preposterous to think so. But strangely enough, that's often how associations are interpreted—as being causal in nature—and complex issues in the social and behavioral sciences are reduced to trivialities because of this misunderstanding. Did long hair and hippiedom have anything to do with the Vietnam conflict? Of course not. Does the rise in the number of crimes committed have anything to do with more efficient and safer cars? Of course not. But they all happen at the same time, creating the illusion of being associated.

Using SPSS to Compute a Correlation Coefficient

Let's use SPSS to compute a correlation coefficient. The data set we are using is an SPSS data file named Chapter 5 Data Set 1.

There are two variables in this data set:

Variable	Definition
Income	Annual income in thousands of dollars
Education	Level of education measured in years

To compute the Pearson correlation coefficient, follow these steps:

1. Open the file named Chapter 5 Data Set 1.

2. Click Analyze → Correlate → Bivariate, and you will see the Bivariate Correlations dialog box, as shown in Figure 5.6.

3. Double-click on the variable named Income to move it to the Variables: box.

4. Double-click on the variable named Education to move it to the Variables: box. You can also hold down the Ctrl key to select more than one variable at a time and then use the "move" arrow in the center of the dialog box to move them both.

5. Click OK.

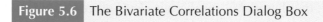

Figure 5.6 The Bivariate Correlations Dialog Box

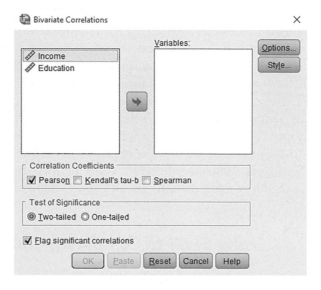

Understanding the SPSS Output

The output in Figure 5.7 shows the correlation coefficient to be equal to .574. Also shown are the sample size, 20, and a measure of the statistical significance of the correlation coefficient (we'll cover the topic of statistical significance in Chapter 9).

Figure 5.7 SPSS Output for the Computation of the Correlation Coefficient

Correlations

		Education	Income
Education	Pearson Correlation	1	.574**
	Sig. (2-tailed)		.008
	N	20	20
Income	Pearson Correlation	.574**	1
	Sig. (2-tailed)	.008	
	N	20	20

. Correlation is significant at the 0.01 level (2-tailed).

The SPSS output shows that the two variables are related to one another and that as level of income increases, so does level of education. Similarly, as level of income decreases, so does level of education. The fact that the correlation is significant means that this relationship is not due to chance.

As for the meaningfulness of the relationship, the coefficient of determination is $.574^2$ or .329 or .33, meaning that 33% of the variance in one variable is accounted for by the other. According to our eyeball strategy, this is a relatively weak relationship. Once again, remember that low levels of income do not cause low levels of education nor does not finishing high school mean that someone is destined to a life of low income. That's causality, not association, and correlations speak only to association.

Creating a Scatterplot (or Scattergram or Whatever)

You can draw a scatterplot by hand, but it's good to know how to have SPSS do it for you as well. Let's take the same data that we just used to produce the correlation matrix in Figure 5.7 and use it to create a scatterplot. Be sure that the data set named Chapter 5 Data Set 1 is on your screen.

1. Click Graphs → Chart Builder → Scatter/Dot, and you will see the Chart Builder dialog box shown in Figure 5.8.

2. Double-click on the first Scatter/Dot example.

3. Highlight and drag the variable named Income to the *y*-axis.

4. Highlight and drag the variable named Education to the *x*-axis.

5. Click OK, and you'll have a very nice, simple, and easy-to-understand scatterplot like the one you see in Figure 5.9.

| **Figure 5.8** | The Chart Builder Dialog Box |

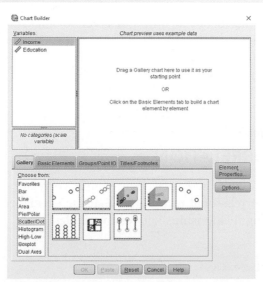

Computing a Scatterplot

| Figure 5.9 | A Simple Scatterplot |

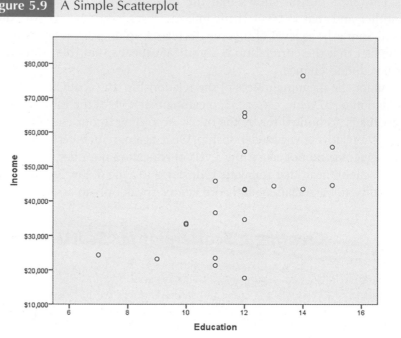

OTHER COOL CORRELATIONS

Other
Important
Correlation
Coefficients

There are different ways in which variables can be assessed. For example, nominal-level variables are categorical in nature; examples are race (black or white) and political affiliation (Independent or Republican). Or, if you are measuring income and age, you are measuring interval-level variables, because the underlying continuum on which they are based has equally appearing intervals. As you continue your studies, you're likely to come across correlations between data that occur at different levels of measurement. And to compute these correlations, you need some specialized techniques. Table 5.4 summarizes what these different techniques are and how they differ from one another.

PARTING WAYS: A BIT ABOUT PARTIAL CORRELATION

Okay, now you have the basics about simple correlation, but there are many other correlational techniques that are specialized tools to use when exploring relationships between variables.

| Table 5.4 | Correlation Coefficient Shopping, Anyone? |

Level of Measurement and Examples			
Variable X	**Variable Y**	**Type of Correlation**	**Correlation Being Computed**
Nominal (voting preference, such as Republican or Democrat)	Nominal (sex, such as male or female)	Phi coefficient	The correlation between voting preference and sex
Nominal (social class, such as high, medium, or low)	Ordinal (rank in high school graduating class)	Rank biserial coefficient	The correlation between social class and rank in high school
Nominal (family configuration, such as two-parent or single-parent)	Interval (grade point average)	Point biserial	The correlation between family configuration and grade point average
Ordinal (height converted to rank)	Ordinal (weight converted to rank)	Spearman rank coefficient	The correlation between height and weight
Interval (number of problems solved)	Interval (age in years)	Pearson correlation coefficient	The correlation between number of problems solved and age in years

A common "extra" tool is called **partial correlation**, where the relationship between two variables is explored, but the impact of a third variable is removed from the relationship between the two. Sometimes that third variable is called a *mediating* or a *confounding* variable.

For example, let's say that we are exploring the relationship between level of depression and incidence of chronic disease and we find that, on the whole, the relationship is positive. In other words, the more chronic disease is evident, the higher the likelihood that depression is present as well (and of course vice versa). Now remember, one variable does not "cause" the other, and the presence of one does not mean that the other will be present as well. The positive correlation is just an assessment of the relationship between these two variables, the key idea being that they share some variance in common.

And that's exactly the point—it's what they share in common that we want to control and, in some cases, remove from the relationship.

For example, how about level of family support? Nutritional habits? Severity or length of illness? These and many more variables can all be responsible for the relationship between these two variables, or they may at least account for some of the variance.

And think back a bit. That's exactly the same argument we made when focusing on the relationship between consumption of ice cream and level of crime. Once outside temperature (the mediating or confounding variable) is removed from the equation . . . boom! The relationship between consumption of ice cream and crime level plummets. Let's take a look.

Here are some data on consumption of ice cream and crime rate for 10 cities.

	Consumption of Ice Cream	Crime Rate
Consumption of Ice Cream	1.00	.743
Crime Rate		1.00

So, the correlation between these two variables, consumption of ice cream and crime rate, is .743. This is a pretty healthy relationship, accounting for about 50% of the variance between the two variables ($.743^2 = .55$ or 55%).

Now, we'll add a third variable, average outside temperature. Here are the Pearson correlation coefficients for the set of three variables.

	Consumption of Ice Cream	Crime Rate	Average Outside Temperature
Consumption of Ice Cream	1.00	.743	.704
Crime Rate		1.00	.655
Average Outside Temperature			1.00

As you can see by these values, there's a fairly strong relationship between ice cream consumption and outside temperature and between crime rate and outside temperature. We're interested in the question, "What's the correlation between ice cream consumption and crime rate with the effects of outside temperature removed or *partialed out*."

That's what partial correlation does. It looks at the relationship between two variables (in this case, consumption of ice cream and crime rate) as it removes the influence of a third (in this case, outside temperature).

Using SPSS to Compute Partial Correlations

Let's use some data and SPSS to illustrate the computation of a partial correlation. Here are the raw data.

City	Ice Cream Consumption	Crime Rate	Average Outside Temperature
1	3.4	62	88
2	5.4	98	89
3	6.7	76	65
4	2.3	45	44
5	5.3	94	89
6	4.4	88	62
7	5.1	90	91
8	2.1	68	33
9	3.2	76	46
10	2.2	35	41

1. Click Analyze → Correlate → Partial and you will see the Partial Correlations dialog box, as shown in Figure 5.10.

2. Move Ice_Cream and Crime_Rate to the Variables: box by dragging them or double-clicking on each one.

3. Move the variable named Outside_Temp to the Controlling for: box.

4. Click OK and you will see the SPSS output as shown in Figure 5.11.

Figure 5.10 The Partial Correlations Dialog Box

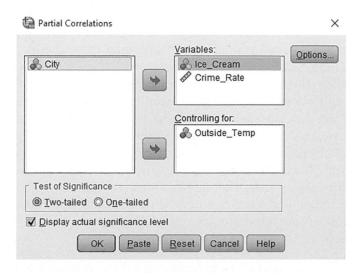

Understanding the SPSS Output

As you can see in Figure 5.11, the correlation between ice cream consumption (Ice_Cream) and crime rate (Crime_Rate) with the influence or moderation of outside temperature (Outside_Temp) removed is .525. This is less than the simple Pearson correlation between ice cream consumption and crime rate (which is .743), which does not consider the influence

Figure 5.11 The Completed Partial Correlation Analysis

Correlations

Control Variables			Ice_Cream	Crime_Rate
Outside_Temp	Ice_Cream	Correlation	1.000	.525
		Significance (2-tailed)	.	.147
		df	0	7
	Crime_Rate	Correlation	.525	1.000
		Significance (2-tailed)	.147	.
		df	7	0

of outside temperature. What seemed to explain 55% of the variance (and was what we call "significant at the .05 level"), with the removal of Outside_Temp as a moderating variable now explains $.525^2 = 0.28 = 28\%$ of the variance (and the relationship is no longer significant).

Our conclusion? Outside temperature accounted for enough of the shared variance between consumption of ice cream and crime rate for us to conclude that the two-variable relationship was significant. But, with the removal of the moderating or confounding variable outside temperature, the relationship was no longer significant. And we don't need to stop selling ice cream to try to reduce crime.

REAL-WORLD STATS

This is a fun one and consistent with the increasing interest in using statistics in various sports in various ways, a discipline informally named *sabermetrics*. The term was coined by Bill James (of *Moneyball* fame).

Stephen Hall and his colleagues examined the link between teams' payrolls and the competitiveness of those teams (for both professional baseball and soccer), and he was one of the first to look at this from an empirical perspective. In other words, until these data were published, most people made decisions based on anecdotal evidence rather than quantitative assessments. Hall looked at data on team payrolls in Major League Baseball and English soccer between 1980 and 2000, and he used a model that allows for the establishment of causality (and not just association) to examine the link.

While in baseball, payroll and performance both increased significantly in the 1990s, there is no evidence that causality runs in the direction from payroll to performance. In comparison, for English soccer, the researchers did show that higher payrolls caused better performance. Pretty cool, isn't it, how association can be explored using special causal models?

Want to know more? Go online or to the library and find . . .

Hall, S., Szymanski, S., & Zimbalist, A. S. (2002). Testing causality between team performance and payroll: The cases of Major League Baseball and English soccer. *Journal of Sports Economics, 3,* 149–168.

Real-World Stats

SUMMARY

The idea of showing how things are related to one another and what they have in common is a very powerful one, and the correlation coefficient is a very useful descriptive statistic (one used in inference as well, as we will show you later). Keep in mind that correlations express a relationship that is only associative and not causal, and you'll be able to understand how this statistic gives us valuable information about relationships between variables and how variables change or remain the same in concert with others. Now it's time to change speeds just a bit and wrap up Part II with a focus on reliability and validity. You need to know about these ideas because you'll be learning about how to determine what differences in outcomes, such as scores and other variables, represent.

Chapter Summary

TIME TO PRACTICE

1. Use these data to answer Questions 1a and 1b. These data are saved as Chapter 5 Data Set 2.

   a. Compute the Pearson product-moment correlation coefficient by hand and show all your work.

   b. Construct a scatterplot for these 10 pairs of values by hand. Based on the scatterplot, would you predict the correlation to be direct or indirect? Why?

Problem 1

Number Correct (out of a possible 20)	Attitude (out of a possible 100)
17	94
13	73
12	59

(Continued)

(Continued)

Number Correct (out of a possible 20)	Attitude (out of a possible 100)
15	80
16	93
14	85
16	66
16	79
18	77
19	91

2. Use these data to answer Questions 2a and 2b. These data are saved as Chapter 5 Data Set 3.

Problem 2

Speed (to complete a 50-yard swim)	Strength (number of pounds bench-pressed)
21.6	135
23.4	213
26.5	243
25.5	167
20.8	120
19.5	134
20.9	209
18.7	176
29.8	156
28.7	177

a. Using either a calculator or a computer, compute the Pearson correlation coefficient.

b. Interpret these data using the general range of very weak to very strong. Also compute the coefficient of determination. How does the subjective analysis compare with the value of r^2?

3. Rank the following correlation coefficients on strength of their relationship (list the weakest first).

    71

    +.36

−.45

.47

−.62

4. For the following set of scores, calculate the Pearson correlation coefficient and interpret the outcome. These data are saved as Chapter 5 Data Set 4.

Achievement Increase Over 12 Months	Classroom Budget Increase Over 12 Months
0.07	0.11
0.03	0.14
0.05	0.13
0.07	0.26
0.02	0.08
0.01	0.03
0.05	0.06
0.04	0.12
0.04	0.11

5. For the following set of data, by hand, correlate minutes of exercise with GPA. What do you conclude given your analysis? These data are saved as Chapter 5 Data Set 5.

Exercise	GPA
25	3.6
30	4.0
20	3.8
60	3.0
45	3.7
90	3.9
60	3.5
0	2.8
15	3.0
10	2.5

Problem 6

6. Use SPSS to determine the correlation between hours of studying and grade point average for these honor students. Why is the correlation so low?

Hours of Studying	GPA
23	3.95
12	3.90
15	4.00
14	3.76
16	3.97
21	3.89
14	3.66
11	3.91
18	3.80
9	3.89

7. The coefficient of determination between two variables is .64. Answer the following questions:

   a. What is the Pearson correlation coefficient?

   b. How strong is the relationship?

   c. How much of the variance in the relationship between these two variables is unaccounted for?

8. Here is a set of 3 variables for each of 20 participants in a study on recovery from a head injury. Create a simple matrix that shows the correlations between each variable. You can do this by hand (and plan on being here for a while) or use SPSS or any other application. These data are saved as Chapter 5 Data Set 6.

Age at Injury	Level of Treatment	12-Month Treatment Score
25	1	78
16	2	66
8	2	78
23	3	89
31	4	87
19	4	90
15	4	98
31	5	76

Age at Injury	Level of Treatment	12-Month Treatment Score
21	1	56
26	1	72
24	5	84
25	5	87
36	4	69
45	4	87
16	4	88
23	1	92
31	2	97
53	2	69
11	3	79
33	2	69

9. Look at Table 5.4 on page 99. What type of correlation coefficient would you use to examine the relationship between sex (defined as male or female) and political affiliation? How about family configuration (two-parent or single-parent) and high school GPA? Explain why you selected the answers you did.

10. When two variables are correlated (such as strength and running speed), they are associated with one another. But if they are associated with one another, then why doesn't one cause the other?

11. Provide three examples of an association between two variables where a causal relationship makes perfect sense conceptually but, since correlations do not imply causality, makes little sense statistically until further examination.

12. Why can't correlations be used as a tool to prove a causal relationship between variables, rather than just an association?

13. When would you use partial correlation?

STUDENT STUDY SITE

⑤SAGE edge™

Get the tools you need to sharpen your study skills! Visit **edge.sagepub.com/salkind6e** to access practice quizzes, eFlashcards, original and curated videos, data sets, journal articles, and more!

6 Just the Truth

An Introduction to Understanding Reliability and Validity

Difficulty Scale ☺ ☺ ☺ (not so hard)

WHAT YOU'LL LEARN ABOUT IN THIS CHAPTER

✦ Defining reliability and validity and understanding why they are important

✦ This is a stats class! What's up with this measurement stuff?

✦ Understanding scales of measurement

✦ Computing and interpreting various types of reliability coefficients

✦ Computing and interpreting various types of validity coefficients

---

AN INTRODUCTION TO RELIABILITY AND VALIDITY

Ask any parent, teacher, pediatrician, or almost anyone in your neighborhood what the five top concerns are about today's children, and there is sure to be a group who identifies obesity as one of those concerns. Sandy Slater and her colleagues developed and tested the reliability and validity of a self-reported questionnaire on home, school, and neighborhood physical activity environments for youth located in low-income urban minority neighborhoods and rural areas. In particular, the researchers looked at such variables as information on the presence of electronic and play equipment in youth participants' bedrooms and homes and outdoor play equipment at schools. They also looked at what people close to the children thought about being active. A total of 205 parent–child pairs

Introduction to Chapter 6

109

completed a 160-item take-home survey on two different occasions, a perfect model for establishing test–retest reliability. The researchers found that 90% of the measures had good reliability and validity. The researchers hope that this survey can be used to help identify opportunities and develop strategies to encourage underserved youth to be more physically active.

Want to know more? Go online or to the library and find . . .

Slater, S., Full, K., Fitzgibbon, M., & Uskali, A. (2015, June 4). Test–retest reliability and validity results of the Youth Physical Activity Supports Questionnaire. *SAGE Open, 5*(2). doi: 10.1177/2158244015586809

What's Up With This Measurement Stuff?

Measurement

An excellent question, and one that you should be asking. After all, you enrolled in a stats class, and up to now, that's been the focus of the material that has been covered. Now it looks like you're faced with a topic that belongs in a tests and measurements class. So, what's this material doing in a stats book?

Well, much of what we have covered so far in *Statistics for People Who (Think They) Hate Statistics* has to do with the collection and description of data. Now we are about to begin the journey toward analyzing and interpreting data. But before we begin learning those skills, we want to make sure that the data are what you think they are—that the data represent what it is you want to know about. In other words, if you're studying poverty, you want to make sure that the measure you use to assess poverty works, and that it works time after time. Or, if you are studying aggression in middle-aged males, you want to make sure that whatever tool you use to assess aggression works, and that it works time after time.

More really good news: Should you continue in your education and want to take a class on tests and measurements, this introductory chapter will give you a real jump on understanding the scope of the area and what topics you'll be studying.

And to make sure that the entire process of collecting data and making sense out of them works, you first have to make sure that what you use to collect data works as well. The fundamental questions that will be answered in this chapter are "How do I know that the test, scale, instrument, etc., I use works every time I use it?" (that's reliability) and "How do I know that the test, scale, instrument, etc., I use measures what it is supposed to?" (that's validity).

Anyone who does research will tell you about the importance of establishing the reliability and validity of your test tool, whether it's a simple observational instrument of consumer behavior or one that measures a complex psychological construct such as attachment. However, there's another very good reason. If the tools that you use to collect data are unreliable or invalid, then the results of any test or any hypothesis, and the conclusions you may reach based on those results, are necessarily inconclusive. If you are not sure that the test does what it is supposed to and that it does so consistently, how do you know that the nonsignificant results you got aren't a function of the lousy test tools rather than an actual rejection of the null hypothesis when it is true (the Type I error friend you will learn about in Chapter 9)? Want a clean test of the null? Make reliability and validity an important part of your research.

You may have noticed a new term at the beginning of this chapter—**dependent variable**. In an experiment, this is the outcome variable, or what the researcher looks at to see whether any change has occurred as a function of the treatment that has taken place. And guess what? The treatment has a name as well—the **independent variable**. For example, if a researcher examined the effect of different reading programs on comprehension, the independent variable would be the reading program, and the dependent or outcome variable would be reading comprehension score. Although these terms will not be used often throughout the remainder of *Statistics for People* . . . , you should have some familiarity with them.

RELIABILITY: DOING IT AGAIN UNTIL YOU GET IT RIGHT

Reliability is pretty easy to understand and figure out. It's simply whether a test, or whatever you use as a measurement tool, measures something consistently. If you administer a test of personality type before a special treatment occurs, will the administration of that same test 4 months later be reliable? That, my friend, is one of the questions. And that is one reason why there are different types of reliability, each of which we will get to after we define reliability just a bit more.

Reliability

Test Scores: Truth or Dare?

When you take a test in this class, you get a score, such as 89 (good for you) or 65 (back to the books!). That test score consists of several elements, including the **observed score** (or what you actually get on the test, such as 89 or 65) and a **true score** (the true, 100% accurate reflection of what you *really* know). We can't directly measure true score because it is a theoretical reflection of the actual amount of the trait or characteristic possessed by the individual.

Nothing about this tests and measurement stuff is clear-cut, and this true score stuff surely qualifies. Here's why. We just defined true score as the real, real, real value associated with some trait or attribute. So far, so good. But there's another point of view as well. Some psychometricians (the people who do tests and measurement for a living) believe that true score has nothing to do with whether the construct of interest is really being reflected. Rather, true score is the *mean* score an individual would get if he or she took a test an infinite number of times, and it represents the theoretical typical level of performance on a given test. Now, one would hope that the typical level of performance would reflect the construct of interest, but that's another question (a validity one at that). The distinction here is that a test is reliable if it consistently produces whatever score a person would get on average, regardless of what the test is measuring. In fact, a perfectly reliable test might not produce a score that has anything to do with the construct of interest, such as "what you really know."

Why aren't true scores and observed scores the same? Well, they can be if the test (and the accompanying observed score) is a perfect (and we mean absolutely perfect) reflection of what's being measured.

But the Yankees don't always win, the bread sometimes falls on the buttered side, and Murphy's law tells us that the world is not perfect. So, what you see as an observed score may come close to the true score, but rarely are they the same. Rather, the difference as you see is the amount of error that is introduced.

Observed Score = True Score + Error Score

Error? Yes—in all its glory. For example, let's suppose for a moment that someone gets an 89 on a stats test, but his or her true score (which we never really know but only theorize about) is 80. That means the

9-point difference (that's the **error score**) is due to error, or the reason why individual test scores vary from being 100% true.

What might be the source of such error? Well, perhaps the room in which the test was taken was so warm that it caused some students to fall asleep. That would certainly have an impact on a test score. Or perhaps the test taker didn't study for the test as much as he or she should have. Ditto. Both of these examples would reflect testing situations or conditions rather than qualities of the trait being measured, right?

Our job is to reduce those errors as much as possible by having, for example, good test-taking conditions and making sure the test takers are encouraged to get enough sleep. Reduce the error and you increase the reliability, because the observed score more closely matches the true score.

The less error, the more reliable—it's that simple.

DIFFERENT TYPES OF RELIABILITY

There are several types of reliability, and we'll cover the four most important and most often used in this section. They are all summarized in Table 6.1.

| Table 6.1 | Different Types of Reliability, When They Are Used, How They Are Computed, and What They Mean |

Type of Reliability	When You Use It	How You Do It	An Example of What You Can Say When You're Done
Test–retest reliability	When you want to know whether a test is reliable over time	Correlate the scores from a test given at Time 1 with the same test given at Time 2.	The Bonzo test of identity formation for adolescents is reliable over time.
Parallel forms reliability	When you want to know if several different forms of a test are reliable or equivalent	Correlate the scores from one form of the test with the scores from a second form of the same test of the same content (but not the exact same test).	The two forms of the Regular Guy test are equivalent to one another and have shown parallel forms reliability.
Internal consistency reliability	When you want to know if the items on a test assess one, and only one, dimension	Correlate each individual item score with the total score.	All of the items on the SMART Test of Creativity assess the same construct.
Interrater reliability	When you want to know whether there is consistency in the rating of some outcome	Examine the percentage of agreement between raters.	The interrater reliability for the Best-Dressed Football Player judging was .91, indicating a high degree of agreement between judges.

Test–Retest Reliability

Test-Retest
Reliability

Test–retest reliability is used when you want to examine whether a test is reliable over time.

For example, let's say that you are developing a test that will examine preferences for different types of vocational programs. You may administer the test in September and then readminister the same test (and it's important for it to be the same test) again in June. Then, the two sets of scores (remember, the same people took it twice) are correlated (just like we did in Chapter 5), and you have a measure of reliability. Test–retest reliability is a must when you are examining differences or changes over time.

You must be very confident that what you are measuring has been measured in a reliable way such that the results you are getting come as close as possible to the individual's score each and every time.

Computing Test–Retest Reliability. Here are some scores from a test at Time 1 and Time 2 for the Mastering Vocational Education (MVE) test, which is under development. Our goal is to compute the Pearson correlation coefficient as a measure of the test–retest reliability of the instrument.

ID	Scores From Time 1	Scores From Time 2
1	54	56
2	67	77
3	67	87
4	83	89
5	87	89
6	89	90
7	84	87
8	90	92
9	98	99
10	65	76

The first and last step in this process is to compute the Pearson product-moment correlation (see Chapter 5 for a refresher on this), which is equal to

$$r_{\text{Time1·Time2}} = .90$$

What does .90 mean as far as test–retest reliability? We'll get to the interpretation of this value shortly.

Parallel Forms Reliability

Parallel forms reliability is used when you want to examine the equivalence or similarity between two different forms of the same test.

For example, let's say that you are doing a study on memory and part of the task is to look at 10 different words, memorize them as best you can, and then recite them back after 20 seconds of study and 10 seconds of rest. Because this study takes place over a 2-day period and involves some training of memory skills, you want to have another set of items that is exactly similar in task demands, but it obviously cannot be the same as far as content. So, you create another list of words that is hopefully similar to the first. In this example, you want the consistency to be high across forms—the same ideas are being tested, just using a different form.

Computing Parallel Forms Reliability. Here are some scores from the I Remember Memory Test (IRMT) on Form A and Form B. Our goal is to compute the Pearson correlation coefficient as a measure of the parallel forms reliability of the instrument.

ID	Scores From Form A	Scores From Form B
1	4	5
2	5	6
3	3	5
4	6	6
5	7	7
6	5	6
7	6	7
8	4	8
9	3	7
10	3	7

The first and last steps in this process are to compute the Pearson product-moment correlation (again, see Chapter 5 for a refresher on this), which is equal to

$$r_{\text{FormA·FormB}} = .13$$

We'll get to the interpretation of this value shortly.

Internal Consistency Reliability

Internal consistency reliability is quite different from the two previous types that we have explored. It is used when you want to know whether the items on a test are consistent with one another in that they represent one—and only one—dimension, construct, or area of interest.

Let's say that you are developing a test of attitudes toward different types of health care and you want to make sure that the set of five items measures just that—and nothing else. You would look at the score for each item (for a group of test takers) and see if the individual score correlates with the total score. You would expect that people who scored high on certain items (e.g., "I like my HMO") would score low on others (e.g., "I don't like spending money on health care") and that this would be consistent across all the people who took the test.

Cronbach's alpha (or α) is a special measure of reliability known as internal consistency. The more consistently individual item scores vary with the total score on the test, the higher the value of Cronbach's alpha. And the higher the value, the more confidence you can have that this test is internally consistent, or measures one thing, and that one thing is the sum of what each item evaluates.

For example, here's a five-item test that has lots of internal consistency:

1. $4 + 4 = ?$

2. $5 - ? = 3$

3. $6 + 2 = ?$

4. $8 - ? = 3$

5. $1 + 1 = ?$

All of the items seem to measure the same thing, regardless of what that same thing is (which is a validity question—stay tuned).

Now, here's a five-item test that doesn't quite come up to speed as far as being internally consistent:

1. $4 + 4 = ?$

2. Who is the fattest of the three little pigs?

3. $6 + 2 = ?$

4. $8 - ? = 3$

5. So, just what did the wolf want?

It's obvious why. These questions are inconsistent with one another— the key criterion for internal consistency.

Computing Cronbach's Alpha

Here are some sample data for 10 people on this five-item attitude test (the I♥HMO test) where scores are between 1 (*strongly disagree*) and 5 (*strongly agree*) on each item.

 When you compute Cronbach's alpha (named after Lee Cronbach), you are actually correlating the score for each item with the total score for each individual, and then comparing that with the variability present for all individual item scores. The logic is that any individual test taker with a high total test score should have a high(er) score on each item (such as 5, 5, 3, 5, 3, 4, 4, 2, 4, 5 for a total score of 40) and that any individual test taker with a low(er) total test score should have a low(er) score on each individual item (such as 4, 1, 2, 1, 3, 2, 4, 1, 2, 1).

ID	Item 1	Item 2	Item 3	Item 4	Item 5
1	3	5	1	4	1
2	4	4	3	5	3
3	3	4	4	4	4
4	3	3	5	2	1
5	3	4	5	4	3
6	4	5	5	3	2
7	2	5	5	3	4
8	3	4	4	2	4
9	3	5	4	4	3
10	3	3	2	3	2

And here's the formula to compute Cronbach's alpha:

$$\alpha = \left(\frac{k}{k-1} \right) \left(\frac{s_y^2 - \Sigma s_i^2}{s_y^2} \right), \tag{6.1}$$

where

- k = the number of items;
- s_y^2 = the variance associated with the observed score; and
- Σs_i^2 = the sum of all the variances for each item.

Here's the same set of data with the values (the variance associated with the observed score, or s_y^2, and the sum of all the variances for each item, or Σs_i^2) needed to complete the preceding equation.

ID	Item 1	Item 2	Item 3	Item 4	Item 5	Total Score
1	3	5	1	4	1	14
2	4	4	3	5	3	19
3	3	4	4	4	4	19
4	3	3	5	2	1	14
5	3	4	5	4	3	19
6	4	5	5	3	2	19
7	2	5	5	3	4	19
8	3	4	4	2	4	17
9	3	5	4	4	3	19
10	3	3	2	3	2	13
						$s_y^2 = 6.4$
Item Variance	0.32	0.62	1.96	0.93	1.34	$\Sigma s_i^2 = 5.17$

And when you plug all these figures into the equation and get the following . . .

$$\alpha = \left(\frac{5}{5-1} \right) \left(\frac{6.40 - 5.17}{6.4} \right) = .24 \qquad (6.2)$$

you find that coefficient alpha is .24 and you're done (except for the interpretation—that comes later!).

If we told you that there were many other types of internal consistency validity, you would not be surprised, right? This is especially true for measures of internal consistency. Not only is there coefficient alpha, but there are also split-half reliability, Spearman–Brown, Kuder–Richardson 20 and 21 (KR_{20} and KR_{21}), and still others that basically do the same thing—examine the one-dimensional nature of a test—only in different ways.

Using SPSS to Calculate Cronbach's Alpha

Once you know how to compute Cronbach's alpha by hand and want to move on to using SPSS, the transition is very easy. We are using the data set shown above, the five-item test with 10 people's responses.

1. Enter the data in the Data Editor. Be sure that there is a separate column for each test item.

2. Click Analyze → Scale → Reliability Analysis, and you will see the dialog box shown in Figure 6.1.

3. Move each of the variables (Item 1 through Item 5) to the Items box by double-clicking on one at a time or selecting them all (using the Shift+Click technique) and move them over to the Items: area of the dialog box. Be sure, under Model, that Alpha is selected from the drop-down menu.

4. Click OK. SPSS will conduct the analysis and produce the output you see in Figure 6.2.

Figure 6.1 The Reliability Analysis Dialog Box

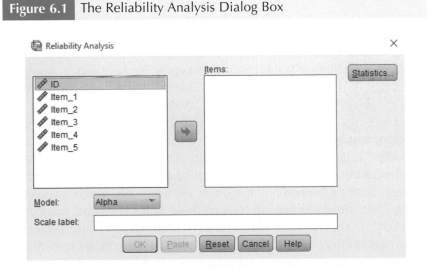

Understanding the SPSS Output

For the data set that you see on page 118, for five items across 10 individuals, the value of Chronbach's alpha is .239. As you can see, that is uncannily close to what was done manually. The output does not reveal a whole lot more than this. As you will learn later, this is a relatively low degree of reliability.

Computing Internal Consistency Estimates

Figure 6.2	SPSS Output for the Reliability Analysis

Scale: ALL VARIABLES

Case Processing Summary

		N	%
Cases	Valid	10	100.0
	Excluded[a]	0	.0
	Total	10	100.0

a. Listwise deletion based on all variables in the procedure.

Reliability Statistics

Cronbach's Alpha	N of Items
.239	5

Interrater Reliability

Interrater reliability is the measure that tells you how much two raters agree on their judgments of some outcome.

For example, let's say you are interested in a particular type of social interaction during a transaction between a banker and a potential checking-account customer, and you observe both people in real time (you're observing behind a one-way mirror) to see if the new and improved customer relations course that the banker took resulted in increased smiling and pleasant types of behavior toward the potential customer. Your job is to note every 10 seconds whether the banker is demonstrating one of the three different behaviors he has been taught—smiling, leaning forward in his chair, or using his hands to make a point. Each time you see any one of those behaviors, you mark it on your scoring sheet as an X. If you observe nothing, you score a dash (–).

As part of this process, and to be sure that what you are recording is a reliable measure, you will want to find out what the level of agreement is between observers as to the occurrence of these behaviors. The more similar the ratings are, the higher the level of interrater agreement and interrater reliability.

Computing Interrater Reliability

In this example, the really important variable is whether a customer-friendly act occurred within a set of 10-second time frames across

2 minutes (or twelve 10-second periods). So, what we are looking at is the rating consistency across a 2-minute period broken down into twelve 10-second periods. An *X* on the scoring sheet means that the behavior occurred, and a dash (–) means it did not.

For a total of 12 periods (and 12 possible agreements), there are 7 where both Dave and Maureen agreed that friendly behavior did take place (periods 1, 3, 4, 5, 7, 8, and 12) and 3 where they agreed it did not (periods 2, 6, and 9), for a total of 10 agreements and 2 disagreements.

	Time Period	1	2	3	4	5	6	7	8	9	10	11	12
Rater 1	Dave	X	–	X	X	X	–	X	X	–	–	X	X
Rater 2	Maureen	X	–	X	X	X	–	X	X	–	X	–	X

Interrater reliability is computed using the following simple formula:

$$Interrater\ reliability = \frac{Number\ of\ agreements}{Number\ of\ possible\ agreements}$$

And when we plug in the numbers as you see here . . .

$$Interrater\ reliability = \frac{10}{12} = .833$$

Computing
Interrater
Reliability

the resulting interrater reliability coefficient is .833.

HOW BIG IS BIG? FINALLY: INTERPRETING RELIABILITY COEFFICIENTS

Okay, now we get down to business, and guess what? Remember all you learned about interpreting the value of the correlation coefficient in Chapter 5? It's almost the same as when you interpret reliability coefficients as well, with a (little) bit of a difference.

We want only two things, and here they are . . .

- Reliability coefficients to be positive and not to be negative
- Reliability coefficients that are as large as possible (between .00 and +1.00)

So, the interpretation is left to you. The interrater reliability coefficient of .833 that we just computed is certainly high and represents a strong level of agreement between the two sets of observations. The earlier Chronbach value of .239—not so strong.

And If You Can't Establish Reliability . . . Then What?

The road to establishing the reliability of a test is not a smooth one at all—it takes a good deal of work. What if the test is not reliable?

Here are a few things to keep in mind. Remember that reliability is a function of how much error contributes to the observed score. Lower that error, and you increase the reliability.

- Make sure that the instructions are standardized and clear across all settings in which the test is administered.
- Increase the number of items or observations because the larger the sample from the universe of behaviors you are investigating, the more likely the sample is representative and reliable. This is especially true for achievement tests.
- Delete unclear items because some people will respond in one way and others will respond in a different fashion, regardless of their knowledge or ability level or individual traits.
- For an achievement test especially (such as a spelling or history test), moderate the easiness and difficulty of the test. Any test that is too difficult or too easy does not reflect an accurate picture of the test taker's performance.
- Minimize the effects of external events and standardize directions. For example, if a particularly important event, such as Mardi Gras or graduation, occurs near the time of testing, you might postpone an assessment.

Just One More Thing

The first step in creating an instrument that has sound *psychometric* (how's that for a big word?) properties is to establish its reliability (and we just spent some good time on that). Why? Well, if a test or measurement instrument is not reliable, is not consistent, and does not do the same thing time after time after time, it does not matter what it measures (and that's the validity question), right?

On the Kids Are Cool at Spelling (KACAS) test of introductory spelling, the first three items could be . . .

$$16 + 12 = ?$$

$$21 + 13 = ?$$

$$41 + 33 = ?$$

This is surely a highly reliable test, but surely not a valid one. Now that we have reliability well understood, let's move on to an introduction to validity.

VALIDITY: WHOA! WHAT IS THE TRUTH?

Validity is, most simply, the property of an assessment tool that indicates that the tool does what it says it does. A valid test is a test that measures what it is supposed to. If an achievement test is supposed to measure knowledge of history, then that's what it does. If an intelligence test is supposed to measure whatever intelligence is defined as by the test's creators, then it does just that.

Different Types of Validity

Just as there are different types of reliability, so there are different types of validity, and we'll cover the three most important categories and most often used in this section. They are all summarized in Table 6.2.

Content Validity

Content validity is the property of a test such that the test items sample the universe of items for which the test is designed. Content validity is most often used with achievement tests (e.g., everything from your first-grade spelling test to the Scholastic Aptitude Test [SAT]).

Table 6.2	Different Types of Validity, When They Are Used, How They Are Computed, and What They Mean

Type of Validity	When You Use It	How You Do It	An Example of What You Can Say When You're Done
Content validity	When you want to know whether a sample of items truly reflects an entire universe of items in a certain topic	Ask Mr. or Ms. Expert to make a judgment that the test items reflect the universe of items in the topic being measured.	My weekly quiz in my stats class fairly assesses the chapter's content.
Criterion validity	When you want to know whether test scores are systematically related to other criteria that indicate that the test taker is competent in a certain area	Correlate the scores from the test with some other measure that is already valid and that assesses the same set of abilities.	The EATS test (of culinary skills) has been shown to be correlated with being a fine chef 2 years after culinary school (an example of predictive validity).
Construct validity	When you want to know whether a test measures some underlying psychological construct	Correlate the set of test scores with some theorized outcome that reflects the construct for which the test is being designed.	It's true—men who participate in body contact and physically dangerous sports score higher on the TEST(osterone) test of aggression.

Establishing Content Validity. Establishing content validity is actually very easy. All you need to do is locate your local (and cooperative) content expert. For example, if I were designing a test of introductory physics, I would go to the local physics expert (perhaps the teacher at the local high school or a professor at the university who teaches physics), and I would say, "Hey, Albert or Alberta, do you think this set of 100 multiple-choice questions accurately reflects all the possible topics and ideas that I would expect the students in my introductory class to understand?"

I would probably tell Albert or Alberta what the topics were, and then he or she would look at the items and provide a judgment as to whether they met the criterion I established—a representation of the entire universe of all items that are introductory in physics. If the answer is yes, I'm done (at least for now). If the answer is no, it's back to the drawing board and either the creation of new items or the refinement of existing ones (until the content is deemed correct by the expert).

Criterion Validity

Criterion validity assesses whether a test reflects a set of abilities in a current or future setting. If the criterion is taking place in the here and now, we talk about **concurrent criterion validity**. If the criterion is taking place in the future, we talk about **predictive validity**. For criterion validity to be present, one need not establish both concurrent and predictive validity but only the one that works for the purposes of the test.

Establishing Concurrent Validity

For example, you've been hired by the Universal Culinary Institute to design an instrument that measures culinary skills. Some part of culinary training has to do with straight knowledge (for example, what's a roux?). And that's left to the achievement test side of things.

So, you develop a test that you think does a good job of measuring culinary skills, and now you want to establish the level of concurrent validity. To do this, you design the COOK scale (whose score can range from 1 to 100 by your design), a set of 5-point items across a set of criteria (presentation, food safety, cleanliness, etc.) that each judge will use. As a criterion (and that's the key here), you have another set of judges rank each student from 1 to 10 on overall cooking ability. Then, you simply correlate the COOK scores with the judges' rankings. If the validity coefficient (a simple correlation) is high, you're in business—if not, it's back to the drawing board.

Establishing Predictive Validity

Let's say that the cooking school has been percolating (heh-heh) along just fine for 10 years and you are interested not only in how well people cook (and that's the concurrent validity part of this exercise that you just established) but in the test's predictive validity as well. Now, the criterion changes from a here-and-now score (the one that judges give) to one that looks to the future.

Here, we are interested in developing a test that *predicts* success as a chef 10 years down the line. To establish the predictive validity of the COOK test, you go back and locate graduates of the program who have been out cooking for 10 years and administer the test to them. The criterion that is used here is their level of success, and you use as measures (a) whether they own their own restaurant and (b) whether it has been in business for more than 1 year (given that the failure rate for new restaurants is more than 80% within the first year). The rationale here is that if a restaurant is in business for more than 1 year, then the chef must be doing something right.

To complete this exercise, you correlate the COOK score with a value of 1 (if the restaurant is in business for more than a year and owned by the graduate) with the previous (10 years earlier) COOK score. A high correlation coefficient indicates predictive validity, and a low correlation coefficient indicates a lack thereof.

Construct Validity

Construct validity is the most interesting and the most difficult of all the validities to establish because it is based on some underlying construct, or idea, behind a test or measurement tool.

You may remember from your extensive studies in Psych 1 that a construct is a group of interrelated variables. For example, aggression is a construct (consisting of such variables as inappropriate touching, violence, lack of successful social interaction, etc.), as are intelligence, mother–infant attachment, and hope. And keep in mind that these constructs are generated from some theoretical position that the researcher assumes. For example, he or she might propose that aggressive men have more trouble with law enforcement officials than nonaggressive men.

Establishing Construct Validity

So, you have the FIGHT test (of aggression), an observational tool that consists of a series of items. It is an outgrowth of your theoretical view about what the construct of aggression consists of. You know from the criminology literature that males who are aggressive do certain types of things more than others—for example, they get into more arguments, they are more physically aggressive (pushing and such), they commit more crimes of violence against others, and they have fewer successful interpersonal relationships. The FIGHT scale includes items that describe different behaviors, some that are theoretically related to aggressive behaviors and some that are not.

Once the FIGHT scale is completed, you examine the results to see whether positive scores on the FIGHT scale correlate with the presence of the kinds of behaviors you would predict given your theory (level of involvement in crime, quality of personal relationships, etc.) and don't correlate with the kinds of behaviors that should not be related (such as handedness or preferences for certain types of food). And, if the correlation is high for the items that you predict should correlate and low for the items that should not, then you can conclude that there is something about the FIGHT scale (probably the items you designed to assess elements of aggression) that works. Congratulations.

And If You Can't Establish Validity . . . Then What?

Well, this is a tough one, especially because there are so many different types of validity.

In general, if you don't have the validity evidence you want, it's because your test is not doing what it should. If it's an achievement test and a satisfactory level of content validity is what you seek, then you probably have to rewrite the questions on your test to make sure they are more consistent with the opinion of the expert you consulted.

If you are concerned with criterion validity, then you probably need to reexamine the nature of the items on the test and answer the question *How well do I expect these responses to these questions to relate to the criterion?* And, of course, this assumes that the criterion you are using makes sense.

And finally, if it is construct validity that you are seeking and can't seem to find, better take a close look at the theoretical rationale that underlies the test you developed. Perhaps our definition and model of aggression are wrong, or perhaps the definition and conceptualization of intelligence need some critical rethinking. Your aim is to establish consistency between the theory and the test items based on the theory.

A LAST FRIENDLY WORD

This measurement stuff is pretty cool. It's intellectually interesting, and, in these times of accountability, everyone wants to know about the progress of students, stockbrokers, professors(!), social welfare agency programs, and more.

Because of this strong and growing interest in accountability and the measurement of outcomes, there's a great temptation for under-graduate students working on their honors thesis or semester project or graduate students working on their thesis or dissertation to design an instrument for their final project.

But be aware that what sounds like a good idea might lead to a disaster. The process of establishing the reliability and validity of any instrument can take years of intensive work. And what can make matters even worse is when the naive or unsuspecting individual wants to create a new instrument to test a new hypothesis. That means that on top of everything else that comes with testing a new hypothesis, there is also the work of making sure the instrument works as it should.

 If you are doing original research of your own, such as for your thesis or dissertation requirement, be sure to find a measure that has already had reliability and validity evidence well established. That way, you can get on with the main task of testing your hypotheses and not fool with the huge task of instrument development—a career in and of itself. Want a good start? Try the Buros Center for Testing, available online at http://buros.com.

VALIDITY AND RELIABILITY: REALLY CLOSE COUSINS

Let's step back for a moment and recall one of the reasons that you're even reading this chapter.

It was assigned to you.

No, really. This chapter is important because you need to know something about reliability and the validity of the instruments you are using to measure outcomes. Why? If these instruments are not reliable *and* valid, then the results of your experiment will always be in doubt.

As we have mentioned earlier in this chapter, you can have a test that is reliable but not valid. However, you cannot have a valid test without it first being reliable. Why? Well, a test can do whatever it does over and over (that's reliability) but still not do what it is supposed to (that's validity). But if a test does what it is supposed to, then it has to do it consistently to work.

You've read about the relationship between reliability and validity several places in this chapter, but there's a very cool relationship lurking out there that you may read about later in your coursework that you should know about now. This relationship says that the maximum level of validity is equal to the square root of the reliability coefficient. For example, if the reliability coefficient for a test of mechanical aptitude is .87, the validity coefficient can be no larger than .93 (which is the square root of .87). What this means in tech talk is that the validity of a test is constrained by how reliable it is. And that makes perfect sense if we stop to think that a test must do what it does consistently before we are sure it does what it says it does. But the relationship is closer as well. You cannot have a valid instrument without it first being reliable, because in order for something to do what it is supposed to do, it must first do it consistently, right? So, the two work hand in hand.

REAL-WORLD STATS

Here's a classic example of why validity is such an important concept to understand and to make sure is present when doing any kind of research or using the results of research to inform actions by professionals. In this case, the study has to do with the assessment of attention-deficit/hyperactivity disorder (ADHD).

Often, this diagnosis is biased by the subjectivity of symptoms and reports of parents and teachers. The relatively new use of continuous performance tests (which measure sustained and selective attention) increased expectations that the diagnosis of ADHD would be more standardized and accurate, both qualities of any reliable and valid test. In this study, Nathanel Zelnik and his colleagues looked at scores from the Test of Variables of Attention in 230 children who were referred to their ADHD clinic. Among the 179 children with diagnosed ADHD (the criterion group), the Test of Variables of Attention was suggestive of ADHD in 163 participants (91.1% sensitivity), but it was also suggestive for ADHD in 78.4% of the children without ADHD. In sum, it is just not a reliable enough measure to accurately discriminate between the two groups.

Want to know more? Go online or to the library and find . . .

Real-World
Stats

Zelnik, N., Bennett-Back, O., Miari, W., Geoz, H. R., & Fattal-Valevski, A. (2012). Is the test of variables of attention reliable for the diagnosis of attention-deficit hyperactivity disorder (ADHD)? *Journal of Child Neurology, 27,* 703–707.

Yep, this is a stats course, so what's the measurement stuff doing here? Once again, almost any use of statistics revolves around some outcome being measured. Just as you read basic stats to make sense of lots of data, you need basic measurement information to make sense out of how behaviors, test scores, and rankings and ratings are assessed.

Chapter
Summary

TIME TO PRACTICE

1. Go to the library and find five journal articles in your area of interest in which reliability and validity data are reported. Discuss the outcome measures that are used. Identify the type of reliability that was established and the type of validity and comment on whether you think that the levels are acceptable. If not, how can they be improved?

2. Provide an example of when you would want to establish test–retest and parallel forms reliability.

3. You are developing an instrument that measures vocational preferences (what people want to do for a living), and you need to administer the test several times during the year to students who are attending a vocational program. You need to assess the test–retest reliability of the test and the data from two administrations (available as Chapter 6 Data Set 1)—one in the fall and one in the spring. Would you call this a reliable test? Why or why not?

4. How can a test be reliable and not valid? Provide an example. Why is a test not valid unless it is reliable?

5. Here's the situation. You are in charge of the test development program for the office of state employment, and you need at least two forms of the same test to administer on the same day. What kind of reliability will you want to establish? Use the data in Chapter 6 Data Set 2 to compute the reliability coefficient between the first and second form of the test for the 100 people who took it. Did you reach your goal?

6. In general terms, describe what a test would be like if it were reliable but not valid. Now, do the same for a test that is valid but not reliable.

7. When testing any experimental hypothesis, why is it important that the test you use to measure the outcome be both reliable and valid?

8. Describe the differences among content, predictive, and construct validity. Give examples of how each of these is measured.

9. Describe the steps that you would take to establish the construct validity of an observational paper-and-pencil test that assesses "thinking outside the box."

STUDENT STUDY SITE

⑤SAGE edge™

Get the tools you need to sharpen your study skills! Visit **edge.sagepub.com/ salkind6e** to access practice quizzes, eFlashcards, original and curated videos, data sets, journal articles, and more!

PART III

Taking Chances for Fun and Profit

"And this is my statistically significant other."

What do you know so far, and what's next? To begin with, you've got a really solid basis for understanding how to describe the characteristics of a set of scores and how distributions can differ from one another. That's what you learned in Chapters 2, 3, and 4 of *Statistics for People Who (Think They) Hate Statistics*. In Chapter 5, you learned how to describe the relationship between variables using correlations as tools. And, in Chapter 6, you learned about the importance of reliability and validity for understanding the integrity of any test score or other kind of outcome.

Now it's time to bump up the ante a bit and start playing for real. In Part III of *Statistics for People Who (Think They) Hate Statistics,* you will be introduced in Chapter 7 to the importance, and nature, of hypothesis testing, including an in-depth discussion of what a hypothesis is, what different types of hypotheses there are, the function of the hypothesis, and why and how hypotheses are tested.

Then, in Chapter 8, we'll get to the all-important topic of probability, represented by our discussion of the normal curve and the basic principles underlying probability. This is the part of statistics that helps us define how likely it is that some event (such as a specific score on a test) will occur. We'll use the normal curve as a basis for these arguments, and you'll see how any score or occurrence within any distribution has a likelihood associated with it.

After some fun with probability and the normal curve, we'll be ready to start our extended discussion in Part IV regarding the application of hypothesis testing and probability theory to the testing of specific questions regarding relationships between variables. It only gets better from here!

7 Hypotheticals and You

Testing Your Questions

Difficulty Scale ☺ ☺ ☺½ (don't plan on going out tonight)

WHAT YOU'LL LEARN ABOUT IN THIS CHAPTER

✦ Understanding the difference between a sample and a population

✦ Understanding the importance of the null and research hypotheses

✦ Using criteria to judge a good hypothesis

SO YOU WANT TO BE A SCIENTIST . . .

You might have heard the term *hypothesis* used in other classes. You may even have had to formulate one for a research project you did for another class, or you may have read one or two in a journal article. If so, then you probably have a good idea what a hypothesis is. For those of you who are unfamiliar with this often-used term, a **hypothesis** is basically "an educated guess." Its most important role is to reflect the general problem statement or question that was the motivation for asking the research question in the first place.

Introduction to
Chapter 7

That's why taking the care and time to formulate a really precise and clear research question is so important. This research question will guide your creation of a hypothesis, and in turn, the hypothesis will determine the techniques you will use to test it and answer the question that was originally asked.

So, a good hypothesis translates a problem statement or a research question into a form that is more amenable to testing. This form is called a hypothesis. We will talk about what makes a hypothesis a good one later in this chapter. Before that, let's turn our attention to the difference between a sample and a population. This is an important distinction, because hypothesis testing deals with a sample and then the results are generalized to the larger population. We also address the two main types of hypotheses (the null hypothesis and the research hypothesis). But first, let's formally define some simple terms that we have used earlier in *Statistics for People Who (Think They) Hate Statistics*.

SAMPLES AND POPULATIONS

Looking at
Sampling Error

As a good scientist, you would like to be able to say that if Method A is better than Method B in your study, this is true forever and always and for all people in the universe, right? Indeed. And, if you do enough research on the relative merits of Methods A and B and test enough people, you may someday be able to say that.

But don't get too excited, because it's unlikely you will ever be able to speak with such confidence. It takes too much money ($$$) and too much time (all those people!) to do all that research, and besides, it's not even necessary. Instead, you can just select a representative sample from the population and test your hypothesis about the relative merits of Methods A and B.

Given the constraints of never enough time and never enough research funds, with which almost all scientists live, the next best strategy is to take a portion of a larger group of participants and do the research with that smaller group. In this context, the larger group is referred to as a *population*, and the smaller group selected from that population is referred to as a *sample*.

A measure of how well a sample approximates the characteristics of a population is called **sampling error**. Sampling error is basically the difference between the values of the sample statistic and the population parameter. The higher the sampling error, the less precision you have in sampling, and the more difficult it will be to make the case that what you find in the sample indeed reflects what you expected to find in the population. And just as there are measures of variability regarding distributions, so there are measures of the variability of this difference between a sample measure and a population measure. This is often called the *standard error*—it's basically the standard deviation of the difference between those two measures.

Samples should be selected from populations in such a way that the sample matches as closely as possible the characteristics of the population. The goal is to have the sample be as much like the population as possible. The most important implication of ensuring similarity between the two is that the research results based on the sample can be generalized to the population. When the sample accurately represents the population, the results of the study are said to have a high degree of generalizability.

A high degree of generalizability is an important quality of good research because it means that the time and effort (and $$$) that went into the research may have implications for groups of people other than the original participants.

It's easy to equate "big" with "representative." Keep in mind that it is far more important to have an accurately representative sample than it is to have a big sample (people often think that big is better—only true on Thanksgiving, by the way). Having lots and lots of participants in a sample may be very impressive, but if the participants do not represent the larger population, the research will have little value.

THE NULL HYPOTHESIS

Okay. So we have a sample of participants selected from a population, and to begin the test of our research hypothesis, we first formulate the **null hypothesis**.

The Null Hypothesis

The null hypothesis is an interesting little creature. If it could talk, it would say something like "I represent no relationship between the variables that you are studying." In other words, null hypotheses are statements of equality demonstrated by the following real-life null hypotheses taken from a variety of popular social and behavioral science journals. Names have been changed to protect the innocent.

- There will be *no difference* between the average score of 9th graders and the average score of 12th graders on the ABC memory test.
- There is *no difference* between the effectiveness of community-based, long-term care and the effectiveness of in-home, long-term care in promoting the social activity of older adults when measured using the Margolis Scale of Social Activities.
- There is *no relationship* between reaction time and problem-solving ability.
- There is *no difference* between white and black families in the amount of assistance families offer their children in school-related activities.

What these four null hypotheses have in common is that they all contain a statement that two or more things are equal or unrelated (that's the "no difference" and "no relationship" part) to each other.

The Purposes of the Null Hypothesis

What are the basic purposes of the null hypothesis? The null hypothesis acts as both a starting point and a benchmark against which the actual outcomes of a study can be measured.

Let's examine each of these purposes in more detail.

First, the null hypothesis acts as a starting point because it is the state of affairs that is accepted as true in the absence of any other information. For example, let's look at the first null hypothesis we stated earlier:

> There will be no difference between the average score of 9th graders and the average score of 12th graders on the ABC memory test.

Given absolutely no other knowledge of 9th and 12th graders' memory skills, you have no reason to believe that there will be differences between the two groups, right? If you know nothing about the relationship between these variables, the best you can do is guess. And that's taking a chance. You might speculate as to why one group might outperform another, but if you have no evidence a priori (before the fact), then what choice do you have but to assume that they are equal?

This lack of a relationship as a starting point is a hallmark of this whole topic. In other words, until you prove that there is a difference, you have to assume that there is no difference. And a statement of no difference or no relationship is exactly what the null hypothesis is all about. Such a statement assures us (as members of the scientific community) that we are starting on a level playing field with no bias toward one or the other direction as to how the test of our hypothesis will turn out.

Furthermore, if there are any differences between these two groups, you have to assume that these differences are due to the most attractive explanation for differences between any groups on any variable—chance! That's right: Given no other information, chance is always the most likely and attractive explanation for the observed differences between two groups or the relationship between variables. Chance explains what we cannot. You might have thought of chance as the odds of winning that $5,000 nickel jackpot at the slots, but we're talking about chance as all that other "stuff" that clouds the picture and makes it even more difficult to understand the "true" nature of relationships between variables.

For example, you could take a group of soccer players and a group of football players and compare their running speeds, thinking about whether playing soccer or playing football makes athletes faster. But look at all the factors we don't know about that could contribute to differences. Who is to know whether some soccer players practice more, or whether some football players are stronger, or whether both groups are receiving additional training of different types?

What's more, perhaps the way their speed is being measured leaves room for chance; a faulty stopwatch or a windy day can contribute to differences unrelated to *true* running speed. As good researchers, our job is to eliminate chance factors from explaining observed differences and to evaluate other factors that might contribute to group differences, such as intentional training or nutrition programs, and see how they affect speed.

The point is, if we find differences between groups and the differences are not due to training, we have no choice but to attribute the difference to chance. And, by the way, you might find it useful to think of *chance* as being somewhat equivalent to the idea of *error*. When we can control sources of error, the likelihood that we can offer a meaningful explanation for some outcome increases.

The second purpose of the null hypothesis is to provide a benchmark against which observed outcomes can be compared to see if these differences are due to some other factor. The null hypothesis helps to define a range within which any observed differences between groups can be attributed to chance (which is the null hypothesis's contention) or are due to something other than chance (which perhaps would be the result of the manipulation of some variable, such as training in our example of the soccer and football players).

Most research studies have an implied null hypothesis, and you may not find it clearly stated in a research report or journal article. Instead, you'll find the research hypothesis clearly stated, and this is where we now turn our attention.

THE RESEARCH HYPOTHESIS

Whereas a null hypothesis is a statement of no relationship between variables, a **research hypothesis** is a definite statement that a relationship exists between variables. For example, for each of the null hypotheses stated earlier, here is a corresponding research hypothesis. Notice that we said "a" and not "the" corresponding research hypothesis because there certainly could be more than one research hypothesis for any one null hypothesis.

The Research Hypothesis

- The average score of 9th graders *is different* from the average score of 12th graders on the ABC memory test.
- The effectiveness of community-based, long-term care *is different* from the effectiveness of in-home, long-term care in promoting

the social activity of older adults when measured using the Margolis Scale of Social Activities.

- Slower reaction time and problem-solving ability *are positively related.*
- There *is a difference* between white and black families in the amount of assistance families offer to their children in school-related activities.

Each of these four research hypotheses has one thing in common: They are all statements of inequality. They posit a relationship between variables and not an equality, as does the null hypothesis.

The nature of this inequality can take two different forms—a directional or a nondirectional research hypothesis. If the research hypothesis posits no direction to the inequality (such as "different from"), the hypothesis is a nondirectional research hypothesis. If the research hypothesis posits a direction to the inequality (such as "more than" or "less than"), the research hypothesis is a directional research hypothesis.

The Nondirectional Research Hypothesis

A **nondirectional research hypothesis** reflects a difference between groups, but the direction of the difference is not specified.

For example, the research hypothesis . . .

The average score of 9th graders is different from the average score of 12th graders on the ABC memory test

. . . is nondirectional in that the direction of the difference between the two groups is not specified. The hypothesis is a research hypothesis because it states that there is a difference, and it is nondirectional because it says nothing about the direction of that difference.

A nondirectional research hypothesis, like this one, would be represented by the following equation:

$$\text{H}_1: \overline{X}_9 \neq \overline{X}_{12}, \tag{7.1}$$

where

- H_1 represents the symbol for the first (of possibly several) research hypotheses;
- $\overline{X}_9$ represents the average memory score for the sample of 9th graders;

- $\overline{X}_{12}$ represents the average memory score for the sample of 12th graders; and
- ≠ means "is not equal to."

The Directional Research Hypothesis

A **directional research hypothesis** reflects a difference between groups, and the direction of the difference is specified.

For example, the research hypothesis . . .

The average score of 12th graders is greater than the average score of 9th graders on the ABC memory test

. . . is directional because the direction of the difference between the two groups is specified. One is hypothesized to be greater than (not just different from) the other.

Examples of two other directional hypotheses are these:

A is greater than *B* (or *A* > *B*).

B is greater than *A* (or *A* < *B*).

These both represent inequalities (greater than or less than). A directional research hypothesis such as the one described above, where 12th graders are hypothesized to score better than 9th graders, would be represented by the following equation:

$$H_1: \overline{X}_{12} > \overline{X}_9, \qquad (7.2)$$

where

- H_1 represents the symbol for the first (of possibly several) research hypotheses;
- $\overline{X}_9$ represents the average memory score for the sample of 9th graders;
- $\overline{X}_{12}$ represents the average memory score for the sample of 12th graders; and
- > means "is greater than."

What is the purpose of the research hypothesis? It is this hypothesis that is directly tested as an important step in the research process. The results of this test are compared with what you expect by chance alone (reflecting the null hypothesis) to see which of the two is the more attractive explanation for any differences between groups or variables you might observe.

Table 7.1 gives the four null hypotheses and accompanying directional and nondirectional research hypotheses.

 Another way to talk about directional and nondirectional hypotheses is to talk about one- and two-tailed tests. A **one-tailed test** (reflecting a directional hypothesis) posits a difference in a particular direction, such as when we hypothesize that Group 1 will score higher than Group 2. A **two-tailed test** (reflecting a nondirectional hypothesis) posits a difference but in no particular direction. The importance of this distinction begins when you test different types of hypotheses (one- and two-tailed) and establish probability levels for rejecting or not rejecting the null hypothesis. More about this in Chapter 9. Promise.

Table 7.1   Null Hypotheses and Corresponding Research Hypotheses

Null Hypothesis	Nondirectional Research Hypothesis	Directional Research Hypothesis
There will be **no difference** in the average score of 9th graders and the average score of 12th graders on the ABC memory test.	Twelfth graders and 9th graders will **differ** on the ABC memory test.	Twelfth graders will have a **higher** average score on the ABC memory test than will 9th graders.
There is **no difference** between the effectiveness of community-based, long-term care for older adults and the effectiveness of in-home, long-term care for older adults when measured using the Margolis Scale of Social Activities.	The effect of community-based, long-term care for older adults is **different** from the effect of in-home, long-term care for older adults when measured using the Margolis Scale of Social Activities.	Older adults exposed to community-based, long-term care score **higher** on the Margolis Scale of Social Activities than do older adults receiving in-home, long-term care.
There is **no relationship** between reaction time and problem-solving ability.	There **is a relationship** between reaction time and problem-solving ability.	There **is a positive relationship** between reaction time and problem-solving ability.
There is **no difference** between white and black families in the amount of assistance families offer their children in educational activities.	The amount of assistance offered by white families to their children in educational activities is **different** from the amount of support offered by black families to their children in educational activities.	The amount of assistance offered by white families to their children in educational activities is **more** than the amount of support offered by black families to their children in educational activities.

Some Differences Between the Null Hypothesis and the Research Hypothesis

Besides the null hypothesis representing an equality and the research hypothesis representing an inequality, the two types of hypotheses differ in several other important ways.

Questions About Test Direction

First, for a bit of review, the two types of hypotheses differ in that one (the null hypothesis) states that there is no relationship between variables (an equality), whereas the research hypothesis states that there is a relationship between the variables (an inequality). This is the primary difference.

Second, null hypotheses always refer to the population, whereas research hypotheses always refer to the sample. We select a sample of participants from a much larger population. We then try to generalize the results from the sample back to the population. If you remember your basic philosophy and logic (you did take these courses, right?), you'll remember that going from small (as in a sample) to large (as in a population) is a process of inference.

Third, because the entire population cannot be directly tested (again, it is impractical, uneconomical, and often impossible), you can't say with 100% certainty that there is no real difference between segments of the population on some variable. Rather, you have to infer it (indirectly) from the results of the test of the research hypothesis, which is based on the sample. Hence, the null hypothesis must be indirectly tested, and the research hypothesis can be directly tested.

Fourth, the null hypothesis is written with an equal sign, while the research hypothesis is written with a not equal to, greater than, or less than sign.

Fifth, null hypotheses are always written using Greek symbols, and research hypotheses are always written using Roman symbols. Thus, the null hypothesis that the average score for 9th graders is equal to that of 12th graders is represented like this:

$$H_0: \mu_9 = \mu_{12}, \tag{7.3}$$

where

- H_0 represents the null hypothesis;
- μ_9 represents the theoretical average for the population of 9th graders; and
- μ_{12} represents the theoretical average for the population of 12th graders.

The research hypothesis that the average score for a sample of 12th graders is greater than the average score for a sample of 9th graders is shown in Formula 7.2 (on page 139).

Finally, because you cannot directly test the null hypothesis, it is an *implied* hypothesis. But the research hypothesis is explicit and is stated as such. This is another reason why you rarely see null hypotheses stated in research reports and will almost always see a statement (be it in symbols or words) of the research hypothesis.

WHAT MAKES A GOOD HYPOTHESIS?

You now know that hypotheses are educated guesses—a starting point for a lot more to come. As with any guess, some hypotheses are better than others right from the start. We can't stress enough how important it is to ask the question you want answered and to keep in mind that any hypothesis you present is a direct extension of the original question you asked. This question will reflect your personal interests and motivation and your understanding of what research has been done previously. With that in mind, here are criteria you might use to decide whether a hypothesis you read in a research report or one that you formulate is acceptable.

To illustrate, let's use an example of a study that examines the effects of providing afterschool child care to employees who work late on the parents' adjustment to work. Here is a well-written directional research hypothesis:

> Parents who enroll their children in afterschool programs will miss fewer days of work in 1 year and will have a more positive attitude toward work, as measured by the Attitude Toward Work survey, than will parents who do not enroll their children in such programs.

Here are the criteria.

First, a good hypothesis is stated in declarative form and not as a question. While the above hypothesis may have started in the researcher's mind as the question "Are parents better employees . . . ?" it was not posed because hypotheses are most effective when they make a clear and forceful statement.

Second, a good hypothesis posits an expected relationship between variables. The hypothesis in our example clearly describes the expected relationships among afterschool child care, parents' attitude, and absentee rate. These variables are being tested to see if one (enrollment in the afterschool program) has an effect upon the others (absentee rate and attitude).

Notice the word *expected* in the above criterion. Defining an expected relationship is intended to prevent a fishing trip to look for any relationships that may be found (sometimes called the "shotgun" approach), which may be tempting but is not very productive. You do get some-where using the shotgun approach, but because you don't know where you started, you have no idea where you end up.

 In the fishing-trip approach, you throw out your line and take anything that bites. You collect data on as many things as you can, regardless of your interest or even whether collecting the data is a reasonable part of a sci-entific investigation. Or, to use a shotgun analogy, you load up them guns and blast away at anything that moves, and you're bound to hit something. The problem is, you may not want what you hit, and worse, you may miss what you want to hit, and worst of all (if possible), you may not know what you hit! Big data and data min-ing (see Chapter 17) anyone? Good researchers do not want just anything they can catch or shoot. They want specific results. To get them, researchers need their open-ing questions and hypotheses to be clear, forceful, and easily understood.

Third, hypotheses reflect the theory or literature on which they are based. As you read in Chapter 1, the accomplishments of scientists rarely can be attributed to just their own hard work. Their accomplish-ments are always due, in part, to many other researchers who came before them and laid the framework for later explorations. A good hypothesis reflects this, in that it has a substantive link to existing literature and theory. In the aforementioned example, let's assume that there is literature indicating that parents are more comfortable know-ing their children are being cared for in a structured environment and parents can then be more productive at work. Knowing this would allow one to hypothesize that an afterschool program would provide the security parents are looking for. In turn, this allows them to con-centrate on working rather than calling or texting to find out whether Rachel or Gregory got home safely.

Fourth, a hypothesis should be brief and to the point. You want your hypothesis to describe the relationship between variables in a declara-tive form and to be as direct and explicit as possible. The more to the point it is, the easier it will be for others (such as your master's thesis or doctoral dissertation committee members!) to read your research and understand exactly what you are hypothesizing and what the important variables are. In fact, when people read and evaluate research (as you will learn more about later in this chapter), the first thing many of them do is find the hypotheses to get a good idea as to the general purpose of the research and how things will be done. A good hypoth-esis tells you both of these things.

Fifth, good hypotheses are testable hypotheses—and testable hypotheses contain variables that can be measured. This means that you can actually carry out the intent of the question reflected by the hypothesis. You can see from our example hypothesis that the important comparison is between parents who have enrolled their child in an afterschool program and those who have not. Then, such things as attitude and workdays missed will be measured. Identifying these two groups of parents and measuring the variables attitude and absenteeism are both reasonable objectives. Attitude is measured by the Attitude Toward Work survey (a fictitious title, but you get the idea), and absenteeism (the number of days missed at work) is an easily recorded and unambiguous measure almost always available in payroll records (assuming you've been granted access). Think how much harder things would be if the hypothesis were stated as *Parents who enroll their children in afterschool care feel better about their jobs.* Although you might get the same message, the results might be more difficult to interpret given the ambiguous nature of the phrase *feel better.*

In sum, hypotheses should . . .

Creating
Hypotheses

- be stated in declarative form,
- posit a relationship between variables,
- reflect a theory or a body of literature on which they are based,
- be brief and to the point, and
- be testable.

When a hypothesis meets each of these five criteria, you know that it is good enough to continue with a study that will accurately test the general question from which the hypothesis was derived.

REAL-WORLD STATS

You might think that the scientific method and the use of null and research hypotheses is beyond question in the world of scientific research. Well, you would be wrong. Here's just a sampling of some of the articles published in professional journals over the past few years that raise concerns. No one is yet ready to throw out the scientific method as the best approach for testing hypotheses, but it's not a bad idea to every now and then question whether the method always and forever is the best model to use.

Want to know more? Go online or to the library and find . . .

- Jeff Gill from California Polytechnic State University raises a variety of issues that call into question the use of the null hypothesis significance-testing model as the best way to evaluate hypotheses. He focuses on political science (his area) and how the use of the technique is widely misunderstood. Major problems are discussed and some solutions offered. You can find the article by looking for this reference: Gill, J. (1999). The insignificance of null hypothesis testing. *Politics Research Quarterly, 52,* 647–674.

Real-World Stats

• Howard Wainer and Daniel Robinson take these criticisms one step further and suggest that the historical use of such procedures was reasonable but that modifications to significance testing and the interpretations of outcomes would serve modern science well. Basically, they are saying that other tools (such as effect size, which we discuss in Chapter 11) should be used to evaluate outcomes. Read all about it in Wainer, H., & Robinson, D. H. (2003). Shaping up the practice of null hypothesis significance testing. *Educational Researcher, 32,* 22–30.

More Real-World Stats

• Finally, in the really interesting article "A Vast Graveyard of Undead Theories: Publication Bias and Psychological Science's Aversion to the Null," Christopher Ferguson and Moritz Heene raise the very real issue that many journals refuse to publish outcomes where a null result is found (such as no difference between groups). They believe that when such outcomes are not published, false theories are never taken to task and never tested for their truthfulness. Thus, the replicability of science (a very important aspect of the entire scientific process) is compromised. You can find more about this in Ferguson, C. J., & Heene, M. (2012). A vast graveyard of undead theories: Publication bias and psychological science's aversion to the null. *Perspectives on Psychological Science, 7,* 555–561.

Even More Real-World Stats

We hope you get the idea from the above examples that science is not black-and-white, cut-and-dried, or any other metaphor indicating there is only a right way and only a wrong way of doing things. Science is an organic and dynamic process that is always changing in its focus, methods, and potential outcomes.

SUMMARY

A central component of any scientific study is the hypothesis, and the different types of hypotheses (null and research) help form a plan for answering the questions asked by the purpose of our research. The null hypothesis provides a starting point and benchmark for research, and we use it as a comparison as we evaluate the acceptability of the research hypothesis. Now let's move on to how null hypotheses are actually tested.

Chapter Summary

TIME TO PRACTICE

1. Go to the library and select five empirical research articles (those that contain actual data) from your area of interest. For each one, list the following:

    a. What is the null hypothesis (implied or explicitly stated)?

    b. What is the research hypothesis (implied or explicitly stated)?

    c. And what about those articles with no hypothesis clearly stated or implied? Identify those articles and see if you can write a research hypothesis for them.

2. While you're at the library, select two other articles from an area in which you are interested and write a brief description of the sample and how it was selected from the population. Be sure to include some words about whether the researchers did an adequate job of selecting the sample; be able to justify your answer.

3. For the following research questions, create one null hypothesis, one directional research hypothesis, and one nondirectional research hypothesis.

    a. What are the effects of attention span on out-of-seat classroom behavior?

    b. What is the relationship between the quality of a marriage and the quality of the spouses' relationships with their siblings?

    c. What is the best way to treat an eating disorder?

4. Go back to the five hypotheses that you found in question 1 above and evaluate each using the five criteria that were discussed at the end of the chapter.

5. What kinds of problems might using a poorly written or ambiguous research hypothesis introduce?

6. What is the null hypothesis, and what is one of its important purposes? How does it differ from the research hypothesis?

7. What is *chance* in the context of a research hypothesis? And what do we do about chance in our experiments?

8. Why does the null hypothesis presume no relationship between variables?

STUDENT STUDY SITE

⑤SAGE edge™

Get the tools you need to sharpen your study skills! Visit **edge.sagepub.com/ salkind6e** to access practice quizzes, eFlashcards, original and curated videos, data sets, journal articles, and more!

8

Are Your Curves Normal?

Probability and Why It Counts

Difficulty Scale ☺ ☺ ☺ (not too easy and not too hard, but very important)

WHAT YOU'LL LEARN ABOUT IN THIS CHAPTER

✦ Understanding probability and why it is basic to the understanding of statistics

✦ Applying the characteristics of the normal, or bell-shaped, curve

✦ Computing and interpreting z scores and understanding their importance

---

WHY PROBABILITY?

And here you thought this was a statistics class! Ha! Well, as you will learn in this chapter, the study of probability is the basis for the normal curve (much more on that later) and the foundation for inferential statistics.

Introduction to Chapter 8

Why? First, the normal curve provides us with a basis for understanding the probability associated with any possible outcome (such as the odds of getting a certain score on a test or the odds of getting a head on one flip of a coin).

Second, the study of probability is the basis for determining the degree of confidence we have in stating that a particular finding or outcome is "true." Or, better said, that an outcome (like an average score) may not have occurred due to chance alone. For example, let's compare Group A (which participates in 3 hours of extra swim practice each week) and Group B (which has no extra swim practice each week). We find that Group A differs from Group B on a test of fitness, but can we say that the difference is due to the extra practice? Or might it be due to something else? The tools that the study of probability provides allow us to determine the exact mathematical likelihood that

A Basic
Introduction to
Probability

the difference is due to practice (and practice only) versus something else (such as chance).

All that time we spent on hypotheses in the previous chapter was time well spent. Once we put together our understanding of what a null hypothesis and a research hypothesis are with the ideas that are the foundation of probability, we'll be in a position to discuss how likely certain outcomes (formulated by the research hypothesis) are.

THE NORMAL CURVE
(A.K.A. THE BELL-SHAPED CURVE)

What is a normal curve? Well, the **normal curve** (also called a **bell-shaped curve,** or bell curve) is a visual representation of a distribution of scores that has three characteristics. Each of these characteristics is illustrated in Figure 8.1.

The normal curve represents a distribution of values in which the mean, median, and mode are equal to one another. You probably remember from Chapter 4 that if the median and the mean are different, then the distribution is skewed in one direction or the other. The normal curve is not skewed. It's got a nice hump (only one), and that hump is right in the middle.

Second, the normal curve is perfectly symmetrical about the mean. If you fold one half of the curve along its center line, the two halves would lie perfectly on top of each other. They are identical. One half of the curve is a mirror image of the other.

Figure 8.1   The Normal, or Bell-Shaped, Curve

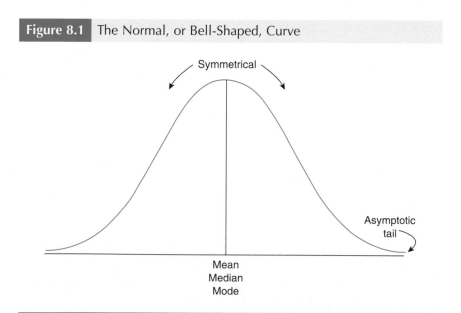

Finally (and get ready for a mouthful), the tails of the normal curve are **asymptotic**—a big word. What it means is that they come closer and closer to the horizontal axis but never touch. See if you have some idea (in advance, because we will talk about it later) why this is so important, because it's really a cornerstone of all this probability stuff.

The normal curve's shape of a bell also gives the graph its other name, the bell-shaped curve.

When your devoted author was knee-high, he always wondered how the tail of a normal curve can approach the horizontal or x-axis yet never touch it. Try this. Place two pencils one inch apart and then move them closer (by half) so they are one-half inch apart, and then closer (one-quarter inch apart), and closer (one-eighth inch apart). They continually get closer, right? But they never (and never will) touch. Same thing with the tails of the curve. The tail slowly approaches the axis on which the curve "rests," but the tail and the axis can never really touch.

Why is this important? As you will learn later in this chapter, the fact that the tails never touch the x-axis means that there is an infinitely small likelihood that a score can be obtained that is very extreme (way out under the left or right tail of the curve). If the tails did touch the x-axis, then the likelihood that a very extreme score could be obtained would be nonexistent.

Hey, That's Not Normal!

We hope your next question is "But there are plenty of sets of scores where the distribution is not normal or bell shaped, right?" Yes (and it's a big *but*).

When we deal with large sets of data (more than 30), and as we take repeated samples of the data from a population, the values in the curve closely approximate the shape of a normal curve. This is very important, because a lot of what we do when we talk about inferring from a sample to a population is based on the assumption that the sample taken from the population is distributed normally. And that's just another way to say that the sample's characteristics continue to approach those characteristics of the population.

And as it turns out, in nature in general, many things are distributed with the characteristics that we call normal. That is, there are lots of events or occurrences right in the middle of the distribution but relatively few on each end, as you can see in Figure 8.2, which shows the distribution of IQ and height in the general population.

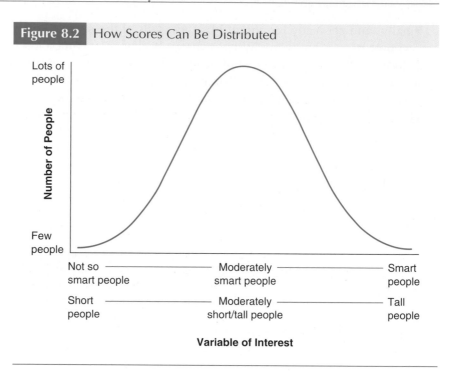

Figure 8.2 How Scores Can Be Distributed

For example, there are very few people who are brilliant and very few who are intellectually or cognitively at the absolute bottom of the group. There are lots who are right in the middle and fewer as we move toward the tails of the curve. Likewise, there are *relatively* few very tall people and relatively few very short people, but lots of people fall right in the middle. In both of these examples, the distribution of intellectual skills and of height approximates a normal distribution.

Consequently, those events that tend to occur in the extremes of the normal curve have a smaller probability associated with each occurrence. We can say with a great deal of confidence that the odds of any one person (whose height we do not know beforehand) being very tall (or very short) are just not very great. But we know that the odds of any one person being average in height, or right around the middle, are pretty good. Those events that tend to occur in the middle of the normal curve have a higher probability of occurring than do those in the extreme. And this is true for height, weight, general intelligence, weight-lifting ability, income, number of Star Wars action figures owned, and on and on . . .

More Normal Curve 101

You already know the three main characteristics that make a curve normal or make it appear bell shaped, but there's more to it than that. Take a look at the curve in Figure 8.3.

Figure 8.3	A Normal Curve Divided Into Different Sections

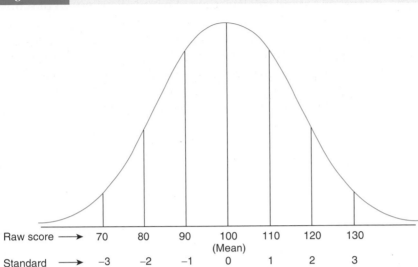

Raw score ⟶ 70   80   90   100   110   120   130
                                        (Mean)

Standard ⟶ −3   −2   −1   0   1   2   3
deviations

The distribution represented here has a mean of 100 and a standard deviation of 10. We've added numbers across the *x*-axis that represent the distance in standard deviations from the mean for this distribution. You can see that the *x*-axis (representing the scores in the distribution) is marked from 70 through 130 in increments of 10 (which is the standard deviation for the distribution), the value of 1 standard deviation. We made up these numbers (100 and 10), so don't go nuts trying to find out where we got them from.

Normal Curves

So, a quick review tells us that this distribution has a mean of 100 and a standard deviation of 10. Each vertical line within the curve separates the curve into a section, and each section is bound by particular scores. For example, the first section to the right of the mean of 100 is bound by the scores 100 and 110, and this section represents 1 standard deviation from the mean (which is 100).

And below each raw score (70, 80, 90, 100, 110, 120, and 130), you'll find a corresponding standard deviation (−3, −2, −1, 0, +1, +2, and +3). As you may have figured out already, each standard deviation in our example is 10 points. So 1 standard deviation from the mean (which is 100) is the mean plus 10 points or 110. Not so hard, is it?

If we extend this argument further, then you should be able to see how the range of scores represented by a normal distribution with a mean of 100 and a standard deviation of 10 is 70 through 130 (which includes −3 to +3 standard deviations).

Now, here's a big fact that is always true about normal distributions, means, and standard deviations: For any distribution of scores (regardless of the value of the mean and standard deviation), if the scores are distributed normally, almost 100% of the scores will fit between −3 and +3 standard deviations from the mean. This is very important, because it applies to all normal distributions. Because the rule does apply (once again, regardless of the value of the mean or standard deviation), distributions can be compared with one another. We'll get to that again later.

With that said, we'll extend our argument a bit more. If the distribution of scores is normal, we can also say that between different points along the x-axis (such as between the mean and 1 standard deviation), a certain percentage of cases will fall. In fact, between the mean (which in this case is 100—got that yet?) and 1 standard deviation above the mean (which is 110), about 34% (actually 34.13%) of all cases in the distribution of scores will fall. This is a fact you can take to the bank because it will always be true.

Want to go further? Take a look at Figure 8.4. Here, you can see the same normal curve in all its glory (the mean equals 100 and the standard deviation equals 10) and the percentage of cases that we would expect to fall within the boundaries defined by the mean and standard deviation.

Figure 8.4 A Normal Curve Divided Into Different Sections

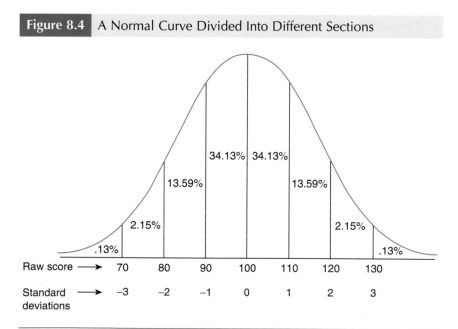

Here's what we can conclude:

The distance between . . .	includes . . .	and the scores that are included (if the mean = 100 and the standard deviation = 10) are . . .
the mean and 1 standard deviation	34.13% of all the cases under the curve	from 100 to 110
1 and 2 standard deviations	13.59% of all the cases under the curve	from 110 to 120
2 and 3 standard deviations	2.15% of all the cases under the curve	from 120 to 130
3 standard deviations and above	0.13% of all the cases under the curve	130 and above

If you add up all the values in either half of the normal curve, guess what you get? That's right, (just about almost) 50%. Why? The distance between the mean and all the scores to the right of the mean underneath the normal curve includes 50% of all the scores.

And because the curve is symmetrical about its central axis (each half is a mirror image of the other), the two halves together represent 100% of all the scores. Not rocket science, but important to point out, nonetheless.

Now, let's extend the same logic to the scores to the left of the mean of 100.

The distance between . . .	includes . . .	and the scores that are included (if the mean = 100 and the standard deviation = 10) are . . .
the mean and −1 standard deviation	34.13% of all the cases under the curve	from 90 to 100
−1 and −2 standard deviations	13.59% of all the cases under the curve	from 80 to 90
−2 and −3 standard deviations	2.15% of all the cases under the curve	from 70 to 80
−3 standard deviations and below	0.13% of all the cases under the curve	70 and below

Now, be sure to keep in mind that we are using a mean of 100 and a standard deviation of 10 only as a particular example. Obviously, not all distributions (not most distributions!) have a mean of 100 and a standard deviation of 10.

All of this is pretty neat, especially when you consider that the values of 34.13% and 13.59% and so on are absolutely independent of the actual values of the mean and the standard deviation. These percentages are due to the shape of the curve and do not depend on the value of any of the scores in the distribution or the value of the mean or standard deviation. In fact, you could draw a normal curve on a piece of cardboard and cut it out, so you had a bell-shaped piece of cardboard. Then if you cut out the area between the mean and +1 standard deviation and weighed it, it would tip the scale at exactly 34.13% of the entire piece of bell-shaped cardboard. (Try it—it's true.)

In our example, this means that (roughly) 68% (34.13% doubled) of the scores fall between the raw score values of 90 and 110. What about the other 32%? Good question. One half (16%, or 13.59% + 2.15% + 0.13%) fall above (to the right of) 1 standard deviation above the mean, and one half fall below (to the left of) 1 standard deviation below the mean. And because the curve slopes, and the amount of area decreases as you move farther away from the mean, it is no surprise that the likelihood that a score will fall more toward the extremes of the distribution is less than the likelihood it will fall toward the middle. That's why the curve has a bump in the middle and is not skewed in either direction. And that's why scores that are farther from the mean have a lower probability of occurring than scores that are closer to the mean.

OUR FAVORITE STANDARD SCORE: THE *Z* SCORE

You have read more than once that distributions differ in their measures of central tendency and variability.

But in the general practice of applying statistics (and using them in research activities), we find ourselves working with distributions that are indeed different, yet we will be required to compare them with one another. And to do such a comparison, we need some kind of a standard.

Say hello to **standard scores**. These are scores that are comparable because they are standardized in units of standard deviations. For example, a standard score of 1 in a distribution with a mean of 50 and a standard deviation of 10 means the same as a standard score of 1 from a distribution with a mean of 100 and a standard deviation of 5; they both represent 1 standard score and are an equivalent distance from their respective means. Also, we can use our knowledge of the normal curve and assign a probability to the occurrence of a value that is 1 standard deviation from the mean. We'll do that later.

Although there are other types of standard scores, the one that you will see most frequently in your study of statistics is called a z **score**. This is the result of dividing the amount that a raw score differs from the mean of the sample scores by the standard deviation, as shown in Formula 8.1:

$$z = \frac{X - \overline{X}}{s},$$ (8.1)

where

- z is the z score;
- X is the individual score;
- $\overline{X}$ is the mean of the distribution; and
- s is the distribution standard deviation.

For example, in Formula 8.2, you can see how the z score is calculated if the mean is 100, the raw score is 110, and the standard deviation is 10.

$$z = \frac{(110 - 100)}{10} = +1.0$$ (8.2)

It's just as easy to compute a raw score given a z score as the other way around. You already know the formula for a z score given the raw score, mean, and standard deviation. But if you know only the z score and the mean and standard deviation, then what's the corresponding raw score? Easy: Just use the formula $X = z(s) + \overline{X}$. You can easily convert raw scores to z scores and back again if necessary. For example, a z score of −0.5 in a distribution with a mean of 50 and an s of 5 would equal a raw score of $X = (-0.5)(5) + 50$, or 47.5.

As you can see in our formula (Equation 8.1), we use $\overline{X}$ and s for the mean and standard distribution, respectively. In some books (and in some lectures), the population mean is represented by the Greek letter *mu*, or μ, and the standard deviation is represented by the Greek letter *sigma*, or σ. One can be strict about when to use what, but for our purposes, we will use letters of the Roman alphabet.

The following data show the original raw scores plus the z scores for a sample of 10 scores that has a mean of 12 and a standard deviation of 2. Any raw score above the mean will have a corresponding z score that is positive, and any raw score below the mean will have a corresponding z score that is negative. For example, a raw score of 15 has a corresponding z score of +1.5, and a raw score of 8 has a corresponding z score of –2.0. And of course, a raw score of 12 (or the mean) has a z score of 0 (because 12 is no distance from the mean).

X	$X - \bar{X}$	z Score
12	0	0.0
15	3	1.5
11	–1	–0.5
13	1	0.5
8	–4	–2.0
14	2	1.0
12	0	0.0
13	1	0.5
12	0	0.0
10	–2	–1.0

Following are just a few observations about these scores, as a little review.

First, those scores below the mean (such as 8 and 10) have negative z scores, and those scores above the mean (such as 13 and 14) have positive z scores.

Second, positive z scores always fall to the right of the mean and are in the upper half of the distribution. And negative z scores always fall to the left of the mean and are in the lower half of the distribution.

Third, when we talk about a score being located 1 standard deviation above the mean, it's the same as saying that the score is 1 z score above the mean. For our purposes, when comparing scores across distributions, z scores and standard deviations are equivalent. In other words, a z score is simply the number of standard deviations from the mean.

Finally (and this is very important), z scores across different distributions are comparable. Here's another table, similar to the one above, that will illustrate this last point. These 10 scores were selected from a set of 100 scores, with the scores having a mean of 59 and a standard deviation of 14.5.

Raw Score	$X - \bar{X}$	z Score
67	8	0.55
54	−5	−0.34
65	6	0.41
33	−26	−1.79
56	−3	−0.21
76	17	1.17
65	6	0.41
33	−26	−1.79
48	−11	−0.76
76	17	1.17

In the first distribution you saw, with a mean of 12 and a standard deviation of 2, a raw score of 12.8 has a corresponding z score of +0.4, which means that a raw score of 12.8 is 0.4 standard deviations from the mean. In the second distribution, with a mean of 59 and a standard deviation of 14.5, a raw score of 64.8 has a corresponding z score of +0.4 as well. A miracle? No—just a good idea that is a huge aid in comparing scores from different sets of data or distributions.

Both raw scores of 12.8 and 64.8, *relative to one another,* are equal distances (and equally distant) from the mean. When these raw scores are represented as standard scores, then they are directly comparable to one another in terms of their relative location in their respective distributions.

What z Scores Represent

You already know that a particular z score not only represents a raw score but also represents a particular location along the x-axis of a distribution. And the more extreme the z score (such as −2.0 or +2.6), the farther it is from the mean.

Because you already know the percentage of area that falls between certain points along the x-axis (such as about 34% between the mean and a standard deviation of +1, for example, or about 14% between a standard deviation of +1 and a standard deviation of +2), we can make the following true statements as well:

- 84% of all the scores fall below a z score of +1 (the 50% that fall below the mean plus the 34% that fall between the mean and the +1 z score).

- 16% of all the scores fall above a z score of +1 (because the total area under the curve has to equal 100%, and 84% of the scores fall below a score of +1.0).

Think about both of these facts for a moment.

Moment passes.

What we are saying is that, given the normal distribution, different areas of the curve are encompassed by different numbers of standard deviations or z scores.

Okay—here it comes. These percentages or areas can also easily be seen as representing *probabilities* of a certain score occurring. For example, here's a big sample question of the kind you can now ask and answer (drum roll, please):

In a distribution with a mean of 100 and a standard deviation of 10, what is the probability that any one score will be 110 or above?

The answer? The probability is 16% or 16 out of 100 or .16. How did we get this?

First, we computed the corresponding z score, which is +1 [(110 − 100)/10]. Then, given the knowledge we already have (see Figure 8.4), we know a z score of 1 represents a location on the x-axis below which 84% (50% plus 34%) of all the scores in the distribution fall. Above that is 16% of the scores or a probability of .16.

In other words, because we already know the areas between the mean and 1, 2, or 3 standard deviations above or below the mean, we can easily figure out the probability that the value of any one z score has of occurring.

Now the method we just went through is fine for z values of 1, 2, and 3. But what if the value of the z score is not a whole number like 2 but is instead 1.23 or −2.01? We need to find a way to be more precise.

How do we do that? Simple—learn calculus and apply it to the curve to compute the area underneath it at almost every possible point along the x-axis, or (and we like this alternative much more) use Table B.1 found in Appendix B (the normal distribution table). This is a listing of all the values (except the very most extreme) for the areas under a curve that correspond to different z scores.

Table B.1 has two columns. The first column, labeled "z Score," is simply the z score that has been computed. The second column, "Area Between the Mean and the z Score," is the exact area underneath the curve that is contained between the two points.

For example (and you should turn to Table B.1 and try this as you read along), if we wanted to know the area between the mean and a

z score of +1, we would find the value 1.00 in the column labeled "z Score" and read across to the second column, where we would find the area between the mean and a z score of 1.00 to be 34.13. Seen that before?

Why aren't there any plus or minus signs in this table (such as −1.00)? Because the curve is symmetrical, it does not matter whether the value of the z score is positive or negative. The area between the mean and 1 standard deviation in any direction is *always* 34.13%.

Here's the next step. Let's say that for a particular z score of 1.38, you want to know the probability associated with that z score. If you wanted to know the percentage of the area between the mean and a z score of 1.38, you would find in Table B.1 the corresponding area for the z score of 1.38, which is 41.62, indicating that more than 41% of all the cases in the distribution fall between a z score of 0 and 1.38. Then we know that about 92% (50% plus 41.62%) will fall at or below a z score of 1.38. Now, you should notice that we did this last example without any raw scores at all. Once you get to this table, they are just no longer needed.

But are we always interested only in the amount of area between the mean and some other z score? What about between two z scores, neither of which is the mean? For example, what if we were interested in knowing the amount of area between a z score of 1.5 and a z score of 2.5, which translates to a probability that a score falls between the two z scores? How can we use the table to compute the answer to such questions? It's easy. Just find the corresponding amount of area each z score encompasses and subtract one from the other. Often, drawing a picture helps, as in Figure 8.5.

For example, let's say that we want to find the area between raw scores of 110 and 125 in a distribution with a mean of 100 and a standard deviation of 10. Here are the steps we would take.

How to Compute a z Score

1. Compute the z score for a raw score of 110, which is (110 − 100)/10, or +1.

2. Compute the z score for a raw score of 125, which is (125 − 100)/10, or +2.5.

3. Using Table B.1 in Appendix B, find the area between the mean and a z score of +1, which is 34.13%.

4. Using Table B.1 in Appendix B, find the area between the mean and a z score of +2.5, which is 49.38%.

5. Because you want to know the distance between the two, subtract the smaller from the larger: 49.38 − 34.13 = 15.25%. Here's the picture that's worth a thousand words, in Figure 8.5.

| Figure 8.5 | Using a Drawing to Figure Out the Difference in Area Between Two *z* Scores |

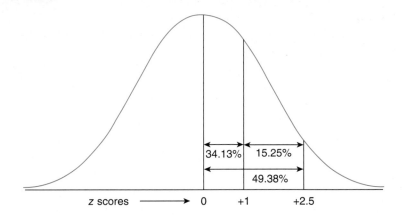

Okay—so we can be pretty confident that the probability of a particular score occurring can be best understood by examining where that score falls in a distribution relative to other scores. In this example, the probability of a score occurring between a *z* score of +1 and a *z* score of +2.5 is about 15%.

Here's another example. In a set of scores with a mean of 100 and a standard deviation of 10, a raw score of 117 has a corresponding *z* score of 1.70. This *z* score corresponds to an area under the curve of 95.54% (50% + 45.54%), meaning that the probability of this score occurring between a score of 0 and a score of 1.70 is 95.54% or 95.5 out of 100 or .955.

Just two things about standard scores. First, even though we are focusing on *z* scores, there are other types of standard scores as well. For example, a *T* score is a type of standard score that is computed by multiplying the *z* score by 10 and adding 50. One advantage of this type of score is that you rarely have a negative *T* score. As with *z* scores, *T* scores allow you to compare standard scores from different distributions.

Second, a *standard* score is a whole different animal from a *standardized* score. A standardized score is one that comes from a distribution with a predefined mean and standard deviation. Standardized scores from tests such as the SAT and GRE (Graduate Record Exam) are used so that comparisons can easily be made between scores from different forms or administrations of the test, which all have the same mean and standard deviation.

What z Scores Really Represent

The name of the statistics game is being able to estimate the probability of an outcome. If we take what we have talked about and done so far in this chapter one step further, we can determine the probability of some event occurring. Then, we will use some criterion to judge whether we think that event is as likely, more likely, or less likely than what we would expect by chance. The research hypothesis presents a statement of the expected event, and we use our statistical tools to evaluate how likely that event is.

That's the 20-second version of what inferential statistics is, but that's a lot. So let's take everything from this paragraph and go through it again with an example.

Let's say that your lifelong friend, trusty Lew, gives you a coin and asks you to determine whether it is a "fair" one—that is, if you flip it 10 times, you should come up with 5 heads and 5 tails.

We would expect 5 heads (or 5 tails) because the probability is .5 of a head or a tail on any one flip (if it's a legit coin). On 10 independent flips (meaning that one flip does not affect another), we should get 5 heads, and so on. Now the question is "How many heads would disqualify the coin as being fake or rigged?"

Let's say the criterion for fairness we will use is that if, in flipping the coin 10 times, we get heads (or heads turn up) less than 5% of the time, we'll say the coin is rigged and call the police on Lew. This 5% criterion is one standard that is used by statisticians. If the probability of the event (be it the number of heads or the score on a test or the difference between the average scores for two groups) occurs in the extreme (and we're saying the extreme is defined as less than 5% of all such occurrences), it's an unlikely, or in this case an unfair, outcome.

Back to the coin and Lew.

Because there are 2 possible outcomes (heads or tails) and we are flipping the coin 10 times, there are 2^{10} or 1,024 possible outcomes, such as 9 heads and 1 tail, 7 heads and 3 tails, 10 heads and 0 tails, and on and on. Here's the distribution of how many heads you can expect, just by chance alone, on 10 flips. For example, the probability associated with getting 6 heads in 10 flips is about 21%.

Number of Heads	Probability
0	0.00
1	0.01
2	0.04
3	0.12

(Continued)

(Continued)

Number of Heads	Probability
4	0.21
5	0.25
6	0.21
7	0.12
8	0.04
9	0.01
10	0.00

So, the likelihood of any particular outcome is known. The likelihood of 6 heads from 10 tosses? About .21, or 21%. Now it's decision time. Just how many heads would one have to get on 10 flips to conclude that the coin is fixed, biased, busted, broken, or loony?

Well, as all good statisticians do, we'll define the criterion as 5%, which we discussed. If the probability of the observed outcome (the results of all our flips) is less than 5%, we'll conclude that it is so unlikely that something other than chance must be responsible—and that something is a bogus coin.

If you look at the table, you can see that 8, 9, or 10 heads all represent outcomes that have less than 5% probability of occurring. So if the result of 10 coin flips were 8, 9, or 10 heads, the conclusion would be that the coin is not a fair one. (Yep—you're right: 0, 1, and 2 qualify for the same decision. Sort of the other side of the coin—groan.)

The same logic applies to our discussion of z scores earlier. Just how extreme a z score would we expect before we could proclaim that an outcome is not due just to chance but to some other factor? If you look at the normal curve table in Appendix B, you'll see that the cutoff point for a z score of 1.65 includes about 45% of the area under the curve. If you add that to the other 50% of the area on the other side of the curve, you come up with a total of 95%. That leaves just 5% above that point on the x-axis. Any score that represents a z score of 1.65 or above is then into pretty thin air—or at least in a location that has a much smaller chance of occurring than others.

Hypothesis Testing and z Scores: The First Step

What we showed you here is that any event can have a probability associated with it. And we use those probability values to decide how unlikely we think an event might be. For example, getting only

1 head and 9 tails in 10 tosses of a coin is highly unlikely. We also said that if an event seems to occur only 5 out of 100 times (5%), we will deem that event to be rather *unlikely* relative to all the other events that could occur.

Hypothesis Testing and z Scores

It's much the same with any outcome related to a research hypothesis. The null hypothesis, which you learned about in Chapter 7, claims that there is no difference between groups (or variables) and that the likelihood of no difference occurring is 100%. We try to test the armor of the null for any chinks that might be there.

In other words, if, through the test of the research hypothesis, we find that the likelihood of an event that occurred is somewhat extreme, then the research hypothesis is a more attractive explanation than the null. So, if we find a z score (and remember that z scores have probabilities of occurrence associated with them as well) that is extreme (how extreme?—less than a 5% chance of occurring), we like to say that the reason for the extreme score is something to do with treatments or relationships and not just chance. We'll go into much greater detail on this point in the following chapter.

Using SPSS to Compute z Scores

SPSS does lots of really cool things, but it's the little treats like the one you'll see here that make the program such a great time-saver. Now that you know how to compute z scores by hand, let's let SPSS do the work.

Using SPSS to Compute z Scores

To have SPSS compute z scores for the set of data you see in the first column in Figure 8.6 (which you also saw earlier in the chapter on page 159), follow these steps.

1. Enter the data in a new SPSS window.
2. Click Analyze → Descriptive Statistics → Descriptives.
3. Double-click on the variable Score to move it to the Variable(s): box.
4. Click Save standardized values as variables in the Descriptives dialog box.
5. Click OK.

You can see in Figure 8.6 how SPSS data computes the corresponding z scores. (Be careful—when SPSS does almost anything, it automatically takes you to an Output window where you will not see the computed z scores! You have to switch back to the Data View.)

Figure 8.6	Having SPSS Compute z Scores for You

	Score	ZScore
1	67	.62153
2	54	-.21145
3	65	.49338
4	33	-1.55703
5	56	-.08330
6	76	1.19821
7	65	.49338
8	33	-1.55703
9	48	-.59590
10	76	1.19821

FAT AND SKINNY FREQUENCY DISTRIBUTIONS

You could certainly surmise by now that distributions can be very different from one another in a variety of ways. In fact, there are four different ways in which they can differ: average value (you know—the mean, median, or mode), variability (range, variance, and standard deviation), skewness, and kurtosis. Those last two are new terms, and we'll define them as we show you what they look like. Let's discuss each of the four characteristics and then illustrate them.

Average Value

We're back once again to measures of central tendency. You can see in Figure 8.7 how three different distributions can differ in their average

Figure 8.7	How Distributions Can Differ in Their Average Score

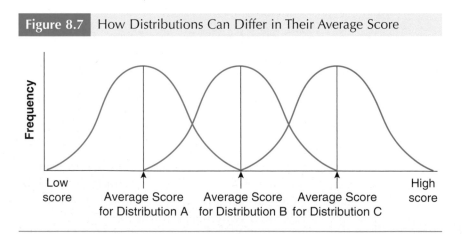

value. Notice that the average for Distribution C is more than the average for Distribution B, which, in turn, is more than the average for Distribution A

Variability

In Figure 8.8, you can see three distributions that all have the same average value but differ in variability. The variability in Distribution A is less than that in Distribution B and, in turn, less than that found in C. Another way to say this is that Distribution C has the largest amount of variability of the three distributions and A has the least.

Figure 8.8 How Distributions Can Differ in Variability

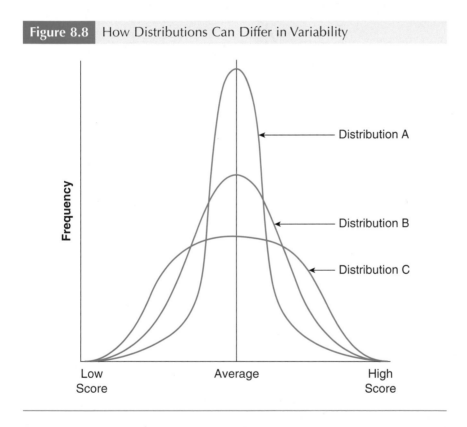

Skewness

Skewness is a measure of the lack of symmetry, or the lopsidedness, of a distribution. In other words, one "tail" of the distribution is longer than another. For example, in Figure 8.9, Distribution A's right tail is longer than its left tail, corresponding to a smaller number of occurrences at the high end of the distribution. This is a positively skewed distribution. This might be the case when you have a test that is very difficult, such that only a few people get scores that are relatively high and many more get scores that are relatively low. Distribution C's right tail is shorter than its left tail, corresponding to a larger number of occurrences at the high end of the distribution. This is a negatively skewed distribution and would be the case for an easy test (lots of high scores and relatively few low scores). And Distribution B—well, it's just right, with equal lengths of tails and no skewness.

Figure 8.9	Degree of Skewness in Different Distributions

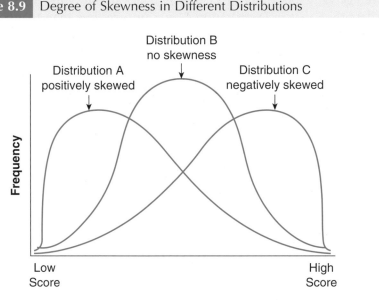

Kurtosis

Even though this sounds like a medical condition, it's the last of the four ways in which we can classify how distributions differ from one another. **Kurtosis** has to do with how flat or peaked a distribution appears, and the terms used to describe this characteristic are relative ones.

For example, the term **platykurtic** refers to a distribution that is relatively flat compared with a normal, or bell-shaped, distribution. The term **leptokurtic** refers to a distribution that is relatively peaked compared with a normal, or bell-shaped, distribution. In Figure 8.10, Distribution A is platykurtic compared with Distribution B. Distribution C is leptokurtic compared with Distribution B. Figure 8.10 looks similar to Figure 8.8 for a good reason—distributions that are platykurtic, for example, are relatively more dispersed than those that are not. Similarly, a distribution that is leptokurtic is less variable or dispersed relative to others.

Figure 8.10 Degrees of Kurtosis in Different Distributions

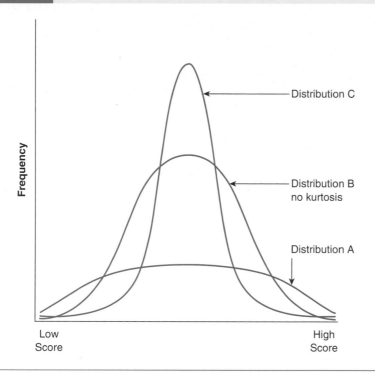

Frequency

Distribution C

Distribution B
no kurtosis

Distribution A

Low
Score

High
Score

While skewness and kurtosis are used mostly as descriptive terms (such as "That distribution is negatively skewed"), there are mathematical indicators of how skewed or kurtotic a distribution is. For example, skewness is computed by subtracting the value of the median from the mean. If the mean of a distribution is 100 and the median is 95, the skewness value is 100 − 95 = 5, and the distribution is positively skewed. If the mean of a distribution is 85 and the median is 90, the skewness value is 85 − 90 = −5, and the distribution is negatively skewed. There's an even more sophisticated formula, which uses the standard deviation of the distribution so that skewness indicators can be compared with one another (see Formula 8.3).

$$Sk = \frac{3(\overline{X} - M)}{s},$$ (8.3)

where

- *Sk* is Pearson's (he's the correlation guy you learned about in Chapter 5) measure of skewness;

(Continued)

(Continued)

- $\overline{X}$ is the mean; and
- M is the median.

Here's an example: The mean of Distribution A is 100, the median is 105, and the standard deviation is 10. For Distribution B, the mean is 120, the median is 116, and the standard deviation is 10. Using Pearson's formula, the skewness of Distribution A is –1.5, and the skewness of Distribution B is 1.2. Distribution A is negatively skewed, and Distribution B is positively skewed. However, Distribution A is more skewed than Distribution B, regardless of the direction.

Let's not leave kurtosis out of this discussion. It, too, can be computed using a fancy formula as follows . . .

$$K = \frac{\Sigma \left(\dfrac{X - \overline{X}}{s} \right)^4}{n} - 3, \qquad (8.4)$$

where

- K = measure of kurtosis;
- Σ = sum;
- X = the individual score;
- $\overline{X}$ = the mean of the sample;
- s = the standard deviation; and
- n = the sample size.

This is a pretty complicated formula that basically looks at how flat or peaked a set of scores is. You can see that if each score is the same, then the numerator is zero and $K = 0$, indicating no skewness. K equals zero when the distribution is normal or *mesokurtic* (now there's a new word to throw around). If the individual scores (the Xs in the formula) differ greatly from the mean (and there is lots of variability), then the curve will probably be quite flat.

REAL-WORLD STATS

More about obesity in children . . .

You have probably heard about all the concerns regarding childhood obesity. These researchers investigated the possible reduction of obesity in children by focusing on physical activity as an intervention. What might this have to do with z scores? A z score was one of their primary outcome or dependent variables: mean body mass index (BMI) z score = 3.24, SD = 0.49.

The participants were invited to participate in a 1-week sports camp, and after the camp, a coach from a local sports club supported the children during participation in a chosen activity for 6 months. Weight, height, body composition, and lifestyle were measured at baseline and after 12 months. The results? Children who participated in the intervention had a significant decrease in BMI z score.

Why did the researchers use z scores? Most probably because the children who were being compared with one another came from different distributions of scores (they had different BMIs), and by using a standard score, those differences (at least in the variability of the scores) were eliminated.

Want to know more? Go online or to the library and find . . .

Nowicka, P., Lanke, J., Pietrobelli, A., Apitzsch, E., & Flodmark, C. E. (2009). Sports camp with six months of support from a local sports club as a treatment for childhood obesity. *Scandinavian Journal of Public Health, 37*, 793–800.

Real-World Stats

SUMMARY

Being able to figure out a z score, and being able to estimate how likely it is to occur in a sample of data, is the first and most important skill for understanding the whole notion of inference. Once we know how likely a test score or a difference between groups is, we can compare that likelihood with what we would expect by chance and then make informed decisions. As we start Part IV of *Statistics for People Who (Think They) Hate Statistics,* we'll apply this model to specific examples of testing questions about the difference.

Chapter Summary

TIME TO PRACTICE

1. What are the characteristics of the normal curve? What human behavior, trait, or characteristic can you think of that is distributed normally? What makes you think it may be distributed normally?

2. To compute a standard score, what three bits of information do you need?

3. Standard scores, such as z scores, allow us to make comparisons across different samples. Why?

4. Why is a z score a standard score, and why can standard scores be used to compare scores from different distributions with one another?

5. The mean of a set of test scores is 50, and the standard deviation is 5. For a raw score of 55, the corresponding z score is +1. What's the z score when the standard deviation is half as much, or 2.5? From this example, what can you conclude the effect of decreasing the amount of variability in a set of scores is on a standard score (given all else is equal, such as the same raw score), and why is this effect important?

6. For the following set of scores, fill in the cells. The mean is 74.13, and the standard deviation is 9.98.

Raw Score	z Score
68.0	?
?	−1.6
82.0	?
?	1.8
69.0	?
?	−0.5
85.0	?
?	1.7
72.0	?

Problem 7

7. For the following set of scores, compute standard scores. Do it using SPSS (easy) and do it manually to keep SPSS honest (not as easy as using SPSS but once you get the hang of it—easy enough). Notice any differences?

18
19
15
20
25
31
17
35
27
22
34
29
40
33
21

8. Questions 8a through 8d are based on a distribution of scores with $\bar{X} = 75$ and standard deviation = 6.38. Draw a small picture to help you see what's required.

 a. What is the probability of a score falling between a raw score of 70 and 80?

 b. What is the probability of a score falling above a raw score of 80?

 c. What is the probability of a score falling between a raw score of 81 and 83?

 d. What is the probability of a score falling below a raw score of 63?

9. Jake needs to score in the top 10% of his class to earn a physical fitness certificate. The class mean is 78, and the standard deviation is 5.5. What raw score does he need to get that valuable piece of paper?

10. Imagine you are in charge of a program in which members are evaluated on five different tests at the end of the program. Why doesn't it make sense to simply compute the average of the five scores as a measure of performance rather than compute a z score for each test for each individual and average those?

11. Who is the better student, relative to his or her classmates? Here's all the information you ever wanted to know . . .

Math		
Class Mean	81	
Class Standard Deviation	2	
Reading		
Class Mean	87	
Class Standard Deviation	10	
Raw Scores		
	Math Score	Reading Score	Average
Noah	85	88	86.5
Talya	87	81	84.0
z Scores		
	Math Score	Reading Score	Average
Noah	_____	_____	_____
Talya	_____	_____	_____

12. Here's an interesting extra-credit question. As you know, one of the defining characteristics of the normal curve is that the tails do not touch the x-axis. Why don't they touch?

STUDENT STUDY SITE

$SAGE edge™

Get the tools you need to sharpen your study skills! Visit **edge.sagepub.com/ salkind6e** to access practice quizzes, eFlashcards, original and curated videos, data sets, journal articles, and more!

PART IV

Significantly Different

Using Inferential Statistics

"Talk about a sum of squares."

You've gotten this far and you're still alive and kicking, so congratulate yourself. At this point, you should have a good understanding of what descriptive statistics is all about, how chance figures as a factor in making decisions about outcomes, and how likely outcomes are to have occurred due to chance or some treatment.

You're an expert on creating and understanding the role that hypotheses play in social and behavioral science research. Now it's time for the rubber to meet the road. Let's see what you're made of in the next part of *Statistics for People Who (Think They) Hate Statistics*. The hard work you've put in will shortly pay off with an understanding of applied problems!

This part of the book deals exclusively with understanding and applying certain types of statistics to answer certain types of research questions, and we'll cover the most common statistical tests. Then, in Part V, we'll even cover a few that are a bit more sophisticated, and we'll show you some of the more useful software packages that can be used to compute the same values that we'll compute using a good old-fashioned calculator.

Let's start with a brief discussion of what the concept of significance is and go through the steps for performing an inferential test. Then we'll go on to examples of specific tests. We have lots of hands-on work ahead of us, so let's get started.

9 Significantly Significant

What It Means for You and Me

Difficulty Scale ☺ ☺ (somewhat thought provoking and key to it all!)

WHAT YOU'LL LEARN ABOUT IN THIS CHAPTER

✦ Understanding the concept of significance and why it is important

✦ Distinguishing between Type I and Type II errors

✦ Understanding how inferential statistics work

✦ Selecting the appropriate statistical test for your purposes

---

THE CONCEPT OF SIGNIFICANCE

Probably no term or concept causes the beginning statistics student more confusion than *statistical significance*. But it doesn't have to be that way for you. Although it's a powerful idea, it is also relatively straightforward and can be understood by anyone in a basic statistics class.

Introduction to Chapter 9

   We need an example of a study to illustrate the points we want to make. Let's take E. Duckett and M. Richards's "Maternal Employment and Young Adolescents' Daily Experiences in Single Mother Families" (paper presented at the Society for Research in Child Development, Kansas City, MO, 1989—a long time ago in a galaxy far away [Kansas City, Missouri . . .]). These two authors examined the attitudes of 436 fifth- through ninth-grade adolescents toward maternal employment. Even though the presentation took place some years ago, it's a perfect example for illustrating many of the important ideas at the heart of this chapter.

Specifically, the two researchers investigated whether differences are present between the attitudes of adolescents whose mothers work and the attitudes of adolescents whose mothers do not work. They also examined some other factors, but for this example, we'll stick with the mothers-who-work and mothers-who-don't-work groups. One more thing. Let's add the word *significant* to our discussion of differences, and we have a research hypothesis something like this:

> *There is a significant difference in attitude toward maternal employment between adolescents whose mothers work and adolescents whose mothers do not work, as measured by a test of emotional state.*

What we mean by the word *significant* is that any difference between the attitudes of the two groups is due to some systematic influence and not due to chance. In this example, that influence is whether or not mothers work. We assume that all of the other factors that might account for any differences between groups were controlled. Thus, the only thing left to account for the differences between adolescents' attitudes is whether or not their mothers work. Right? Yes. Finished? Not quite.

If Only We Were Perfect

Because our world is not a perfect one, we must allow for some leeway in how confident we are that only those factors we identify could cause any difference between groups. In other words, you need to be able to say that although you are pretty sure the difference between the two groups of adolescents is due to maternal employment, you cannot be absolutely, 100%, positively, unequivocally, indisputably (get the picture?) sure. There's always a chance, no matter how small, that you are wrong. The improbable is probable: So said Aristotle, and he was right. No matter how small the chance, there's always a chance.

 And, by the way, the whole notion of the normal curve's tails never, ever really touching the x-axis (as we mentioned in the last chapter) is directly relevant to our discussion here. If the tails did touch, the probability of an event being very extreme in one or the other tail would be absolutely zero. But since they do not touch, there is always a chance, no matter how perfect we might be, that an event can occur—no matter how small and unlikely its probability might be.

Why? Many reasons. For example, you could just be wrong. Maybe during this one experiment, differences between adolescents' attitudes were not due to whether their mothers worked or didn't work but were due to some other factor that was inadvertently not accounted for, such as a speech given by the local Mothers Who Work Club that several students attended. How about if the people in one group were mostly adolescent males and the people in the other group were mostly adolescent females? That could be the source of a difference as well. If you are a good researcher and do your homework, you can account for such differences, but it's always possible that you can't. And as a good researcher, you have to take that possibility into account.

So what do you do? In most scientific endeavors that involve testing hypotheses (such as the group differences example here), there is bound to be a certain amount of error that cannot be controlled—this is the chance factor that we have been talking about in the past few chapters. The level of chance or risk you are willing to take is expressed as a *significance level*, a term that unnecessarily strikes fear in the hearts of even strong men and women.

Significance level (here's the quick-and-dirty definition) is the risk associated with not being 100% confident that what you observe in an experiment is due to the treatment or what was being tested—in our example, whether or not mothers worked. If you read that significant findings occurred at the .05 level (or $p < .05$ in tech talk and what you regularly see in professional journals), the translation is that there is 1 chance in 20 or 5 in 100 (or .05 or 5%) that any differences found were not due to the hypothesized reason (whether mom works) but to some other, unknown reason or reasons (yep—or chance). Your job is to reduce this likelihood as much as possible by removing all of the competing reasons for any differences that you observed. Because you cannot fully eliminate the likelihood (because no one can control every potential factor), you assign some level of probability and report your results with that caveat.

In sum (and in practice), the researcher defines a level of risk that he or she is willing to take. If the results fall within the region that says, "This could not have occurred by chance alone—something else is going on," the researcher knows that the null hypothesis (which states an equality) is not the most attractive explanation for the observed outcomes. Instead, the research hypothesis (that there is an inequality or a difference) is the favored explanation.

Let's take a look at another example, this one hypothetical.

A researcher is interested in seeing whether there is a difference between the academic achievement of children who participated in a preschool program and that of children who did not participate. The null hypothesis is that the two groups are equal to each other on some measure of achievement.

The research hypothesis is that the mean score for the group of children who participated in the program is higher than the mean score for the group of children who did not participate in the program.

As a good researcher, your job is to show (as best you can—and no one is so perfect that he or she can account for everything) that *any* difference that exists between the two groups is due only to the effects of the preschool experience and no other factor or combination of factors. Through a variety of techniques (that you'll learn about in your next statistics or research methods class!), you control or eliminate all the possible sources of difference, such as the influence of parents' education, number of children in the family, and so on. Once these other potential explanatory variables are removed, the only remaining alternative explanation for differences is the effect of the preschool experience itself.

But can you be absolutely (which is pretty darn) sure? No, you cannot. Why? First, because you can never be sure that you are testing a sample that identically reflects the profile of the population. And even if the sample perfectly represents the population, there are always other influences that might affect the outcome and that you inadvertently missed when designing the experiment. There's *always* the possibility of error (another word for chance).

By concluding that the differences in test scores are due to differences in treatment, you accept some risk. This degree of risk is, in effect (drumroll, please), the level of statistical significance at which you are willing to operate.

Statistical significance (here's the formal definition) is the degree of risk you are willing to take that you will reject a null hypothesis when it is actually true. In our preceding example, the null says that there is no difference between the two sample groups (remember, the null is always a statement of equality). In your data, however, you did find a difference. That is, given the evidence you have so far, group membership seems to have an effect on achievement scores. In reality, however, maybe there is no difference. If you reject the null you stated, you would be making an error. The risk you take in making this kind of error (or the level of significance) is also known as a Type I error.

So the next step is to develop a set of steps to test whether our findings indicate that error is responsible for differences or actual differences are responsible.

The World's Most Important Table (for This Semester Only)

Here's what it all boils down to.

A null hypothesis can be true or false. Either there really is no difference between groups, or there really and truly is an inequality

| Table 9.1 | Different Types of Errors |

		Action You Take	
		Accept the Null Hypothesis	**Reject the Null Hypothesis**
True nature of the null hypothesis	The null hypothesis is really true.	1 ☺ Bingo, you accepted a null when it is true and there is really no difference between groups.	2 ☹ Oops—you made a Type I error and rejected a null hypothesis even when there really is no difference between groups. Type I errors are also represented by the Greek letter alpha, or α.
	The null hypothesis is really false.	3 ☹ Uh-oh—you made a Type II error and accepted a false null hypothesis. Type II errors are also represented by the Greek letter beta, or β.	4 ☺ Good job, you rejected the null hypothesis when there really are differences between the two groups. This is also called power, or 1 − β.

(such as the difference between two groups). But remember, you'll never know this true state because the null cannot be tested directly (remember that the null applies only to the population and, for a variety of reasons we have talked about, the population cannot be directly tested).

And, as a crackerjack statistician, you can choose to either reject or accept the null hypothesis. Right? These four different conditions create the table you see in Table 9.1.

Let's look at each cell.

More About Table 9.1

Table 9.1 has four important cells that describe the relationship between the nature of the null (whether it's true or not) and your action (accept or reject the null hypothesis). As you can see, the null can be either true or false, and you can either reject or accept it.

Accepting and Rejecting the Null Hypothesis

The most important thing about understanding this table is the fact that the researcher never really knows the true nature of the null hypothesis and whether there really is or is not a difference between groups. Why? Because the population (which the null represents) is never directly tested. Why? Because it's impractical to do so, and that's why we have inferential statistics.

- ☺ Cell 1 in Table 9.1 represents a situation in which the null hypothesis is really true (there's no difference between groups) and the researcher made the correct decision accepting the null. No problem here. In our example, our results would show that there is no difference between the two groups of children, and we have acted correctly by accepting the null that there is no difference.
- ☹ Oops. Cell 2 represents a serious error. Here, we have rejected the null hypothesis (that there is no difference) when it is really true (and there is no difference between groups). Even though there is no difference between the two groups of children, we will conclude there is, and that's an error—clearly a boo-boo called a **Type I error**, also known as the level of significance.
- ☹ Uh-oh, another type of error. Cell 3 represents a serious error as well. Here, we have accepted the null hypothesis (that there is no difference) when it is really false (and, indeed, there is a difference between groups). We have said that even though there is a difference between the two groups of children, we will conclude there is not—also clearly a boo-boo, this time known as a **Type II error**.
- ☺ Cell 4 represents a situation where the null hypothesis is really false and the researcher made the correct decision in rejecting it. No problem here. In our example, our results show that there is a difference between the two groups of children, and we have acted correctly by rejecting the null that states there is no difference.

So, if .05 is good and .01 is even "better," why not set your Type I level of risk at .000001? For the very good reason that you would be so rigorous in your rejection of false null hypotheses that you might reject a null when it was actually true. Such a stringent Type I error rate allows for little leeway—indeed, the research hypothesis might be true but the associated probability might be .015—still quite rare and probably very useful information, but missed with the too-rigid Type I level of error.

Back to Type I Errors

Evaluating
Levels of
Significance

Let's focus a bit more on Cell 2, where a Type I error was made, because this is the focus of our discussion.

This Type I error, or level of significance, has certain values associated with it that define the risk you are willing to take in any test of the null hypothesis. The conventional levels set are between .01 and .05.

For example, if the level of significance is .01, it means that on any one test of the null hypothesis, there is a 1% chance you will reject the null hypothesis when the null is true and conclude that there is a group difference when there really is no group difference at all.

If the level of significance is .05, it means that on any one test of the null hypothesis, there is a 5% chance you will reject it (and conclude that there is a group difference) when the null is true and there really is no group difference at all. Notice that the level of significance is associated with an independent test of the null, and it is not appropriate to say that "on 100 tests of the null hypothesis, I will make an error on only 5 times, or 5% of the time."

In a research report, statistical significance is usually represented as $p < .05$, read as "the probability of observing that outcome is less than .05" and often expressed in a report or journal article simply as "significant at the .05 level."

With the introduction of fancy-schmancy software such as SPSS and Excel that can do statistical analysis, there's no longer the worry about the imprecision of such statements as "$p < .05$" or "$p < .01$." For example, $p < .05$ can mean anything from .000 to .049999, right? Instead, software such as SPSS and Excel gives you the *exact* probability, such as .013 or .158, of the risk you are willing to take that you will commit a Type I error. So, when you see in a research article the statement that "$p < .05$," it means that the value of p is equal to anything from .00 to .049999999999 . . . (you get the picture). Likewise, when you see "$p > .05$" or "$p =$ n.s." (for nonsignificant), it means that the probability of rejecting a true null exceeds .05 and, in fact, can range from .0500001 to 1.00. So, it's actually terrific when we know the exact probability of an outcome because we can measure more precisely the risk we are willing to take. But what to do if the p value is exactly .05? Well, given what you've already read, if you want to play by the rules, then the outcome is not significant. A result either is, or is not. So, .04999999999 is and .05 is not. But, if SPSS or Excel (or any other program) generates a value of .05, extend the number of decimal places—it may really be .04999999999.

As discussed earlier, there is another kind of error you can make, which, along with the Type I error, is shown in Table 9.1. A Type II error (Cell 3 in the chart) occurs when you inadvertently accept a false null hypothesis. For example, there may really be differences between the populations represented by the sample groups, but you mistakenly conclude there are not.

When talking about the significance of a finding, you might hear the word *power* used. Power is a construct that has to do with how well a statistical test can detect and reject a null hypothesis when it is false. Mathematically, it's calculated by subtracting the value of the Type II error from 1. A more powerful test is always more desirable than a less powerful test, because the more powerful one lets you get to the heart of what's false and what's not.

Ideally, you want to minimize both Type I and Type II errors, but doing so is not always easy or under your control. You have complete control over the Type I error level or the amount of risk that you are willing to take (because you actually set the level itself). Type II errors, however, are not as directly controlled but instead are related to factors such as sample size. Type II errors are particularly sensitive to the number of subjects in a sample, and as that number increases, Type II error decreases. In other words, as the sample characteristics more closely match those of the population (achieved by increasing the sample size), the likelihood that you will accept a false null hypothesis decreases.

SIGNIFICANCE VERSUS MEANINGFULNESS

Statistical
Versus
Practical
Significance

What an interesting situation arises for the researcher when he or she discovers that the results of an experiment indeed are statistically significant! You know technically what statistical significance means—that the research was a technical success and the null hypothesis is not a reasonable explanation for what was observed. Now, if your experimental design and other considerations were well taken care of, statistically significant results are unquestionably the first step toward making a contribution to the literature in your field. However, the value of statistical significance and its importance or meaningfulness must be kept in perspective.

For example, let's take the case where a very large sample of illiterate adults (say, 10,000) is divided into two groups. One group receives intensive training to read using traditional teaching, and the other receives intensive training to read using computers. The average score for Group 1 (which learned in the traditional way) is 75.6 on a reading test, the dependent variable. The average score on the reading test for Group 2 (which learned using the computer) is 75.7. The amount of variance in both groups is about equal. As you can see, the difference in score is only one tenth of 1 point or 0.1 (75.6 vs. 75.7), but when a *t*-test for the significance between independent means is applied, the results are significant at the .01 level, indicating that computers work better than traditional teaching methods. (Chapters 11 and 12 discuss *t*-tests, the kind we would use in such a situation.)

The difference of 0.1 is indeed statistically significant, at the .01 level, but is it meaningful? Does the improvement in test scores (by such a small margin) provide sufficient rationale for the $300,000 it costs to fund the program? Or is the difference negligible enough that it can be ignored, even if it is statistically significant?

Here are some conclusions about the importance of statistical significance that we can reach, given this and the countless other possible examples:

- Statistical significance, in and of itself, is not very meaningful unless the study that is conducted has a sound conceptual base that lends some meaning to the significance of the outcome.
- Statistical significance cannot be interpreted independently of the context within which the outcomes occur. For example, if you are the superintendent in a school system, are you willing to retain children in Grade 1 if the retention program significantly raises their standardized test scores by one half point?
- Although statistical significance is important as a concept, it is not the end-all and certainly should not be the only goal of scientific research. That is the reason why we set out to *test* hypotheses rather than *prove* them. If our study is designed correctly, then even null results tell you something very important. If a particular treatment does not work, this is important information that others need to know about. If your study is designed well, then you should know why the treatment does not work, and the next person down the line can design his or her study taking into account the valuable information you have provided.

Researchers treat the reporting of statistical significance in many different ways in their written reports. Some use words such as *significant* (assuming that if something is significant, it is statistically so) or the entire phrase *statistically significant*. But some also use the phrase *marginally significant*, where the probability associated with a finding might be .051 or .053. What to do? You're the boss, if your own data are being analyzed or if you are reviewing someone else's. Use your noodle and consider all the dimensions of the work being done. If .051, within the context of the question being asked and answered, is "good enough," then it is. Whether outside reviewers agree is a source of great debate and a good topic for class discussion.

Almost every discipline has "other" terms for this *significant* versus *meaningful* distinction, but the issue is generally considered to concern the same elements. For example, health care professionals refer to the meaningful part of the equation as "clinical significance" rather than "meaningfulness." It's the same idea—they just use a different term given the setting in which their outcomes occur.

Ever hear of "publication bias"? It's where a preset significance value of .05 is used as the *only* criterion in the serious consideration of a paper for publication. It's not exactly .05 or bust, but in times past and even today, some editorial boards hold up significance values such as .05 or .01 as the holy grail of getting things right. If those values are not reached, then the findings cannot be significant, let alone meaningful, according to some people's judgment. Now, there's something to be said for consistency throughout a field, but today's cool stats tools such as SPSS and Excel allow us to pinpoint the exact probability associated with an outcome rather than an all-or-nothing criterion such as .05, which dooms some meaningful work before it is even discussed. Be sophisticated—make up your own mind based on all the evidence.

AN INTRODUCTION TO INFERENTIAL STATISTICS

Whereas descriptive statistics are used to describe a sample's characteristics, inferential statistics are used to infer something about the population based on the sample's characteristics.

At several points throughout the first half of *Statistics for People Who (Think They) Hate Statistics,* we have emphasized that a hallmark of good scientific research is choosing a sample in such a way that it is representative of the population from which it was selected. The process then becomes an inferential one, in which you infer from the smaller sample to the larger population based on the results of tests (and experiments) conducted using the sample.

Before we start discussing individual inferential tests, let's go through the logic of how the inferential method works.

How Inference Works

Here are the general steps of a research project to see how the process of inference might work. We'll stay with adolescents' attitudes toward mothers working as an example.

Here's the sequence of events that might happen:

1. The researcher selects representative samples of adolescents who have mothers who work and adolescents who have mothers who do not work. These are selected in such a way that the samples represent the populations from which they are drawn.

2. Each adolescent is administered a test to assess his or her attitude. The mean scores for groups are computed and compared using some test.

3. A conclusion is reached as to whether the difference between the scores is the result of chance (meaning some factor other than moms working is responsible for the difference) or the result of "true" and statistically significant differences between the two groups (meaning the results are due to moms working).

4. A conclusion is reached as to the relationship between maternal employment and adolescents' attitudes in the population from which the sample was originally drawn. In other words, an inference, based on the results of an analysis of the sample data, is made about the population of all adolescents.

How to Select What Test to Use

Step 3 above brings us to ask the question "How do I select the appropriate statistical test to determine whether a difference between groups exists?" Heaven knows, there are plenty of them, and you have to decide which one to use and when to use it.

Well, the best way to learn which test to use is to be an experienced statistician who has taken lots of courses in this area and participated in lots of research. Experience is still the greatest teacher. In fact, there's no way you can really learn what to use and when to use it unless you've had the real-life, applied opportunity to actually use these tools. And as a result of taking this course, you are learning how to use these very tools.

So, for our purposes and to get started, we've created this nice little flowchart (a.k.a. cheat sheet) of sorts that you see in Figure 9.1 on page 189. You have to have some idea of what you're doing, so selecting the correct statistical test does not put the rest of your study on auto-pilot, but it certainly is a good place to get started.

Don't think for a second that Figure 9.1 takes the place of your need to learn about when these different tests are appropriate. The flowchart is here only to help you get started.

 This is really important. We just wrote that selecting the appropriate statistical test is not necessarily an easy thing to do. And the best way to learn how to do it is to do it, and that means practicing and even taking more statistics courses. The simple flowchart we present here works, in general, but use it with caution. When you make a decision, check with your professor or some other person who has been through this stuff and feels more confident than you might (and who also knows more!).

Here's How to Use the Chart

 1. Assume that you're very new to this statistics stuff (which you are) and that you have some idea of what these tests of significance are, but you're pretty lost as far as deciding which one to use when.

2. Answer the question at the top of the flowchart.

3. Proceed down the chart by answering each of the questions until you get to the end of the chart. That's the statistical test you should use. This is not rocket science, and with some practice (which you will get throughout this part of *Statistics for People . . .*), you'll be able to quickly and reliably select the appropriate test. Each of the remaining chapters in this part of the book will begin with a chart like the one you see in Figure 9.1 and take you through the specific steps to get to the test statistic you should use.

 Does the cute flowchart in Figure 9.1 contain all the statistical tests there are? Not by a long shot. There are hundreds, but the ones in Figure 9.1 are the ones used most often. And if you are going to become familiar with the research in your own field, you are bound to run into these.

AN INTRODUCTION TO TESTS OF SIGNIFICANCE

Tests of
Significance

What inferential statistics does best is allow decisions to be made about populations based on information about samples. One of the most useful tools for doing this is a test of statistical significance that can be applied to different types of situations, depending on the nature of the question being asked and the form of the null hypothesis.

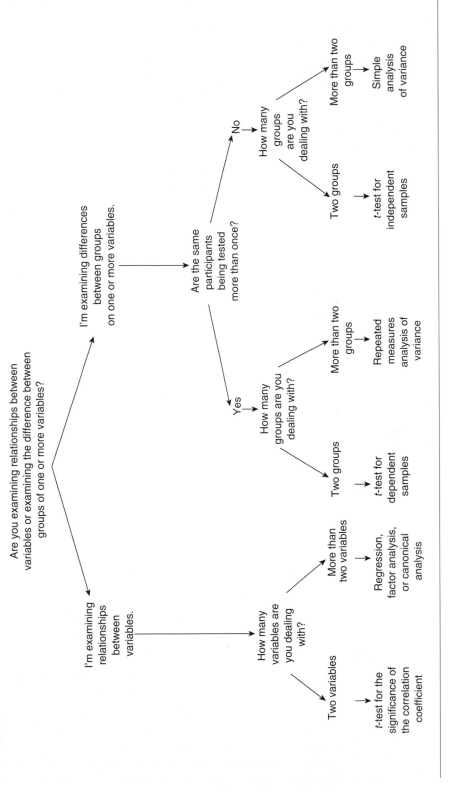

For example, do you want to look at the difference between two groups, such as whether boys score significantly differently than girls on some test? Or the relationship between two variables, such as number of children in a family and average score on intelligence tests? The two cases call for different approaches, but both will result in a test of a null hypothesis using a specific test of statistical significance.

How a Test of Significance Works: The Plan

Tests of significance are based on the fact that each type of null hypothesis has associated with it a particular type of statistic. And each of the statistics has associated with it a special distribution that you compare with the data you obtain from a sample. A comparison between the characteristics of your sample and the characteristics of the test distribution allows you to conclude whether the sample characteristics are different from what you would expect by chance.

Here are the general steps to take in the application of a statistical test to any null hypothesis. These steps will serve as a model for each of the chapters in Part IV.

1. *A statement of the null hypothesis.* Do you remember that the null hypothesis is a statement of equality? The null hypothesis is the "true" state of affairs given no other information on which to make a judgment.

2. *Setting the level of risk (or the level of significance or Type I error) associated with the null hypothesis.* With any research hypothesis comes a certain degree of risk that you are wrong. The smaller this error is (such as .01 compared with .05), the less risk you are willing to take. No test of a hypothesis is completely risk-free because you never really know the "true" relationship between variables. Remember that it is traditional to set the Type I error rate at .01 or .05; SPSS and Excel and other programs specify the exact level.

3. *Selection of the appropriate test statistic.* Each null hypothesis has associated with it a particular test statistic. You can learn what test is related to what type of question in this part of *Statistics for People . . . .*

4. *Computation of the test statistic value.* The **test statistic value** (also called the **obtained value**) is the result or product of a specific statistical test. For example, there are test statistics for the significance of the difference between the averages of two groups, for the significance of the difference of a correlation coefficient from zero, and for the significance of the difference between two proportions. You'll actually compute the test statistic and come up with a numerical value.

5. *Determination of the value needed for rejection of the null hypothesis using the appropriate table of critical values for the particular statistic.* Each test statistic (along with group size and the risk you are willing to take) has a **critical value** associated with it. This is the value you would expect the test statistic to yield if the null hypothesis is indeed true.

6. *Comparison of the obtained value with the critical value.* This is the crucial step. Here, the value you obtained from the test statistic (the one you computed) is compared with the value (the critical value) you would expect to find by chance alone.

7. *If the obtained value is more extreme than the critical value, the null hypothesis cannot be accepted.* That is, the null hypothesis's statement of equality (reflecting chance) is not the most attractive explanation for differences that were found. Here is where the real beauty of the inferential method shines through. Only if your obtained value is more extreme than what would happen by chance (meaning that the result of the test statistic is not a result of some chance fluctuation) can you say that any differences you obtained are not due to chance and that the equality stated by the null hypothesis is not the most attractive explanation for any differences you might have found. Instead, the differences must be due to the treatment. What if the two values are equal (and you were about to ask your instructor that question, right?)? Nope—due to chance.

8. *If the obtained value does not exceed the critical value, the null hypothesis is the most attractive explanation.* If you cannot show that the difference you obtained is due to something other than chance (such as the treatment), then the difference must be due to chance or something you have no control over. In other words, the null is the best explanation.

Here's the Picture That's Worth a Thousand Words

What you see in Figure 9.2 represents the eight steps we just went through. This is a visual representation of what happens when the obtained and critical values are compared. In this example, the significance level is set at .05, or 5%. It could have been set at .01, or 1%.

In examining Figure 9.2, note the following:

1. The entire curve represents all the possible outcomes based on a specific null hypothesis, such as the difference between two groups or the significance of a correlation coefficient.

| Figure 9.2 | Comparing Obtained Values With Critical Values and Making Decisions About Rejecting or Accepting the Null Hypothesis |

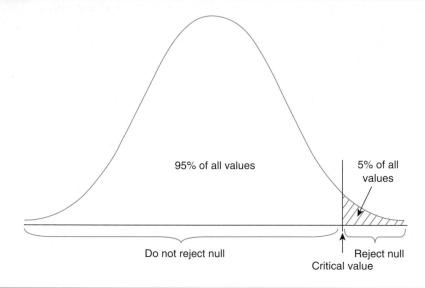

95% of all values

5% of all values

Do not reject null

Reject null
Critical value

2. The critical value is the point beyond which the obtained outcomes are judged to be so rare that the conclusion is that the obtained outcome is not due to chance but to some other factor. In this example, we define *rare* as having a less than 5% chance of occurring.

3. If the outcome representing the obtained value falls to the left of the critical value (it is less extreme), the conclusion is that the null hypothesis is the most attractive explanation for any differences that are observed. In other words, the obtained value falls in the region (95% of the area under the curve) where we expect only outcomes due to chance to occur.

4. If the obtained value falls to the right of the critical value (it is more extreme), the conclusion is that the research hypothesis is the most attractive explanation for any differences that are observed. In other words, the obtained value falls in the region (5% of the area under the curve) where we would expect only outcomes due to something other than chance to occur.

BE EVEN MORE CONFIDENT

You now know that probabilities can be associated with outcomes—that's been an ongoing theme for the past two chapters. Now we are going to say the same thing in a bit different way and introduce a new idea called confidence intervals.

A **confidence interval** (or c.i.) is the best estimate of the range of a population value (or population parameter) that we can come up with given the sample value (or sample statistic representing the population parameter). Say we knew the mean spelling score for a sample of 20 third graders (of all the third graders in a school district). How much confidence could we have that the population mean will fall between two scores? A 95% confidence interval would be correct 95% of the time.

You already know that the probability of a raw score falling within ±1.96 z scores or standard deviations is 95%, right? (See page 162 in Chapter 8 if you need some review.) Or the probability of a raw score falling within ±2.56 z scores or standard deviations is 99%. If we use the positive or negative raw scores equivalent to those z scores, we have ourselves a confidence interval.

Let's fool around with some real numbers.

Let's say that the mean spelling score for a random sample of 100 sixth graders is 64 (out of 75 words) and the standard deviation is 5. What confidence can we have in predicting the population mean for the average spelling score for the entire population of sixth graders?

The 95% confidence interval is equal to . . .

$$64 \pm 1.96(5)$$

. . . or a range from 54.2 to 73.8, so at the least you can say with 95% confidence that the population mean for the average spelling score for all sixth graders falls between those two scores.

Want to be more confident? The 99% confidence interval would be computed as follows . . .

$$64 \pm 2.56(5)$$

or a range from 51.2 to 76.8, so you can conclude with 99% confidence that the population mean falls between those two scores.

Do keep in mind that some people use the standard error of the mean to compute confidence intervals and that the use of either the standard deviation or the standard error of the mean is correct. See what your instructor prefers you do.

 While we are using the standard deviation to compute confidence intervals, many people choose to use the standard error of the mean or SEM (see Chapter 10). The standard error of the mean is the standard deviation of all

(Continued)

(Continued)

the sample means that could, in theory, be selected from the population. Remember that both the standard deviation and the standard error of the mean are "errors" in measurement that surround a certain "true" point (which in our case would be the true mean and the true amount of variability). The use of the SEM is a bit more complex, but it is an alternative way of computing, and understanding, confidence intervals.

Why does the confidence interval itself get larger as the probability of your being correct increases (from, say, 95% to 99%)? Because the larger range of the confidence interval (in this case 19.6 [73.8–54.2] for a 95% confidence interval versus 25.6 [76.8–51.2] for a 99% confidence interval) allows you to encompass a larger number of possible outcomes and you can thereby be more confident. Ha! Isn't this stuff cool?

REAL-WORLD STATS

It's really interesting how different disciplines can learn from one another when they share, and it's a shame that this does not happen more often. This is one of the reasons why interdisciplinary studies are so vital—they create an environment where new and old ideas can be used in new and old settings.

One such discussion took place in a medical journal that devotes itself to articles on anesthesia. The focus was a discussion of the relative merits of statistical versus clinical significance, and Drs. Timothy Houle and David Stump pointed out that many large clinical trials obtain a high level of statistical significance with minuscule differences between groups (just as we talked about earlier in the chapter), making the results clinically irrelevant. However, the authors pointed out that with proper marketing, billions can be made from results of dubious clinical importance. This is really a caveat emptor or buyer-beware state of affairs. Clearly, there are a few very good lessons here about whether the significance of an outcome is really meaningful or not. How to know? Look at the substance behind the results and the context within which the outcomes are found.

Want to know more? Go online or to the library and find . . .

Real-World Stats

Houle, T. T., & Stump, D. A. (2008). Statistical significance versus clinical significance. *Seminars in Cardiothoracic and Vascular Anesthesia, 12,* 5–6.

SUMMARY

So, now you know exactly how the concept of significance works, and all that is left is to apply it to a variety of research questions. That's what we'll start with in the next chapter and continue with throughout this part of the book.

Chapter
Summary

TIME TO PRACTICE

1. Why is significance an important construct in the study and use of inferential statistics?

2. What is statistical significance?

3. What does the (idea of the) critical value represent?

4. Given the following information and setting the level of significance at .05 for decision making, would your decision be to reject or fail to reject the null hypothesis? Provide an explanation for your conclusions.

    a. The null hypothesis that there is no relationship between the type of music a person listens to and his propensity for crime ($p < .05$)

    b. The null hypothesis that there is no relationship between the amount of coffee consumption and GPA ($p = .62$)

    c. The research hypothesis that a negative relationship exists between the number of hours worked and level of job satisfaction ($p = .51$)

5. What's wrong with the following statements?

    a. A Type I error of .05 means that 5 times out of 100 experiments, we will reject a true null hypothesis.

    b. It is possible to set the Type I error rate to zero.

    c. The smaller the Type I error rate, the better the results.

6. Why is it "harder" to find a significant outcome (all other things being equal) when the research hypothesis is being tested at the .01 rather than the .05 level of significance?

7. Why should we think in terms of "failing to reject" the null, rather than just accepting it?

8. What is the difference between *significance* and *meaningfulness*?

9. Here's more exploration of the significance versus meaningfulness debate:

Problem 8

    a. Provide an example of a finding that may be both statistically significant and meaningful.

    b. Now provide an example of a finding that may be statistically significant but not meaningful.

10. What does chance have to do with testing the research hypothesis for significance?

11. In Figure 9.2 (p. 192), there is a striped area on the right side of the illustration.

    a. What does that entire striped area represent?

    b. If the striped area were a larger portion underneath the curve, what would that represent?

STUDENT STUDY SITE

$SAGE edge™

Get the tools you need to sharpen your study skills! Visit **edge.sagepub.com/salkind6e** to access practice quizzes, eFlashcards, original and curated videos, data sets, journal articles, and more!

10 Only the Lonely

The One-Sample Z-Test

Difficulty Scale ☺ ☺ ☺ (not too hard—this is the first chapter of this kind, but you know more than enough to master it)

WHAT YOU'LL LEARN ABOUT IN THIS CHAPTER

✦ Deciding when the Z-test for one sample is appropriate to use

✦ Computing the observed z value

✦ Interpreting the z value

✦ Understanding what the z value means

✦ Understanding what effect size is and how to interpret it

INTRODUCTION TO THE ONE-SAMPLE Z-TEST

Lack of sleep can cause all kinds of problems, from grouchiness to fatigue and, in rare cases, even death. So, you can imagine health care professionals' interest in seeing that their patients get enough sleep. This is especially the case for patients who are ill and have a real need for the healing and rejuvenating qualities that sleep brings. Dr. Joseph Cappelleri and his colleagues looked at the sleep difficulties of patients with a particular illness, fibromyalgia, to evaluate the usefulness of the Medical Outcomes Study (MOS) Sleep Scale as a measure of sleep problems. Although other analyses were completed, including one that compared a treatment group and a control group with one another, the important analysis (for our discussion) was the comparison of participants' MOS scores with national MOS norms. Such a comparison between a sample's score (the MOS score for participants in this study) and a population's score (the norms) necessitates the use

Introduction to Chapter 10

The One-Sample Z-Test

of a one-sample Z-test. And the researchers' findings? MOS Sleep Scale scores were statistically insignificant ($p < .05$). In other words, the null hypothesis that the sample average and the population average were equal could not be accepted.

So why use the one-sample Z-test? Cappelleri and his colleagues wanted to know whether the sample values were different from population (national) values collected using the same measure. The researchers were, in effect, comparing a sample statistic with a population parameter and seeing if they could conclude that the sample was (or was not) representative of the population.

Want to know more? Check out . . .

Cappelleri, J. C., Bushmakin, A. G., McDermott, A. M., Dukes, E., Sadosky, A., Petrie, C. D., & Martin, S. (2009). Measurement properties of the Medical Outcomes Study Sleep Scale in patients with fibromyalgia. *Sleep Medicine, 10*, 766–770.

THE PATH TO WISDOM AND KNOWLEDGE

Here's how you can use Figure 10.1, the flowchart introduced in Chapter 9, to select the appropriate test statistic, the **one-sample Z-test**. Follow along the highlighted sequence of steps in Figure 10.1. Now this is pretty easy (and they are not all this easy) because this is the only inferential procedure in all of Part IV of *Statistics for People* . . . where we have only one group. Plus, there's lots of stuff here that will take you back to Chapter 8 and standard scores, and because you're an expert on those . . .

1. We are examining differences between a sample and a population.

2. There is only one group being tested.

3. The appropriate test statistic is a one-sample Z-test.

COMPUTING THE Z-TEST STATISTIC

Performing a One-Sample Z-Test

The formula used for computing the value for the one-sample Z-test is shown in Formula 10.1. Remember that we are testing whether a sample mean belongs to or represents a population mean. The difference between the sample mean ($\overline{X}$) and the population mean (μ) makes up the numerator for the Z-test value. The denominator, an error term, is called the *standard error of the mean* and is the value we would expect

Figure 10.1 Determining That a One-Sample Z-Test Is the Correct Statistic

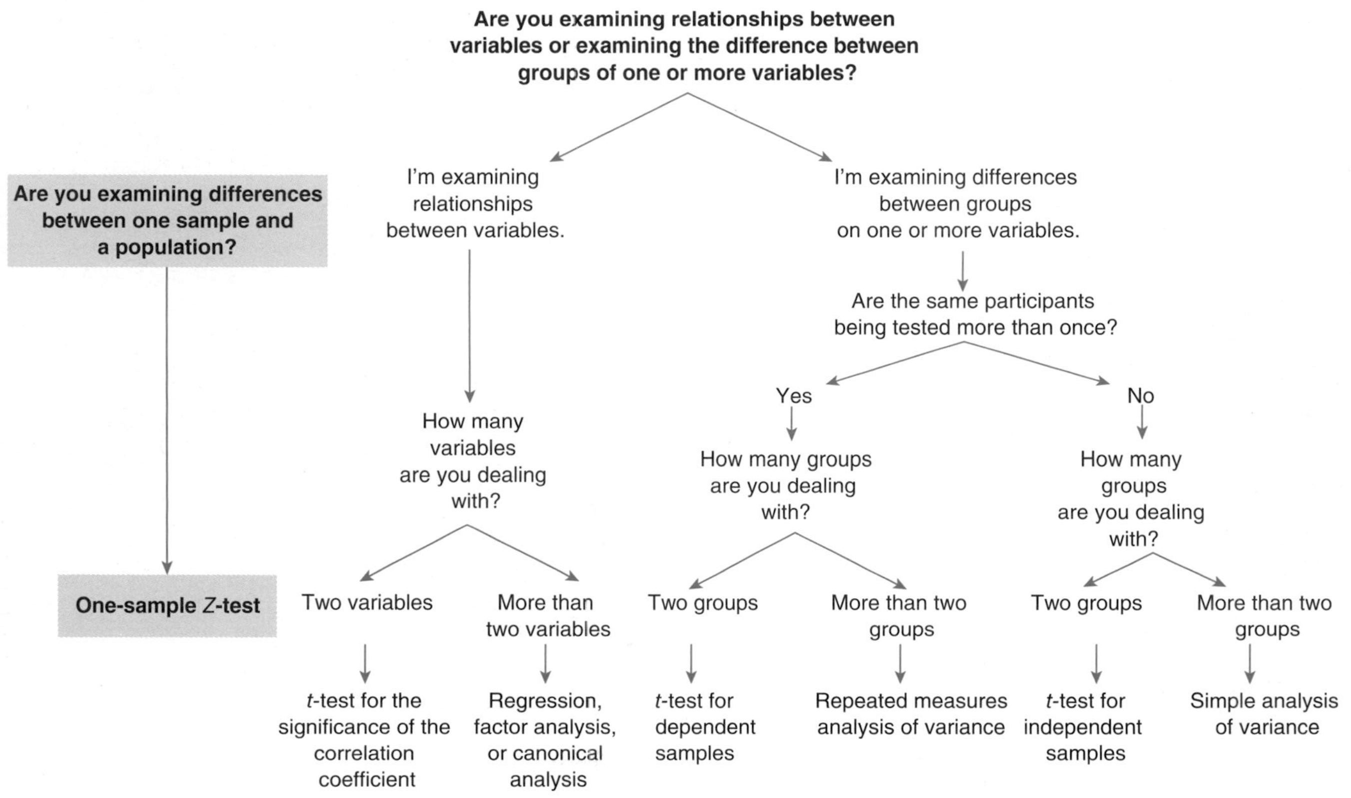

Are you examining relationships between variables or examining the difference between groups of one or more variables?

I'm examining relationships between variables.

I'm examining differences between groups on one or more variables.

How many variables are you dealing with?

Two variables → t-test for the significance of the correlation coefficient

More than two variables → Regression, factor analysis, or canonical analysis

Are the same participants being tested more than once?

Yes → How many groups are you dealing with?

Two groups → t-test for dependent samples

More than two groups → Repeated measures analysis of variance

No → How many groups are you dealing with?

Two groups → t-test for independent samples

More than two groups → Simple analysis of variance

Are you examining differences between one sample and a population?

One-sample Z-test

199

by chance, given all the variability that surrounds the selection of all possible sample means from a population. Using this standard error of the mean (and the key term here is *standard*) allows us once again (as we showed in Chapter 9) to use the table of z scores to determine the probability of an outcome.

$$z = \frac{\bar{X} - \mu}{SEM},$$ (10.1)

where

- $\bar{X}$ is the mean of the sample;
- μ is the population average; and
- *SEM* is the standard error of the mean.

Now, to compute the standard error of the mean, which you need in Formula 10.1, use Formula 10.2:

$$SEM = \frac{\sigma}{\sqrt{n}},$$ (10.2)

where

- σ is the standard deviation for the population; and
- n is the size of the sample.

Populations and Samples

The standard error of the mean is the standard deviation of all the possible means selected from the population. It's the best *estimate* we can come up with, given that it is impossible to compute *all* the possible means. If our sample selection were perfect, the difference between the sample and the population averages would be zero, right? Right. If the sampling from a population were not done correctly (randomly and representatively), however, then the standard deviation of all the means of all these samples could be huge, right? Right. So we try to select the perfect sample, but no matter how diligent we are in our efforts, there's always some error. The standard error of the mean reflects what that value would be for the entire population of all mean values. And yes, Virginia, this is the standard error of the mean. There can be (and are) standard errors for other measures as well.

Time for an example.

Dr. McDonald thinks that his group of earth science students is particularly special (in a good way), and he is interested in knowing whether their class average falls within the boundaries of the average score for the larger group of students who have taken earth science over the past 20 years. Because he's kept good records, he knows the means and standard deviations for both his group of 36 students and the larger group of 1,000 past enrollees. Here are the data.

	Size	Mean	Standard Deviation
Sample	36	100	5.0
Population	1,000	99	2.5

Here are the famous eight steps and the computation of the Z-test statistic.

1. State the null and research hypotheses.

    The null hypothesis states that the sample average is equal to the population average. If the null is not rejected, it means that the sample is representative of the population. If the null is rejected in favor of the research hypothesis, it means that the sample average is different from the population average.

    The null hypothesis is . . .

    $$H_0 : \overline{X} = \mu \qquad (10.3)$$

    The research hypothesis in this example is . . .

    $$H_1 : \overline{X} \neq \mu \qquad (10.4)$$

2. Set the level of risk (or the level of significance or Type I error) associated with the null hypothesis.

    The level of risk or Type I error or level of significance (any other names?) here is .05, but this is totally at the discretion of the researcher.

3. Select the appropriate test statistic.

    Using the flowchart shown in Figure 10.1, we determine that the appropriate test is a one-sample Z-test.

4. Compute the test statistic value (called the obtained value).

    Now's your chance to plug in values and do some computation. The formula for the z value was shown in Formula 10.1. The specific values are plugged in (first for *SEM* in Formula 10.5 and then for z in Formula 10.6). With the values plugged in, we get the following results:

    $$SEM = \frac{2.5}{\sqrt{36}} = 0.42 \qquad (10.5)$$

$$z = \frac{100-99}{0.42} = 2.38 \qquad\qquad (10.6)$$

The z value for a comparison of the sample mean to this population mean, given Dr. McDonald's data, is 2.38.

5. Determine the value needed for rejection of the null hypothesis using the appropriate table of critical values for the particular statistic.

Here's where we go to Table B.1 in Appendix B, which lists the probabilities associated with specific z values, which are the critical values for the rejection of the null hypothesis. This is exactly the same thing we did with several examples in Chapter 9.

We can use the values in Table B.1 to see if two means "belong" to one another by comparing what we would expect by chance (the tabled or critical value) with what we observe (the obtained value).

From our work in Chapter 9, we know that a z value of +1.96 has associated with it a probability of .025, and if we consider that the sample mean could be bigger, or smaller, than the population mean, we need to consider both ends of the distribution (and a range of ±1.96) and a total Type I error rate of .05.

6. Compare the obtained value and the critical value.

The obtained z value is 2.38. So, for a test of this null hypothesis at the .05 level with 36 participants, the critical value is ±1.96. This value represents the value at which chance is the most attractive explanation of why the sample mean and the population mean differ. A result beyond that critical value in either direction (remember that the research hypothesis is nondirectional and this is a two-tailed test) means that we need to provide an explanation as to why the sample and the population means differ.

7. and 8. Decision time!

If the obtained value is more extreme than the critical value (remember Figure 9.2), the null hypothesis cannot be accepted. If the obtained value does not exceed the critical value, the null hypothesis is the most attractive explanation. In this case, the obtained value (2.38) does exceed the critical value (1.96), and it is absolutely extreme enough for us to say that the sample of 36 students in Dr. McDonald's class is different from the previous 1,000 students who have also taken the course. If the obtained value were less than 1.96, it would mean that there is no difference between the test performance of the sample and that of the 1,000 students who have taken the test over the past 20 years. In this case, the 36 students would have performed basically at the same level as the previous 1,000.

And the final step? Why, of course. We wonder why this group of students differs? Perhaps McDonald is right in that they are smarter, but they may also be better users of technology or more motivated. Perhaps they just studied harder. All these are questions to be tested some other time.

So How Do I Interpret z = 2.38, p < .05?

A Z Distribution Table

- z represents the test statistic that was used.
- 2.38 is the obtained value, calculated using the formulas we showed you earlier in the chapter.
- $p < .05$ (the really important part of this little phrase) indicates that the probability is less than 5% that on any one test of the null hypothesis, the sample and the population averages will differ.

USING SPSS TO PERFORM A Z-TEST

We're going to take a bit of a new direction here in that SPSS does not offer a one-sample Z-test but it does offer a one-sample t-test. The results are basically the same, and looking at the one-sample t-test will illustrate how SPSS can be useful—our purpose here. The main difference between this and the Z-test is that SPSS uses a distribution of t scores to evaluate the result.

The real difference between a Z- and a t-test is that for a t-test, the population's standard deviation is not known while for a Z-test, it is known. Another difference is that the tests use different distributions of critical values to evaluate the outcomes (which makes sense given that they're using different test statistics).

In the following example, we are going to use the SPSS one-sample t-test to evaluate whether one score (13) on a test is characteristic of the entire sample. Here's the entire sample:

12

9

7
10
11
15
16
8
9
12

1. After the data are entered, click Analyze → Compare Means → One-Sample T test and you will see the One Sample T test dialog box as shown in Figure 10.2.

2. Double-click on the Score variable to move it to the Test Variable(s): box.

3. Enter a Test Value of 13.

4. Click OK and you will see the output in Figure 10.3.

Figure 10.2   The One-Sample T Test Dialog Box

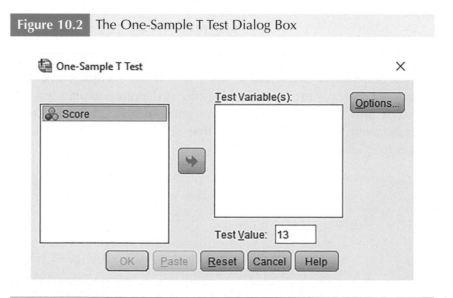

Understanding the SPSS Output

Figure 10.3 shows you the following:

1. For a sample of size 10, the average score is 10.9 and the standard deviation is 2.92.

| Figure 10.3 | The Output From a One-Sample t-Test |

→ **T-Test**

One-Sample Statistics

	N	Mean	Std. Deviation	Std. Error Mean
Score	10	10.9000	2.92309	.92436

One-Sample Test

	Test Value = 13					
					95% Confidence Interval of the Difference	
	t	df	Sig. (2-tailed)	Mean Difference	Lower	Upper
Score	-2.272	9	.049	-2.10000	-4.1911	-.0089

2. The resultant *t* value of −2.72 is significant at the .05 level (barely, but it made it!).

3. The results indicate that that a test value of 13 is significantly different from the values in the sample.

SPECIAL EFFECTS: ARE THOSE DIFFERENCES FOR REAL?

Okay, time for you to entertain a whole new idea, that of effect size, and learn how to use it to make your analysis of any inferential test all that much more interesting and valuable.

Effect Size

In general, using various inferential tools, you may find differences between samples and populations, two samples or more, and so on, but the $64,000 question is not only whether that difference is (statistically) significant but also whether it is *meaningful*. That is, does enough of a separation exist between the distributions that represent each sample or group you test that the difference is really a difference worth discussing! Hmm . . . Welcome to the world of effect size.

Effect size is a measure of how different two groups are from one another—it's a measure of the magnitude of the treatment. Kind of like, How big is big? And what's especially interesting about computing effect size is that sample size is not taken into account. Calculating effect size, and making a judgment about it, adds a whole new dimension to understanding significant outcomes. Another interesting note about effect size is that many different inferential tests use different formulas to compute the effect size (as you will see through the next few chapters), but the same metric (called Cohen's *d*, and we'll get to that shortly) tends to be used regardless. It's like if you use a ruler or a yardstick to measure two different sizes of lumber—you're still dealing with inches.

As an example, let's take the data from Dr. McDonald and the earth science test. Here are the means and standard deviations again.

	Size	Mean	Standard Deviation
Sample	36	100	5.0
Population	1,000	99	2.5

Here's the formula for computing Cohen's *d* for the effect size for a one-sample *Z*-test:

$$d = \frac{\overline{X} - \mu}{\sigma},$$
(10.7)

where

$\overline{X}$ is the sample mean;
μ is the population mean; and
σ is the population standard deviation.

If we substitute Dr. McDonald's values in Formula 10.7, we get this:

$$d = \frac{100 - 99}{2.5} = .4$$

We know from our previous calculations that the obtained *z*-score of 2.38 is significant, meaning that, indeed, the performance of Dr. McDonald's class is different from that of the population. Now we have figured out the effect size (.4), so let's turn our attention to what this statistically significant outcome might mean regarding the size of the effects.

Understanding Effect Size

The great pooh-bah of effect size was Jacob Cohen, who wrote some of the most influential and important articles on this topic. He authored a very important and influential book (your stats teacher has it on his or her shelf!) that instructs researchers in how to figure out the effect size for a variety of different questions that are asked about differences and relationships between variables. And the book also gives some guidelines as to what different sizes of effects might represent for understanding differences. You remember that for our example, the effect size is .4.

What does this mean? One of the very cool things that Cohen (and others) figured out was just what a small, medium, and large effect size is. They used the following guidelines:

- A small effect size ranges from 0 to .2.
- A medium effect size ranges from .2 to .5.
- A large effect size is any value above .5.

Our example, with an effect size of .4, is categorized as medium. But what does it *really* mean?

Effect size gives us an idea about the relative positions of one group to another. For example, if the effect size is zero, that means that both groups tend to be very similar and overlap entirely—there is no difference between the two distributions of scores. On the other hand, an effect size of 1 means that the two groups overlap about 45% (having that much in common). And, as you might expect, as the effect size gets larger, it reflects an increasing lack of overlap between the two groups.

Jacob Cohen's book *Statistical Power Analysis for the Behavioral Sciences,* first published in 1967 with the latest edition (1988, Lawrence Erlbaum) available in reprint from Taylor and Francis, is a must for anyone who wants to go beyond the very general information that is presented here. It is full of tables and techniques for allowing you to understand how a statistically significant finding is only half the story—the other half is the magnitude of that effect.

REAL-WORLD STATS

Been to the doctor lately? Had your test results explained to you? Know anything about the use of electronic medical records? In this study, Noel Brewer and his colleagues compared the usefulness of tables and bar graphs for reporting the results of medical tests.

Using a Z-test, the researchers found that participants required less viewing time when using bar graphs rather than tables. The researchers attributed this difference to superior performance of bar graphs in communicating essential information (and you well remember from Chapter 4, where we stressed that a picture, such as a bar graph, is well worth a thousand words). Also, not very surprisingly, when participants viewed both formats, those with experience with bar graphs preferred bar graphs, and those with experience with tables found bar graphs equally easy to use. Next time you visit your doc and he or she shows you a table, say you want to see the results as a bar graph. Now that's stats applied to real-world, everyday occurrences!

Real-World
Stats

Want to know more? Go online or to the library and find . . .

Brewer, N. T., Gilkey, M. B., Lillie, S. E., Hesse, B. W., & Sheridan, S. L. (2012). Tables or bar graphs? Presenting test results in electronic medical records. *Medical Decision Making, 32,* 545–553.

SUMMARY

Chapter
Summary

The one-sample Z-test is a simple example of an inferential test, and that's why we started off this long section of the book with an explanation of what this test does and how it is applied. The (very) good news is that most (if not all) of the steps we take as we move on to more complex analytic tools are exactly the same as those you saw here. In fact, in the next chapter we move on to a very common inferential test that is an extension of the Z-test we covered here, the simple *t*-test between the means of two different groups.

TIME TO PRACTICE

1. When is it appropriate to use the one-sample Z-test?

2. What does the *z* in Z-test represent? What similarity does it have to a simple *z* or standard score?

3. For the following situations, write out in words a research hypothesis:

   a. Bob wants to know whether the weight loss for his group on the chocolate-only diet is representative of weight loss in a large population of middle-aged men.

   b. The health department is charged with finding out whether the rate of flu per thousand citizens during this past flu season is comparable to the average rate during the past 50 seasons.

   c. Blair is almost sure that his monthly costs for the past year are not representative of his average monthly costs over the past 20 years.

4. Flu cases this past flu season in the Remulak school system (*n* = 500) were 15 per week. For the entire state, the weekly average was 16, and the standard deviation was 15.1. Are the kids in Remulak as sick as the kids throughout the state?

5. The night-shift workers in three of Super Bo's specialty stores stock about 500 products in about 3 hours. How does this rate compare with the stocking done in the other 97 stores in the chain, which average about 496 products stocked in 3 hours? Are the stockers at the specialty stores doing a "better than average" job? Here's the info that you need:

	Size	Average Number of Products Stocked	Standard Deviation
Specialty stores	3	500	12.56
All stores	100	496	22.13

6. A major research study investigated how representative a treatment group's decrease in symptoms was when a certain drug was administered as compared with the response of the entire population. It turns out that the test of the research hypothesis resulted in a Z-test score of 1.67. What conclusion might the researchers put forth? Hint: Notice that the Type I error rate or significance level is *not* stated (as perhaps it should be). What do you make of all this?

7. Millman's golfing group is terrific for a group of amateurs. Are they ready to turn pro? Here's the data. (Hint: Remember that the lower the score (in golf), the better!)

	Size	Average Score	Standard Deviation
Millman's Group	9	82	2.6
The Pros	500	71	3.1

8. Here's a list of units of toys sold by T&K over a 12-month period during 2015. Were the sales of 31,456 for one month in 2016 significantly different from monthly sales in 2015?

Units Sold in 2015	
January	34,518
February	29,540
March	34,889
April	26,764
May	31,429
June	29,962
July	31,084
August	30,506
September	28,546
October	29,560
November	29,304
December	25,852

STUDENT STUDY SITE

⑤SAGE edge™

Get the tools you need to sharpen your study skills! Visit **edge.sagepub.com/ salkind6e** to access practice quizzes, eFlashcards, original and curated videos, data sets, journal articles, and more!

11

t(ea) for Two

Tests Between the Means of Different Groups

Difficulty Scale ☺ ☺ ☺ (A little longer than the previous chapter, but basically the same kind of procedures and very similar questions. Not too hard, but you have to pay attention.)

WHAT YOU'LL LEARN ABOUT IN THIS CHAPTER

+ Using the *t*-test for independent means when appropriate

+ Computing the observed *t* value

+ Interpreting the *t* value and understanding what it means

+ Computing the effect size for a *t*-test for independent means

INTRODUCTION TO THE *T*-TEST FOR INDEPENDENT SAMPLES

Even though eating disorders are recognized for their seriousness, little research has been done that compares the prevalence and intensity of symptoms across different cultures. John P. Sjostedt, John F. Schumaker, and S. S. Nathawat undertook this comparison with groups of 297 Australian and 249 Indian university students. Each student was tested on the Eating Attitudes Test and the Goldfarb Fear of Fat Scale. The groups' scores were compared with one another. On a comparison of means between the Indian and the Australian participants, Indian students scored higher on both of the tests. The results for the Eating Attitudes Test were $t_{(544)} = -4.19$, $p < .0001$, and the results for the Goldfarb Fear of Fat Scale were $t_{(544)} = -7.64$, $p < .0001$.

Now just what does all this mean? Read on.

Why was the *t*-test for independent means used? Sjostedt and his colleagues were interested in finding out whether there was a difference

Introduction to Chapter 11

in the average scores of one (or more) variable(s) between the two groups that were independent of one another. By *independent*, we mean that the two groups were not related in any way. Each participant in the study was tested only once. The researchers applied a *t*-test for independent means, arriving at the conclusion that for each of the outcome variables, the differences between the two groups were significant at or beyond the .0001 level. Such a small Type I error means that there is very little chance that the difference in scores between the two groups was due to something other than group membership, in this case representing nationality, culture, or ethnicity.

Want to know more? Go online or to the library and find . .

Sjostedt, J. P., Schumaker, J. F., & Nathawat, S. S. (1998). Eating disorders among Indian and Australian university students. *Journal of Social Psychology, 138*(3), 351–357.

THE PATH TO WISDOM AND KNOWLEDGE

Here's how you can use Figure 11.1, the flowchart introduced in Chapter 9, to select the appropriate test statistic, the *t*-test for independent means. Follow along the highlighted sequence of steps in Figure 11.1.

1. The differences between the groups of Australian and Indian students are being explored.

2. Participants are being tested only once.

3. There are two groups.

4. The appropriate test statistic is the *t*-test for independent samples.

Selecting the
Dependent
t-Test Between
Means

Almost every statistical test has certain assumptions that underlie the use of the test. For example, the *t*-test makes the major assumption that the amount of variability in each of the two groups is equal. This is the *homogeneity of variance* assumption. Although this assumption can be violated with no adverse consequences if the sample size is big enough, small samples and a violation of this assumption can lead to ambiguous results and conclusions. Don't knock yourself out worrying about these assumptions because they are beyond the scope of this book. However, you should know that while such assumptions are rarely violated, they do exist.

Figure 11.1 Determining That a *t*-Test Is the Correct Statistic

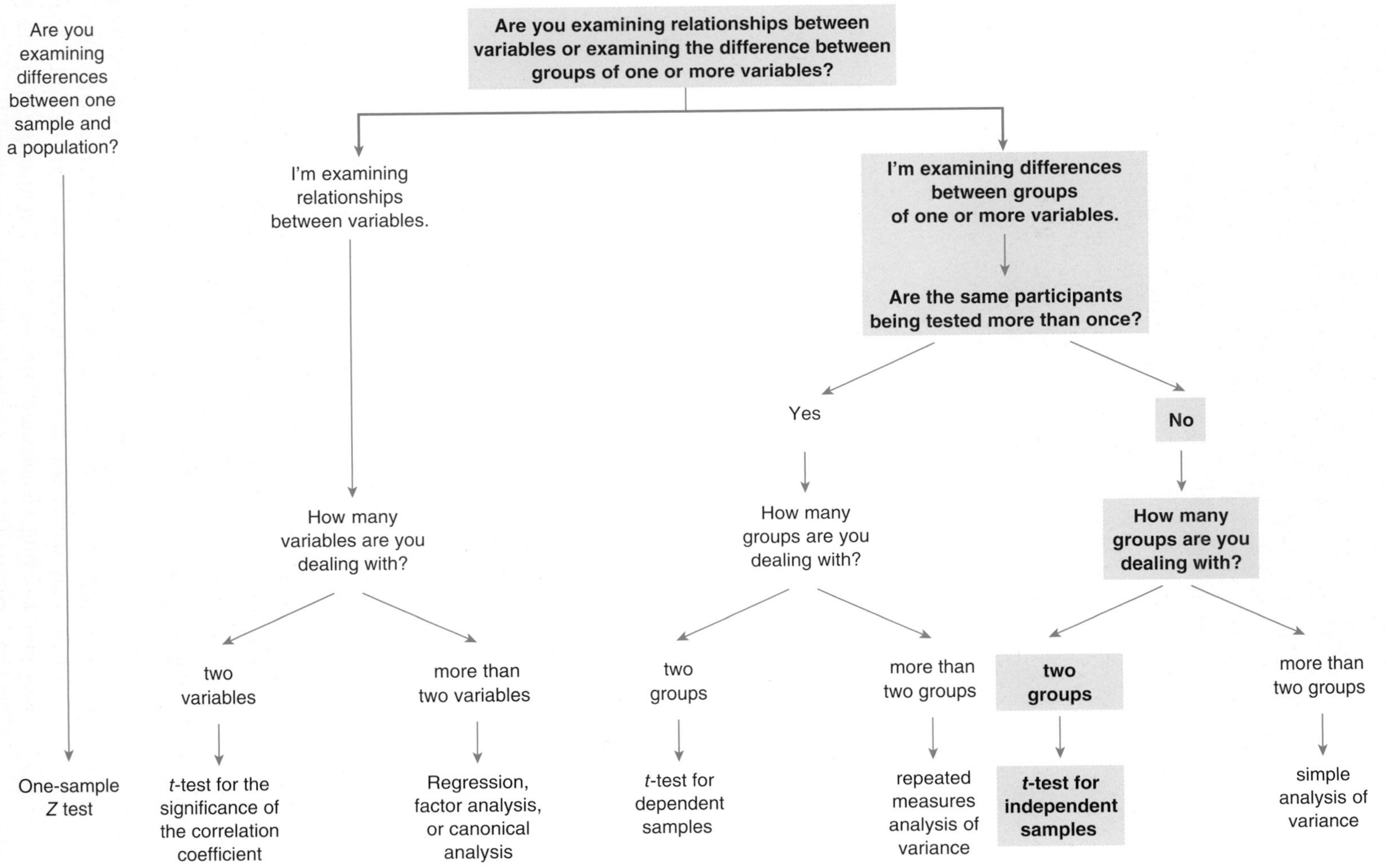

Are you examining differences between one sample and a population?

Are you examining relationships between variables or examining the difference between groups of one or more variables?

I'm examining relationships between variables.

I'm examining differences between groups of one or more variables.

Are the same participants being tested more than once?

Yes

No

How many variables are you dealing with?

two variables

t-test for the significance of the correlation coefficient

more than two variables

Regression, factor analysis, or canonical analysis

How many groups are you dealing with?

two groups

t-test for dependent samples

more than two groups

repeated measures analysis of variance

How many groups are you dealing with?

two groups

***t*-test for independent samples**

more than two groups

simple analysis of variance

One-sample Z test

As we mentioned earlier, there are hundreds of statistical tests. The only inferential one that uses one sample that we cover in this book is the one-sample Z-test (see Chapter 10). But there is also the one-sample t-test, which compares the mean score of a sample with another score, and sometimes that score is, indeed, the population mean, just as with the one-sample Z-test. In any case, you can use the one-sample Z-test or one-sample t-test to test the same hypothesis, and you will reach the same conclusions (although you will be using different values and tables to do so). We discussed this a little bit in Chapter 10 when we talked about using SPSS to compare one sample with a population.

COMPUTING THE *T*-TEST STATISTIC

Computing the t-Test for Independent Samples

The formula for computing the t value for the t-test for independent means is shown in Formula 11.1. The difference between the means makes up the numerator; the amount of variation within and between each of the two groups makes up the denominator.

$$t = \frac{\overline{X}_1 - \overline{X}_2}{\sqrt{\left[\dfrac{(n_1 - 1)s_1^2 + (n_2 - 1)s_2^2}{n_1 + n_2 - 2}\right]\left[\dfrac{n_1 + n_2}{n_1 n_2}\right]}}, \qquad (11.1)$$

where

- $\overline{X}_1$ is the mean for Group 1;
- $\overline{X}_2$ is the mean for Group 2;
- n_1 is the number of participants in Group 1;
- n_2 is the number of participants in Group 2;
- s_1^2 is the variance for Group 1; and
- s_2^2 is the variance for Group 2.

Finding the Critical t-Test Value

This is a bigger formula than we've seen before, but there's really nothing new here at all. It's just a matter of plugging in the correct values.

Time for an Example

Here are some data reflecting the number of words remembered following a program designed to help Alzheimer's patients remember the order of daily tasks. Group 1 was taught using visuals, and Group 2 was taught using visuals and intense verbal rehearsal. We'll use the data to compute the test statistic in the following example.

Group 1			Group 2		
7	5	5	5	3	4
3	4	7	4	2	3
3	6	1	4	5	2
2	10	9	5	4	7
3	10	2	5	4	6
8	5	5	7	6	2
8	1	2	8	7	8
5	1	12	8	7	9
8	4	15	9	5	7
5	3	4	8	6	6

Here are the famous eight steps and the computation of the *t*-test statistic.

1. State the null and research hypotheses.

   As represented by Formula 11.2, the null hypothesis states that there is no difference between the means for Group 1 and Group 2. For our purposes, the research hypothesis (shown as Formula 11.3) states that there is a difference between the means of the two groups. The research hypothesis is two tailed and nondirectional because it posits a difference, but in no particular direction.

   The null hypothesis is

   Computing the t value

   $$H_0: \mu_1 = \mu_2 \tag{11.2}$$

   The research hypothesis is

   $$H_1: \overline{X}_1 \neq \overline{X}_2 \tag{11.3}$$

2. Set the level of risk (or the level of significance or Type I error) associated with the null hypothesis.

   The level of risk or Type I error or level of significance (any other names?) here is .05, but this is totally the decision of the researcher.

3. Select the appropriate test statistic.

   Using the flowchart shown in Figure 11.1, we determined that the appropriate test is a *t*-test for independent means. It is not a *t*-test for dependent means (a common mistake beginning students make) because the groups are independent of one another.

4. Compute the test statistic value (called the obtained value).

Now's your chance to plug in values and do some computation. The formula for the t value was shown in Formula 11.1. When the specific values are plugged in, we get the equation shown in Formula 11.4. (We already computed the mean and standard deviation.)

$$t = \frac{5.43 - 5.53}{\sqrt{\left[\dfrac{(30-1)3.42^2 + (30-1)2.06^2}{30+30-2}\right]\left[\dfrac{30+30}{30\times30}\right]}} \qquad (11.4)$$

With the numbers plugged in, Formula 11.5 shows how we get the final value of −.137. The value is negative because a larger value (the mean of Group 2, which is 5.53) is being subtracted from a smaller number (the mean of Group 1, which is 5.43). Remember, though, that because the test is nondirectional—the research hypothesis is that any difference exists—the sign of the difference is meaningless.

$$t = \frac{-0.1}{\sqrt{\left(\dfrac{339.20 + 123.06}{58}\right)\left(\dfrac{60}{900}\right)}} = -0.137 \qquad (11.5)$$

When a nondirectional test is discussed, you may find that the t value is represented as an absolute value looking like this, $|t|$ or $t = |0.137|$, which ignores the sign of the value altogether. Your teacher may even express the t value as such to emphasize that the sign is relevant for a one-directional test but surely not for a nondirectional one.

5. Determine the value needed for rejection of the null hypothesis using the appropriate table of critical values for the particular statistic.

Here's where we go to Table B.2 in Appendix B, which lists the critical values for the t-test.

We can use this distribution to see whether two independent means differ from one another by comparing what we would expect by chance (the tabled or critical value) with what we observe (the obtained value).

Our first task is to determine the **degrees of freedom** (df), which approximates the sample size. For this particular test statistic, the degrees of freedom is $n_1 - 1 + n_2 - 1$ or $n_1 + n_2 - 2$ (putting the terms in either order results in the same value). So for each group, add the size of the two samples and subtract 2. In this example, $30 + 30 - 2 = 58$. This is the degrees of freedom for this test statistic and not necessarily for any other.

The idea of degrees of freedom means pretty much the same thing no matter what statistical test you use. But the way that the degrees of freedom is computed for specific tests can differ from teacher to teacher and from book to book. We tell you that the correct degrees of freedom for the above test is computed as $n_1 - 1 + n_2 - 1$. However, some teachers believe that you should use the smaller of the two n's (a more conservative alternative you may want to consider).

Using this number (58), the level of risk you are willing to take (earlier defined as .05), and a two-tailed test (because there is no direction to the research hypothesis), you can use the *t*-test table to look up the critical value. At the .05 level, with 58 degrees of freedom for a two-tailed test, the value needed for rejection of the null hypothesis is . . . Oops! There's no 58 degrees of freedom in the table! What do you do? Well, if you select the value that corresponds to 55, you're being conservative in that you are using a value for a sample smaller than what you have (and the critical *t* value will be larger).

If you go for 60 degrees of freedom (the closest to your value of 58), you will be closer to the size of the population, but you'll be a bit liberal in that 60 is larger than 58. Although statisticians differ in their viewpoint as to what to do in this situation, let's always go with the value that's closer to the actual sample size. So the value needed to reject the null hypothesis with 58 degrees of freedom at the .05 level of significance is 2.001.

6. Compare the obtained value and the critical value.

   The obtained value is −0.14 (−0.137 rounded to the nearest hundredth), and the critical value for rejection of the null hypothesis that Group 1 and Group 2 performed differently is 2.001. The critical value of 2.001 represents the value at which chance is the most attractive explanation for any of the observed differences between the two groups, given 30 participants in each group and the willingness to take a .05 level of risk.

7. and 8. Decision time!

   Now comes our decision. If the obtained value is more extreme than the critical value (remember Figure 9.2), the null hypothesis cannot be accepted. If the obtained value does not exceed the critical value, the null hypothesis is the most attractive explanation. In this case, the obtained value (−0.14) does not exceed the critical value (2.001)—it is not extreme enough for us to say that the difference between Groups 1 and 2 occurred due to anything other than chance. If the value were greater than 2.001, that would be just like getting 8, 9, or 10 heads in a coin toss—too extreme a result for us to believe that mere chance is at work. In the case of the coin, the cause would be an unfair coin; in this example, it would be that there is a better way to teach memory skills to these older people.

So, to what can we attribute the small difference between the two groups? If we stick with our current argument, then we could say the difference is due to anything from sampling error to rounding error to simple variability in participants' scores. Most important, we're pretty sure (but, of course, not 100% sure—that's what level of significance and Type I errors are all about, right?) that the difference is not due to anything in particular that one group or the other experienced to make its scores better.

So How Do I Interpret $t_{(58)} = -0.14$, $p > .05$

- t represents the test statistic that was used.
- 58 is the number of degrees of freedom.
- −0.14 is the obtained value, calculated using the formula we showed you earlier in the chapter.
- $p > .05$ (the really important part of this little phrase) indicates that the probability is greater than 5% that on any one test of the null hypothesis, the two groups do not differ because of the way they were taught. Note that $p > .05$ can also appear as $p = $ n.s. for nonsignificant.

THE EFFECT SIZE AND T(EA) FOR TWO

You learned in Chapter 10 that effect size is a measure of how different two groups are from one another—it's a measure of the magnitude of the treatment. Kind of like, How big is big?

And what's especially interesting about computing effect size is that sample size is not taken into account. Calculating effect size, and making a judgment about it, adds a whole new dimension to understanding significant outcomes.

Let's take the following example. A researcher tests the question of whether participation in community-sponsored services (such as card games, field trips, etc.) increases the quality of life (as rated from 1 to 10) for older Americans. The researcher implements the treatment over a 6-month period and then, at the end of the treatment period, measures quality of life in the two groups (each consisting of 50 participants over the age of 80 where one group got the services and one group did not). Here are the results.

	No Community Services	Community Services
Mean	6.90	7.46
Standard Deviation	1.03	1.53

And the verdict is that the difference is significant at the .034 level (which is $p < .05$, right?). So there's a significant difference, but what about the magnitude of the difference?

Computing and Understanding the Effect Size

As we showed you in Chapter 10, the most direct and simple way to compute effect size is to simply divide the difference between the means by any one of the standard deviations. Danger, Will Robinson—this does assume that the standard deviations (and the amount of variance) between groups are equal to one another.

For our example, we'll do this . . .

$$ES = \frac{\overline{X}_1 - \overline{X}_2}{s}, \tag{11.6}$$

where

- *ES* is the effect size;
- $\overline{X}_1$ is the mean for Group 1;
- $\overline{X}_2$ is the mean for Group 2; and
- *s* is the standard deviation from either group.

So, in our example,

$$ES = \frac{7.46 - 6.90}{1.53} = .366 \tag{11.7}$$

So, the effect size for this example is .37.

You saw from our guidelines in Chapter 10 (page 206) that an effect size of .37 is categorized as medium. In addition to the difference between the two means being statistically significant, one might conclude that the difference also is *meaningful* in that the effect size is not negligible. Now, how meaningful you wish to make it in your interpretation of the results depends upon many factors, including the context within which the research question is being asked.

So, you really want to be cool about this effect size thing. You can do it the simple way, as we just showed you (by subtracting means from one another and dividing by either standard deviation), or you can really wow that good-looking classmate who sits next to you. The grown-up formula for the effect size uses the pooled variance in the denominator of the *ES* equation that you saw previously. The pooled standard deviation is sort of an average of the standard deviation from Group 1 and the standard deviation from Group 2. Here's the formula:

$$ES = \frac{\overline{X}_1 - \overline{X}_2}{\sqrt{\dfrac{\sigma_1^2 + \sigma_2^2}{2}}},$$

(11.8)

where

- *ES* is the effect size;
- $\overline{X}_1$ is the mean of Group 1;
- $\overline{X}_2$ is the mean of Group 2;
- σ_1^2 is the variance of Group 1; and
- σ_2^2 is the variance of Group 2.

If we applied this formula to the same numbers we showed you previously, you'd get a whopping effect size of .43—not very different from .37, which we got using the more direct method shown earlier (and still in the same category of medium size). But this is a more precise method and one that is well worth knowing about.

Two Very Cool Effect Size Calculators

Why not take the A train and just go right to http://www.uccs.edu/~lbecker/, where statistician Lee Becker from the University of California

An Effect Size Calculator

Figure 11.2 The Very Cool Effect Size Calculator

Group 1	Group 2
M_1 7.4	M_2 6.9
SD_1 1.53	SD_2 1.03
Compute	Reset
Cohen's d	**effect-size r**
0.38338	0.18826

Source: Lee Becker, http://www.uccs.edu/~lbecker/

developed an effect size calculator? Or ditto for the one located at http://www.psychometrica.de/effect_size.html created by Drs. Wolfgang & Alexandra Lenhard? With these calculators, you just plug in the values, click Compute, and the program does the rest, as you see in Figure 11.2.

USING SPSS TO PERFORM A *T*-TEST

SPSS is willing and ready to help you perform these inferential tests. Here's how to perform the one that we just discussed and interpret the output. We are using the data set named Chapter 11 Data Set 1. From your examination of the data, you can see how the grouping variable (Group 1 or Group 2) is in column 1 and the test variable (Memory) is in column 2.

1. Enter the data in the Data Editor or download the file. Be sure that there is one column for each group and that you have no more than two groups represented in that column.

2. Click Analyze → Compare Means → Independent-Samples T Test, and you will see the Independent-Samples T Test dialog box shown in Figure 11.3.

3. Click on the variable named Group and drag it to move it to the Grouping Variable(s) box.

4. Click on the variable named Memory_Test and drag it to place it in the Test Variable(s) box.

5. SPSS will not allow you to continue until you define the grouping variable. This basically means telling SPSS how many levels of the group variable there are (wouldn't you think that a program this smart could figure that out?). In any case, click Define Groups and enter the values 1 for Group 1 and 2 for Group 2, as shown in Figure 11.4. The name of the grouping variable (in this case, Group) has to be highlighted before you can define it.

6. Click Continue and then click OK, and SPSS will conduct the analysis and produce the output you see in Figure 11.5.

Notice how SPSS uses a capital *T* to represent this test while we have been using a small *t*? This difference is strictly a matter of personal preference and, more often than not, reflects what people were taught way back when. What's important for you to know is that there is a difference in letter only—it's the same exact test.

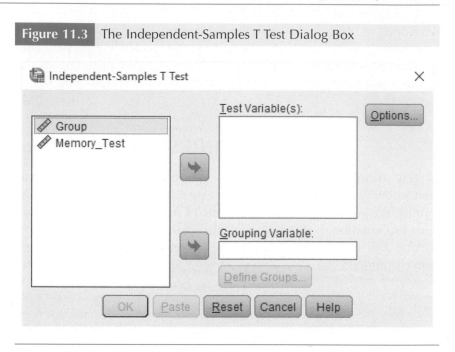

Figure 11.3 The Independent-Samples T Test Dialog Box

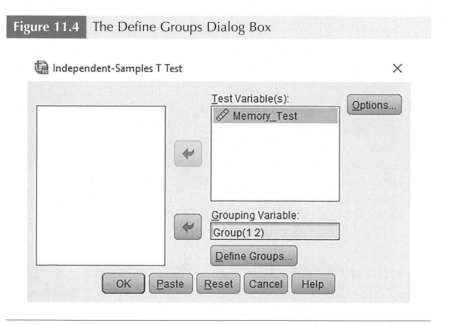

Figure 11.4 The Define Groups Dialog Box

Understanding the SPSS Output

There's a ton of SPSS output from this analysis, and for our purposes, we'll deal only with selected output shown in Figure 11.5. There are three things to note.

Figure 11.5	Copy of SPSS Output for a *t*-Test Between Independent Means

Group Statistics

	Group	N	Mean	Std. Deviation	Std. Error Mean
Memory_Test	1	30	5.43	3.421	.625
	2	30	5.53	2.063	.377

Independent Samples Test

		Levene's Test for Equality of Variances		t-test for Equality of Means						95% Confidence Interval of the Difference	
		F	Sig.	t	df	Sig. (2-tailed)	Mean Difference	Std. Error Difference	Lower	Upper
Memory_Test	Equal variances assumed	4.994	.029	-.137	58	.891	-.100	.729	-1.560	1.360
	Equal variances not assumed			-.137	47.635	.892	-.100	.729	-1.567	1.367

1. The obtained *t* value is −0.137, exactly what we got when we computed the value by hand earlier in this chapter (−0.14, rounded from 0.1368).

2. The number of degrees of freedom is 58 (which you already know is computed using the formula $n_1 + n_2 - 2$).

3. The significance of this finding is .891, or *p* = .891, which means that on one test of this null hypothesis, the likelihood of rejecting the hypothesis when it is true is pretty high (89 out of 100)! So the Type I error is certainly greater than .05, the same conclusion we reached earlier when we went through this process manually. No difference!

REAL-WORLD STATS

There's nothing like being prepared, so say Boy and Girl Scouts. But what's the best way to teach preparedness? In this study by Serkan Celik from Kirikkale University in Turkey, online versus face-to-face first-aid instruction courses were compared with one another to test the effectiveness of the modes of delivery. Interestingly, with the same instructor teaching in both modes, the online form resulted in higher achievement scores at the end of the course. What did Celik do for his analysis? Independent *t*-tests, of course. And in fact, he used SPSS to produce the resulting *t* values showing that there was a difference between pre- and posttest scores.

Want to know more? Go online or to the library and find . . .

Celik, S. (2013). A media comparison study on first aid instruction. *Health Education Journal, 72*, 95–101.

Real-World
Stats

SUMMARY

Chapter
Summary

The *t*-test is your first introduction to performing a real statistical test between two groups and trying to understand this whole matter of significance from an applied point of view. Be sure that you understand what is in this chapter before you move on. And be sure you can do by hand the few calculations that were asked for. Next, we move on to using another form of the same test, only this time, two measures are taken from one group of participants rather than one measure taken from two separate groups.

TIME TO PRACTICE

Problem 1

1. Using the data in the file named Chapter 11 Data Set 2, test the research hypothesis at the .05 level of significance that boys raise their hand in class more often than girls. Do this practice problem by hand, using a calculator. What is your conclusion regarding the research hypothesis? Remember to first decide whether this is a one- or two-tailed test.

Problem 2

2. Using the same data set (Chapter 11 Data Set 2), test the research hypothesis at the .01 level of significance that there is a difference between boys and girls in the number of times they raise their hand in class. Do this practice problem by hand, using a calculator. What is your conclusion regarding the research hypothesis? You used the same data for this problem as for Question 1, but you have a different hypothesis (one is directional and the other is nondirectional). How do the results differ and why?

3. Time for some tedious, by-hand practice just to see if you can get the numbers right. Using the following information, calculate the *t*-test statistics by hand.

   a. $X_1 = 62$ $X_2 = 60$ $n_1 = 10$ $n_2 = 10$ $s_1 = 2.45$ $s_2 = 3.16$

   b. $X_1 = 158$ $X_2 = 157.4$ $n_1 = 22$ $n_2 = 26$ $s_1 = 2.06$ $s_2 = 2.59$

   c. $X_1 = 200$ $X_2 = 198$ $n_1 = 17$ $n_2 = 17$ $s_1 = 2.45$ $s_2 = 2.35$

4. Using the results you got from Question 3, and a level of significance of .05, what are the two-tailed critical values associated with each result? Would the null hypothesis be rejected?

Problem 5

5. Use the following data and SPSS or some other computer application such as Excel or Google Sheets and write a brief paragraph about whether the in-home counseling is equally effective as the out-of-home treatment for two separate groups. Here are the data. The outcome variable is level of anxiety after treatment on a scale from 1 to 10.

In-Home Treatment	Out-of-Home Treatment
3	7
4	6
1	7
1	8
1	7
3	6
3	5
6	6
5	4
1	2
4	5
5	4
4	3
4	6
3	7
6	5
7	4
7	3
7	8
8	7

6. Using the data in the file named Chapter 11 Data Set 3, test the null hypothesis that urban and rural residents both have the same attitude toward gun control. Use SPSS to complete the analysis for this problem.

7. Here's a good one to think about. A public health researcher (Dr. L) tested the hypothesis that providing new car buyers with child safety seats will also act as an incentive for parents to take other measures to protect their children (such as driving more safely, childproofing the home, etc.). Dr. L counted all the occurrences of safe behaviors in the cars and homes of the parents who accepted the seats versus those who did not. The findings? A significant difference at the .013 level. Another researcher (Dr. R) did exactly the same study, and for our purposes, let's assume that everything was

the same—same type of sample, same outcome measures, same car seats, and so on. Dr. R's results were marginally significant (remember that from Chapter 9?) at the .051 level. Whose results do you trust more and why?

8. Here are the results of three experiments where the means of the two groups being compared are exactly the same but the standard deviation is quite different from experiment to experiment. Compute the effect size using Formula 11.6 on page 219 and then discuss why this size changes as a function of the change in variability.

Experiment 1	Group 1 Mean	78.6	Effect Size _____
	Group 2 Mean	73.4	
	Standard Deviation	2.0	
Experiment 2	Group 1 Mean	78.6	Effect Size _____
	Group 2 Mean	73.4	
	Standard Deviation	4.0	
Experiment 3	Group 1 Mean	78.6	Effect Size _____
	Group 2 Mean	73.4	
	Standard Deviation	8.0	

9. Using the data in Chapter 11 Data Set 4 and SPSS, test the null hypothesis that there is no difference in group means between the number of words spelled correctly for two groups of fourth graders. What is your conclusion?

10. For this question, you have to do two analyses. Using Chapter 11 Data Set 5, compute the t score for the difference between two groups on the test variable (named score). Then, use Chapter 11 Data Set 6 and do the same. Notice that there is a difference between the two t values, even though the mean scores for each group are the same. What is the source of this difference, and why—with the same sample size—did the t value differ?

11. Here's an interesting (and hypothetical) situation. The average score for one group on an aptitude test is 89.5 and the average score for the second group is 89.2. Both groups have sample sizes of about 1,500 and the differences between the means is significant, but the effect size is very small (let's say .1). What do you make of the fact that there can be a statistically significant difference between the two groups without their being a meaningful effect size?

STUDENT STUDY SITE

⑤SAGE edge™

Get the tools you need to sharpen your study skills! Visit **edge.sagepub.com/ salkind6e** to access practice quizzes, eFlashcards, original and curated videos, data sets, journal articles, and more!

12

t(ea) for Two (Again)

Tests Between the Means of Related Groups

Difficulty Scale ☺ ☺ ☺ (not too hard—this is the first one of this kind, but you know more than enough to master it)

WHAT YOU'LL LEARN ABOUT IN THIS CHAPTER

✦ Understanding when the t-test for dependent means is appropriate to use

✦ Computing the observed t value

✦ Interpreting the t value and understanding what it means

✦ Computing the effect size for a t-test for dependent means

INTRODUCTION TO THE T-TEST FOR DEPENDENT SAMPLES

How best to educate children is clearly one of the most vexing questions that faces any society. Because children are so different from one another, a balance needs to be found between meeting the basic needs of all and ensuring that special children (on either end of the continuum) get the opportunities they need. An obvious and important part of education is reading, and three professors at the University of Alabama studied the effects of resource classrooms and regular classrooms on the reading achievement of children with learning disabilities. Renitta Goldman, Gary L. Sapp, and Ann Shumate Foster found that, in general, 1 year of daily instruction in both settings resulted in no difference in overall reading achievement scores. On one specific comparison between the pretest and the posttest of the resource group, they found that $t_{(34)} = 1.23$, $p > .05$. At the beginning of the program, reading achievement scores for children in the resource room were 85.8. At the

Introduction to Chapter 12

229

Computing
the t-Test for
Dependent
Samples

end of the program, reading achievement scores for children in the resource room were 88.5—a difference, but not a significant one.

So why a test of dependent means? A *t*-test for dependent means indicates that a single group of the same subjects is being studied under two conditions. In this example, the conditions are before the start of the experiment and after its conclusion. Primarily, the *t*-test for dependent means was used because the same children were tested at two times, before the start of the 1-year program and at the end of the 1-year program. As you can see from the aforementioned result, there was no difference between the scores at the beginning and the end of the program. The relatively small *t* value (1.23) is not nearly extreme enough to fall outside the region where we would reject the null hypothesis. In other words, there is far too little change for us to say that this difference occurred due to something other than chance. The small difference of 2.7 (88.5 − 85.8) is probably due to sampling error or variability within the groups.

Want to know more? Check out . . .

Goldman, R., Sapp, G. L., & Foster, A. S. (1998). Reading achievement by learning disabled students in resource and regular classes. *Perceptual and Motor Skills, 86,* 192–194.

THE PATH TO WISDOM AND KNOWLEDGE

Here's how you can use the flowchart to select the appropriate test statistic, the *t*-test for dependent means. Follow along the highlighted sequence of steps in Figure 12.1.

1. The difference between the students' scores on the pretest and on the posttest is the focus.

2. Participants are being tested more than once.

3. There are two groups of scores.

4. The appropriate test statistic is *t*-test for dependent means.

Selecting the
Dependent
t-Test Between
Means

There's another way that statisticians sometimes talk about dependent tests—as *repeated measures*. Dependent tests are often called "repeated measures" both because the measures are repeated across time or conditions or some factor and because they are repeated across the same cases, be each case a person, place, or thing.

Figure 12.1 Determining That a *t*-Test for Dependent Means Is the Correct Test Statistic

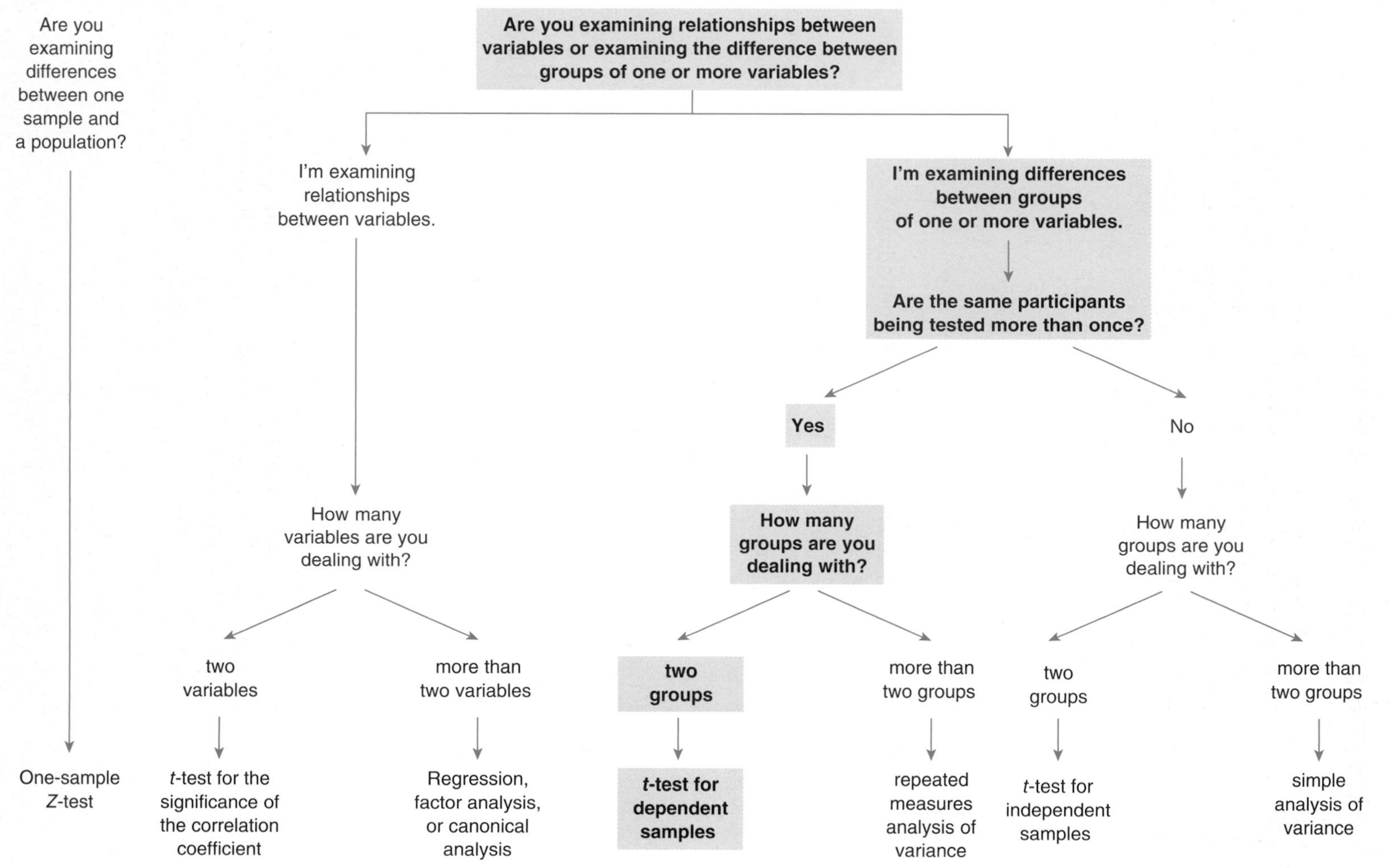

Are you examining differences between one sample and a population?

Are you examining relationships between variables or examining the difference between groups of one or more variables?

I'm examining relationships between variables.

I'm examining differences between groups of one or more variables.

How many variables are you dealing with?

Are the same participants being tested more than once?

two variables

more than two variables

Yes

No

t-test for the significance of the correlation coefficient

Regression, factor analysis, or canonical analysis

How many groups are you dealing with?

How many groups are you dealing with?

two groups

more than two groups

two groups

more than two groups

t-test for dependent samples

repeated measures analysis of variance

t-test for independent samples

simple analysis of variance

One-sample Z-test

COMPUTING THE *T*-TEST STATISTIC

Degrees of
Freedom

The *t*-test for dependent means involves a comparison of means from each group of scores and focuses on the differences between the scores. As you can see in Formula 12.1, the sum of the differences between the two tests forms the numerator and reflects the difference between groups of scores.

$$t = \frac{\Sigma D}{\sqrt{\dfrac{n\Sigma D^2 - (\Sigma D)^2}{n-1}}}, \qquad (12.1)$$

where

- *D* the difference between each individual's score from point 1 to point 2;
- ΣD is the sum of all the differences between groups of scores;
- ΣD^2 is the sum of the differences squared between groups of scores; and
- *n* is the number of pairs of observations.

Coming up, you'll see some data used to illustrate how the *t* value for a dependent *t*-test is computed. Just as in the example discussed earlier in this chapter, there is a pretest and a posttest, and for illustration's sake, assume that these data are before and after scores from a reading program.

Here are the famous eight steps and the computation of the *t*-test statistic:

1. State the null and research hypotheses.

The null hypothesis states that there is no difference between the means for the pretest and the posttest scores on reading achievement. The research hypothesis is one tailed and directional because it posits that the posttest score will be higher than the pretest score.
The null hypothesis is

$$H_0: \mu_{posttest} = \mu_{pretest} \qquad (12.2)$$

The research hypothesis is

$$H_1: \bar{X}_{posttest} > \bar{X}_{pretest} \qquad (12.3)$$

Pretest	Posttest	Difference	D^2
3	7	4	16
5	8	3	9
4	6	2	4
6	7	1	1
5	8	3	9
5	9	4	16
4	6	2	4
5	6	1	1
3	7	4	16
6	8	2	4
7	8	1	1
8	7	−1	1
7	9	2	4
6	10	4	16
7	9	2	4
8	9	1	1
8	8	0	0
9	8	−1	1
9	4	−5	25
8	4	−4	16
7	5	−2	4
7	6	−1	1
6	9	3	9
7	8	1	1
8	12	4	16
Sum	158	188	30	180
Mean	6.32	7.52	1.2	7.2

2. Set the level of risk (or the level of significance or Type I error) associated with the null hypothesis.

The level of risk or Type I error or level of significance here is .05, but this is totally the decision of the researcher.

3. Select the appropriate test statistic.

Using the flowchart shown in Figure 12.1, we determined that the appropriate test is a *t*-test for dependent means. It is not a *t*-test for independent means because the groups are not independent of each other. In fact, they're not groups of participants but groups of scores for the same participants. The groups are dependent on one another. Other names for the *t*-test for dependent means is the *t*-test for paired samples or the *t*-test for correlated samples. You'll see in Chapter 15 that there is a very close relationship between a test of the significance of the correlation between these two sets of scores (pre- and posttest) and the *t* value we are computing here.

4. Compute the test statistic value (called the obtained value).

Now's your chance to plug in values and do some computation. The formula for the *t* value was shown previously. When the specific values are plugged in, we get the equation shown in Formula 12.4. (We already computed the means and standard deviations for the pretest and posttest scores.)

$$t = \frac{30}{\sqrt{\dfrac{(25 \times 180) - 30^2}{25 - 1}}}, \tag{12.4}$$

With the numbers plugged in, we have the following equation with a final obtained *t* value of 2.45. The mean score for pretest performance was 6.32, and the mean score for posttest performance was 7.52.

$$t = \frac{30}{\sqrt{150}} = 2.45 \tag{12.5}$$

5. Determine the value needed for rejection of the null hypothesis using the appropriate table of critical values for the particular statistic.

Here's where we go to Appendix B and Table B.2, which lists the critical values for the test. Once again, we have a *t*-test, and we'll use the same table we used in Chapter 11 to find the critical value for rejection of the null hypothesis.

Our first task is to determine the degrees of freedom (*df*), which approximates the sample size. For this particular test statistic, the degrees of freedom is $n - 1$, where *n* equals the number of pairs of observations, or $25 - 1 = 24$. This is the degrees of freedom for this test statistic only and not necessarily for any other.

Using this number (24), the level of risk you are willing to take (earlier defined as .05), and a one-tailed test (because there is a direction to the research hypothesis—the posttest score will be larger than the pretest score), the value needed for rejection of the null hypothesis is 1.711.

6. Compare the obtained value and the critical value.

The obtained value is 2.45, larger than the critical value needed for rejection of the null hypothesis.

7. and 8. Time for a decision!

Now comes our decision. If the obtained value is more extreme than the critical value, the null hypothesis cannot be accepted. If the obtained value does not exceed the critical value, the null hypothesis is the most attractive explanation. In this case, the obtained value does exceed the critical value—it is extreme enough for us to say that the difference between the pretest and the posttest occurred due to something other than chance. And if we did our experiment correctly, then what could the factor be that affected the outcome? Easy—the introduction of the daily reading program. We know the difference is due to a particular factor. The difference between the pretest and the posttest groups could not have occurred by chance but instead is due to the treatment.

So How Do I Interpret $t_{(24)}$ = 2.45, p < .05?

- *t* represents the test statistic that was used.
- 24 is the number of degrees of freedom.
- 2.45 is the obtained value using the formula we showed you earlier in the chapter.
- $p < .05$ (the really important part of this little phrase) indicates that the probability is less than 5% on any one test of the null hypothesis that the average of posttest scores is greater than the average of pretest scores due to chance alone—there's something else going on. Because we defined .05 as our criterion for the research hypothesis being more attractive than the null hypothesis, our conclusion is that there is a significant difference between the two sets of scores. That's the something else.

Quick Calcs
Again

USING SPSS TO PERFORM A *T*-TEST

SPSS is willing and ready to help you perform these inferential tests. Here's how to perform the one that we just did and understand the output. We are using the data set named Chapter 12 Data Set 1, which was also used in the earlier example.

1. Enter the data in the Data Editor or download the file. Be sure that there is a separate column for pretest and posttest scores. Unlike in a *t*-test for independent means, there are no groups to identify. In Figure 12.2, you can see that the cell entries are labeled pretest and posttest.

2. Click Analyze → Compare Means → Paired-Samples T Test, and you will see the dialog box shown in Figure 12.3.

3. Drag the variable named Posttest to the Variable1 space in the Paired Variables: box, as you see in Figure 12.4.

4. Drag the variable named Pretest to the Variable2 space in the Paired Variables: box.

5. Click OK.

6. SPSS will conduct the analysis and produce the output you see in Figure 12.5.

Computing the
t Value

Why drag Posttest first?

SPSS works in the following way. It subtracts Variable2 from Variable1 because the research hypothesis is directional and "says" that the posttest scores will be greater than the pretest scores; thus, we want to subtract pretest scores from posttest scores and, hence, Posttest has to be identified as Variable1.

Figure 12.2   Data From Chapter 12 Data Set 1

Pretest	Posttest
3	7
5	8
4	6
6	7
5	8
5	9
4	6
5	6
3	7
6	8

Figure 12.3 Paired-Samples *t*-Test Dialog Box

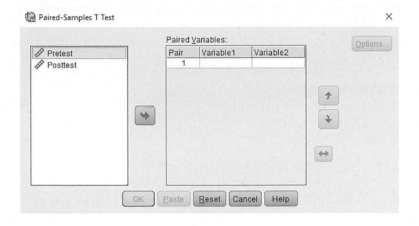

Figure 12.4 Dragging Variables for the Paired *t*-Test Analysis

Figure 12.5 SPSS Output for a *t*-Test Between Dependent Means

➔ **T-Test**

[DataSet2] C:\Textbook Stuff\Stat for People 6e\Data Sets\SPSS\Chapter 12 Data Set 1.sav

Paired Samples Statistics

		Mean	N	Std. Deviation	Std. Error Mean
Pair 1	Posttest	7.52	25	1.828	.366
	Pretest	6.32	25	1.725	.345

Paired Samples Correlations

		N	Correlation	Sig.
Pair 1	Posttest & Pretest	25	.051	.810

Paired Samples Test

		Paired Differences							
					95% Confidence Interval of the Difference				
		Mean	Std. Deviation	Std. Error Mean	Lower	Upper	t	df	Sig. (2-tailed)
Pair 1	Posttest - Pretest	1.200	2.449	.490	.189	2.211	2.449	24	.022

Earlier versions of SPSS may not allow you to define the order of which mean is subtracted from which other mean. So, if you are using an older version of SPSS, then you will get a negative *t* value if the average of the first variable is smaller than the average of the second variable (which, in this case, would have been a *t* value of –2.449 instead of 2.449). As long as you keep the research hypothesis foremost in mind when you interpret the results, you should be fine.

Just for fun, we're going to throw a little caveat bone in the mix here. Sometimes, and only *sometimes*, researchers will use a test of dependent means when the two measures (which are usually on the same participant) are actually on different participants who are very closely matched together on all relevant characteristics (be it age, gender, social class, aggressiveness, workout speed, ice cream preference—you get the picture). In this case, even though they are different participants, they are considered dependent because they are matched. We're just fine-tuning the meaning of the word *dependent* and how it applies to such situations.

Understanding the SPSS Output

Here we go . . .

1. The posttest score (7.52) is larger than the pretest score (6.32). At least this far into the analysis, it appears that the results are supporting the research hypothesis that the children scored higher on the posttest than the pretest.

2. The difference between the means of the pretest and posttest groups is 1.2, where the posttest mean was subtracted from the pretest mean. And the exact probability that a *t* score of 2.449 was obtained by chance is .022—very unlikely.

3. But also notice that the output you see in Figure 12.5 shows this to be the probability associated with a two-tailed (or nondirectional) test. We conducted a one-tailed test, so what to do? Read on! Using Table B.2 in Appendix B (see page 403), we find that for a one-tailed test, with 24 degrees of freedom at the .05 level of significance, the critical value for rejection of the null hypothesis is 1.711. So, although SPSS will give us the specific obtained *t* value, it will not give us the probability of that value for a one-tailed test. It does for two-tailed, but not for one-tailed. For that,

we have to rely on our own skills and use the table as we did here, or else we can use a software program that can do one-tailed tests (see Chapter 20 for more about this).

Believe it or not, way back in the olden days, when your author and perhaps your instructor were graduate students, there were only huge mainframe computers and not a hint of such marvels as we have today on our desktops. In other words, everything that was done in our statistics class was done by hand. The great benefit of doing the calculations by hand is, first, it helps you to better understand the process. Second, should you be without a computer, you can still do the analysis. So, if the computer does not spit out everything you need, use some creativity. As long as you know the basic formula for the critical value and have the appropriate tables, you'll do fine.

THE EFFECT SIZE FOR *T*(EA) FOR TWO (AGAIN)

Guess what, folks? The computation of the effect size for a test between dependent means is the formula and procedure for the computation of the effect size for a test of the difference between independent means.

You can refer back to Chapter 11 (page 221) for all the details, but here's the formula and the effect size. And, as we are sure you realize, we got these values from the output in Figure 12.5.

$$ES = \frac{7.52 - 6.32}{1.828} = .656 \qquad (12.6)$$

So, for this analysis, the effect size (according to the guidelines proposed by Cohen and discussed in Chapter 11) is quite large (over .5 qualifies as "large"). Not only is the difference significant, but the bigness of the difference is for real and perhaps meaningful given the context of the research questions being asked.

REAL-WORLD STATS

You may be part of the "sandwich" generation whose adults are not only taking care of their elderly parents but also raising children. This situation (and the aging of any population such as is taking place in the United States) speaks to the importance of evaluating the elderly (however this cohort is defined) and the use of the tools we've discussed. The purpose of this study was to determine whether the life satisfaction of a sample of Thai elderly depended on their daily living practice. Matched pairs of elderly people were tested to see who perceived themselves as life satisfied or life dissatisfied, and scores of 85% and above on the life satisfaction instrument were used as a criterion to identify the elderly with life satisfaction. A two-tailed dependent *t*-test (because the samples were matched—in effect the researchers considered the paired participants to be the same participants) was used to examine differences in mean scores both overall and for each domain of daily living practices. The life satisfaction group of elderly had significantly higher scores than their dissatisfied counterparts. One of the most interesting questions is how these results might apply to other samples of elderly participants in different cultures.

Real-World
Stats

Want to know more? Go online or to the library and find . . .

Othaganont, P., Sinthuvorakan, C., & Jensupakarn, P. (2002). Daily living practice of the life-satisfied Thai elderly. *Journal of Transcultural Nursing, 13,* 24–29.

SUMMARY

Chapter
Summary

That's it for means. You've just learned how to compare data from independent (Chapter 11) and dependent (Chapter 12) groups, and now it's time to move on to another class of significance tests that deals with more than two groups (be they independent or dependent). This class of techniques, called analysis of variance, is very powerful and popular and will be a valuable tool in your war chest!

TIME TO PRACTICE

1. What is the difference between a test of independent means and a test of dependent means, and when is each appropriate?

2. In the following examples, indicate whether you would perform a *t*-test of independent means or dependent means.

   a. Two groups were exposed to different levels of treatment for ankle sprains. Which treatment was most effective?

   b. A researcher in nursing wanted to know if the recovery of patients was quicker when some received additional in-home care whereas others received the standard amount.

   c. A group of adolescent boys was offered interpersonal skills counseling and then tested in September and May to see if there was any impact on family harmony.

   d. One group of adult men was given instructions in reducing their high blood pressure whereas another was not given any instructions.

   e. One group of men was provided access to an exercise program and tested two times over a 6-month period for heart health.

3. For Chapter 12 Data Set 2, compute the *t* value manually and write a conclusion as to whether there was a change in tons of paper used as a function of the recycling program in 25 different districts. (Hint: *Before* and *after* become the two levels of treatment.) Test the hypothesis at the .01 level.

Problem 3

4. Here are the data from a study in which adolescents were given counseling at the beginning of the school year to see if it had a positive impact on their tolerance for adolescents who were ethnically different from them. Assessments were made right before the treatment and then 6 months later. Did the program work? The outcome variable is scored on an attitude-toward-others test with possible scores ranging from 0 to 50; the higher the score, the more tolerance. Use SPSS or some other computer application to complete this analysis.

Before Treatment	After Treatment
45	46
46	44
32	47
34	42
33	45
21	32
23	36
41	43
27	24
38	41
41	38
47	31
41	22
32	36
22	36
34	27
36	41
19	44
23	32
22	32

5. For Chapter 12 Data Set 3, compute the *t* value and write a conclusion as to whether there is a difference in families' satisfaction level, measured on a scale of 1 to 15, in their use of service centers following a social service intervention. Do this exercise using SPSS and report the exact probability of the outcome.

Problem 5

6. Do this exercise the good old-fashioned way—by hand. A famous brand-name manufacturer wants to know whether people prefer Nibbles or Wribbles. Consumers get a chance to sample each type of cracker and indicate their like or dislike on a scale from 1 to 10. Which do they like the most?

Nibbles Rating	Wribbles Rating
9	4
3	7
1	6
6	8
5	7
7	7
8	8
3	6
10	7
3	8
5	9
2	8
9	7
6	3
2	6
5	7
8	6
1	5
6	5
3	6

7. Take a look at Chapter 12 Data Set 4. Does shift matter when it comes to stress (the higher the stress score, the more stress the worker feels)?

Problem 8

8. Chapter 12 Data Set 5 provides you with data for two groups of adults, tested in the fall and spring during a resistance weight-lifting class. The outcome variable is bone density (the higher the score, the denser the bones) on a scale from 1 to 10. Does lifting weights work? How about the effect size?

STUDENT STUDY SITE

⑤SAGE edge™

Get the tools you need to sharpen your study skills! Visit **edge.sagepub.com/salkind6e** to access practice quizzes, eFlashcards, original and curated videos, data sets, journal articles, and more!

13

Two Groups Too Many?

Try Analysis of Variance

Difficulty Scale ☺ (longer and harder than the others, but a very interesting and useful procedure—worth the work!)

WHAT YOU'LL LEARN ABOUT IN THIS CHAPTER

✦ Deciding when it is appropriate to use analysis of variance

✦ Computing and interpreting the F statistic

✦ Using SPSS to complete an analysis of variance

✦ Computing the effect size for a one-way analysis of variance

INTRODUCTION TO ANALYSIS OF VARIANCE

One of the up-and-coming fields in the area of psychology is the psychology of sports. Although the field focuses mostly on enhancing performance, it gives special attention to many aspects of sports. One area of interest is what psychological skills are necessary to be a successful athlete. With this question in mind, Marious Goudas, Yiannis Theodorakis, and Georgios Karamousalidis have tested the usefulness of the Athletic Coping Skills Inventory.

Introduction to
Chapter 13

As part of their research, they used a simple **analysis of variance** (or ANOVA) to test the hypothesis that number of years of experience in sports is related to coping skill (or an athlete's score on the Athletic Coping Skills Inventory). ANOVA was used because more than two levels of the same variable were being tested and these groups were compared on their average performance. In particular, Group 1 included athletes with 6 years of experience or fewer, Group 2 included athletes with 7 to 10 years of experience, and Group 3 included athletes with more than 10 years of experience.

The test statistic for ANOVA is the *F*-test (named for R. A. Fisher, the creator of the statistic), and the results of this study on the Peaking Under Pressure subscale of the test showed that $F_{(2,110)} = 13.08, p < .01$. The means of the three groups' scores on this subscale differed from one another. In other words, any difference in test score was due to number of years of experience in athletics rather than some chance occurrence of scores.

Want to know more? Go online or to the library and find ?. . .

Goudas, M., Theodorakis, Y., & Karamousalidis, G. (1998). Psychological skills in basketball: Preliminary study for development of a Greek form of the Athletic Coping Skills Inventory-28. *Perceptual and Motor Skills, 86,* 59–65.

THE PATH TO WISDOM AND KNOWLEDGE

Selecting a One-Way ANOVA

Here's how you can use the flowchart shown in Figure 13.1 to select ANOVA as the appropriate test statistic. Follow along the highlighted sequence of steps.

1. We are testing for a difference between scores of different groups, in this case, the difference between the peaking scores of athletes.

2. The athletes are not being tested more than once.

3. There are three groups (fewer than 6 years, 7–10 years, and more than 10 years of experience).

4. The appropriate test statistic is simple analysis of variance.

DIFFERENT FLAVORS OF ANOVA

ANOVA comes in many flavors. The simplest kind, and the focus of this chapter, is the **simple analysis of variance,** used when one factor or one treatment variable (such as group membership) is being explored and this factor has more than two levels. Simple ANOVA is also called **one-way analysis of variance** because there is only one grouping dimension. The technique is called *analysis of variance* because the variance due to differences in performance is separated into (a) variance that's due to differences between individuals *between* groups and (b) variance due to differences *within* groups. Then, the two types of variance are compared with one another. The between-groups variance is due to treatment differences, while the within-group variance is due to differences between individuals within each group.

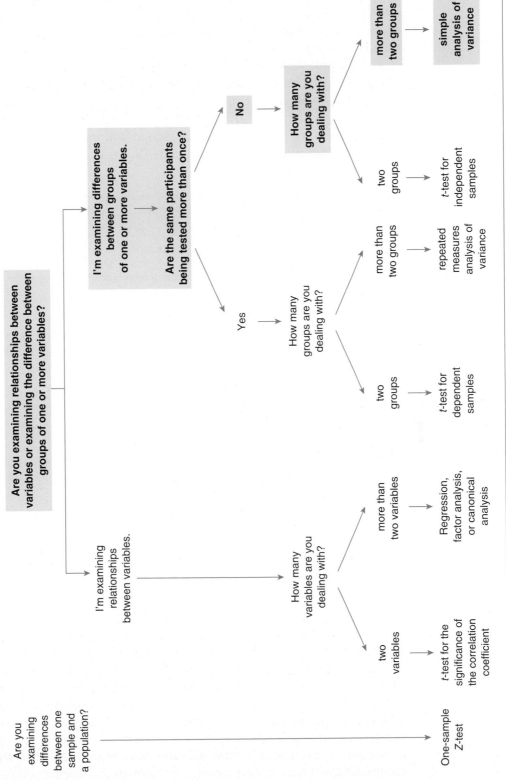

245

In fact, ANOVA is, in many ways, similar to a *t*-test (an ANOVA with two groups is, in effect, a *t*-test!). In both ANOVA and *t*-tests, differences between means are computed. But with ANOVA, there are more than two means.

For example, let's say we were investigating the effects on language development of being in preschool for 5, 10, or 20 hours per week. The group to which the children belong is the treatment variable, or the grouping or between-groups factor. Language development is the outcome measure. The experimental design will look something like this, with three levels of the one variable (hours of participation).

Group 1 (5 hours per week)	Group 2 (10 hours per week)	Group 3 (20 hours per week)
Language development test score	Language development test score	Language development test score

The more complex type of ANOVA, **factorial design**, is used to explore more than one treatment factor. Here's an example where the effect of number of hours of preschool participation is being examined, but the effects of gender differences are being examined as well. The experimental design may look like this:

Gender	Number of Hours of Preschool Participation		
	Group 1 (5 hours per week)	Group 2 (10 hours per week)	Group 3 (20 hours per week)
Male	Language development test score	Language development test score	Language development test score
Female	Language development test score	Language development test score	Language development test score

This factorial design is described as a 3 × 2 factorial design. The 3 indicates that there are three levels of one grouping factor (Group 1, Group 2, and Group 3). The 2 indicates that there are two levels of the other grouping factor (male and female). In combination, there are 6 possibilities (males who spend 5 hours per week in preschool, females who spend 5 hours per week in preschool, males who spend 10 hours per week in preschool, etc.).

These factorial designs follow the same basic logic and principles of simple ANOVA; they are just more ambitious in that they can test the influence of more than one factor at a time as well as a combination of factors. Don't worry—you'll learn all about factorial designs in the next chapter.

COMPUTING THE *F*-TEST STATISTIC

Simple ANOVA involves testing the difference between the means of more than two groups on one factor or dimension. For example, you might want to know whether four groups of people (20, 25, 30, and 35 years of age) differ in their attitude toward public support of private schools. Or you might be interested in determining whether five groups of children from different grades (2nd, 4th, 6th, 8th, and 10th) differ in the level of parental participation in school activities.

ANOVA:
Computing the
F-Test Statistic

Any analysis where . . .

- there is only one dimension or treatment,
- there are more than two levels of the grouping factor, and
- one is looking at differences across groups in average scores

. . . requires that simple ANOVA be used.

The formula for the computation of the *F* value, which is the test statistic needed to evaluate the hypothesis that there are overall differences between groups, is shown in Formula 13.1. It is simple at this level, but it takes a bit more arithmetic to compute than some of the test statistics you have worked with in earlier chapters.

One-Way
ANOVA

$$F = \frac{MS_{between}}{MS_{within}},$$ (13.1)

where

$MS_{between}$ is the variance between groups; and
MS_{within} is the variance within groups.

The logic behind this ratio goes something like this. If there were absolutely no variability within each group (all the scores were the same), then any difference between groups would be meaningful, right? Probably so. The ANOVA formula (which is a ratio) compares the amount of variability between groups (which is due to the grouping factor) with the amount of variability within groups (which is due to chance). If that ratio is 1, then the amount of variability due to within-group differences is

(Continued)

(Continued)

equal to the amount of variability due to between-groups differences, and any difference between groups is not significant. As the average difference between groups gets larger (and the numerator of the ratio increases in value), the *F* value increases as well. As the *F* value increases, it becomes more extreme in relation to the distribution of all *F* values and is more likely due to something other than chance. Whew!

Here are some data and some preliminary calculations to illustrate how the *F* value is computed. For our example, let's assume these are three groups of preschoolers and their language scores.

Group 1 Scores	Group 2 Scores	Group 3 Scores
87	87	89
86	85	91
76	99	96
56	85	87
78	79	89
98	81	90
77	82	89
66	78	96
75	85	96
67	91	93

The F-test
Statistic

Here are the famous eight steps and the computation of the *F*-test statistic:

1. State the null and research hypotheses.

The null hypothesis, shown in Formula 13.2, states that there is no difference among the means of the three different groups. ANOVA, also called the *F*-test (because it produces an *F* statistic or an *F* ratio), looks for an overall difference among groups.

Note that this test does not look at pairwise differences, such as the difference between Group 1 and Group 2. For that, we have to use another technique, which we will discuss later in the chapter.

$$H_0: \mu_1 = \mu_2 = \mu_3 \tag{13.2}$$

The research hypothesis, shown in Formula 13.3, states that there is an overall difference among the means of the three groups. Note that there is no direction to the difference because all *F*-tests are nondirectional.

$$H_1: X_1 \neq X_2 \neq X_3 \qquad (13.3)$$

2. Set the level of risk (or the level of significance or Type I error) associated with the null hypothesis.

   The level of risk or Type I error or level of significance (any other names?) is .05 here. Once again, the level of significance used is totally at the discretion of the researcher.

3. Select the appropriate test statistic.

   Using the flowchart shown in Figure 13.1, we determine that the appropriate test is a simple ANOVA.

4. Compute the test statistic value (called the obtained value).

Now's your chance to plug in values and do some computation. There's a good deal of computation to do.

- The *F* ratio is a ratio of the variability among groups to the variability within groups. To compute these values, we first have to compute what is called the sum of squares for each source of variability—between groups, within groups, and the total.
- The between-groups sum of squares is equal to the sum of the differences between the mean of all scores and the mean of each group's score, which is then squared. This gives us an idea of how different each group's mean is from the overall mean.
- The within-group sum of squares is equal to the sum of the differences between each individual score in a group and the mean of each group, which is then squared. This gives us an idea how different each score in a group is from the mean of that group.
- The total sum of squares is equal to the sum of the between-groups and within-group sum of squares.

Okay—let's figure out these values.

Up to now, we've talked about one- and two-tailed tests. There's no such thing when talking about ANOVA! Because more than two groups are being tested, and because the *F*-test is an *omnibus* robust (how's that for some cool words?) test (meaning that ANOVA of any flavor tests for an overall difference between means), talking about the specific direction of differences does not make any sense.

Figure 13.2 Computing the Important Values for a One-Way ANOVA

Group	Test Score	X^2	Group	Test Score	X^2	Group	Test Score	X^2
1	87	7,569	2	87	7,569	3	89	7,921
1	86	7,396	2	85	7,225	3	91	8,281
1	76	5,776	2	99	9,801	3	96	9,216
1	56	3,136	2	85	7,225	3	87	7,569
1	78	6,084	2	79	6,241	3	89	7,921
1	98	9,604	2	81	6,561	3	90	8,100
1	77	5,929	2	82	6,724	3	89	7,921
1	66	4,356	2	78	6,084	3	96	9,216
1	75	5,625	2	85	7,225	3	96	9,216
1	67	4,489	2	91	8,281	3	93	8,649
n	10			10			10	
ΣX	766			852			916	
$\overline{X}$	76.60			85.20			91.60	
$\Sigma(X^2)$	59,964			72,936			84,010	
$(\Sigma X)^2/n$	58,675.60			72,590.40			83,905.60	

$$N = 30.00$$
$$\Sigma\Sigma X = 2{,}534.00$$
$$(\Sigma\Sigma X)^2/N = 214{,}038.53$$
$$\Sigma\Sigma(X^2) = 216{,}910$$
$$\Sigma(\Sigma X)^2/n = 215{,}171.60$$

Figure 13.2 shows the practice data you saw previously with all the calculations you need to compute the between-groups, within-group, and total sum of squares. First, let's look at what we have in this expanded table. We'll start with the left-hand column:

- n is the number of participants in each group (such as 10).
- ΣX is the sum of the scores in each group (such as 766).
- $\overline{X}$ is the mean of each group (such as 76.60).
- $\Sigma(X^2)$ is the sum of each score squared (such as 59,964).
- $(\Sigma X)^2/n$ is the sum of the scores in each group squared and then divided by the size of the group (such as 58,675.60).

Second, let's look at the right-most column:

- n is the total number of participants (such as 30).
- $\Sigma\Sigma X$ is the sum of all the scores across groups.
- $(\Sigma\Sigma X)^2/n$ is the sum of all the scores across groups squared and divided by n.
- $\Sigma\Sigma(X^2)$ is the sum of all the sums of squared scores.
- $\Sigma(\Sigma X)^2/n$ is the sum of the sum of each group's scores squared divided by n.

That is a load of computation to carry out, but we are almost finished!

First, we compute the sum of squares for each source of variability. Here are the calculations:

Between sum of squares	$\Sigma(\Sigma X)^2/n - (\Sigma\Sigma X)^2/n$ or 215,171.60 – 214,038.53	1,133.07
Within sum of squares	$\Sigma\Sigma(X^2) - \Sigma(\Sigma X)^2/n$ or 216,910 – 215,171.6	1,738.40
Total sum of squares	$\Sigma\Sigma(X^2) - (\Sigma\Sigma X)^2/n$ or 216,910 – 214,038.53	2,871.47

Second, we need to compute the mean sum of squares, which is simply an average sum of squares. These are the variance estimates that we need to eventually compute the all-important F ratio.

We do that by dividing each sum of squares by the appropriate number of degrees of freedom (df). Remember, degrees of freedom is an approximation of the sample or group size. We need two sets of degrees of freedom for ANOVA. For the between-groups estimate, it is $k - 1$, where k equals the number of groups (in this case, there are 3 groups and 2 degrees of freedom). For the within-group estimate, we need $N - k$, where N equals the total sample size (which means that the number of degrees of freedom is 30 – 3, or 27). Then the F ratio is simply a ratio of the mean sum of squares due to between-groups differences to the mean sum of squares due to within-group differences, or 566.54/64.39 or 8.799. This is the obtained F value.

Here's a **source table** of the variance estimates used to compute the F ratio. This is how most F tables appear in professional journals and manuscripts.

Source	Sum of Squares	df	Mean Sum of Squares	F
Between groups	1,133.07	2	566.54	8.799
Within groups	1,738.40	27	64.39	
Total	2,871.47	29		

All that trouble for one little F ratio? Yes, but as we have said earlier, it's essential to do these procedures at least once by hand. It gives you the important appreciation of where the numbers come from and some insight into what they mean.

 Because you already know about *t*-tests, you might be wondering how a *t* value (which is always used for the test between the difference of the means for two groups) and an *F* value (which is always used for more than two groups) are related. Interestingly enough, an *F* value for two groups is equal to a *t* value for two groups squared, or $F = t^2$. Handy trivia question, right? But also useful if you know one and need to know the other.

5. Determine the value needed for rejection of the null hypothesis using the appropriate table of critical values for the particular statistic.

As we have done before, we have to compare the obtained and critical values. We now need to turn to the table that lists the critical values for the F-test, Table B.3 in Appendix B. Our first task is to determine the degrees of freedom for the numerator, which is $k - 1$, or $3 - 1 = 2$. Then we determine the degrees of freedom for the denominator, which is $N - k$, or $30 - 3 = 27$. Together, they are represented as $F_{(2,27)}$.

The obtained value is 8.80, or $F_{(2,27)} = 8.80$. The critical value at the .05 level with 2 degrees of freedom in the numerator (represented by columns in Table B.3) and 27 degrees of freedom in the denominator (represented by rows in Table B.3) is 3.36. So at the .05 level, with 2 and 27 degrees of freedom for an omnibus test among the means of the three groups, the value needed for rejection of the null hypothesis is 3.36.

6. Compare the obtained value and the critical value.

The obtained value is 8.80, and the critical value for rejection of the null hypothesis at the .05 level that the three groups are different from one another (without concern for where the difference lies) is 3.36.

7. and 8. Decision time.

Now comes our decision. If the obtained value is more extreme than the critical value, the null hypothesis cannot be accepted. If the obtained value does not exceed the critical value, the null hypothesis is the most attractive explanation. In this case, the obtained value does exceed the critical value—it is extreme enough for us to say that the difference among the three groups is not due to chance. And if we did our experiment correctly, then what factor do we think affected the outcome? Easy—the number of hours of preschool. We know the difference is due to a particular factor because the difference among the groups could not have occurred by chance but instead is due to the treatment.

So How Do I Interpret $F_{(2,27)} = 8.80, p < .05$?

- F represents the test statistic that was used.
- 2 and 27 are the numbers of degrees of freedom for the between-groups and within-group estimates, respectively.
- 8.80 is the obtained value using the formula we showed you earlier in the chapter.
- $p < .05$ (the really important part of this little phrase) indicates that the probability is less than 5% on any one test of the null hypothesis that the average scores of each group's language skills differ due to chance alone rather than due to the effect of the treatment. Because we defined .05 as our criterion for the research hypothesis being more attractive than the null hypothesis, our conclusion is that there is a significant difference among the three sets of scores.

(Really Important) Tech Talk

Imagine this scenario. You're a high-powered researcher at an advertising company, and you want to see whether the use of color in advertising communications makes a difference to sales. You've decided to test this at the .05 level. So you put together a brochure that

(Continued)

(Continued)

is all black-and-white, one that is 25% color, the next 50%, then 75%, and finally 100% color—for five levels. You do an ANOVA and find out that a difference in sales exists. But because ANOVA is an omnibus test, you don't know where the source of the significant difference lies. So you take two groups (or pairs) at a time (such as 25% color and 75% color) and test them against each other. In fact, you test every combination of two against each other. Kosher? No way. This is called performing multiple *t*-tests, and it is actually against the law in some jurisdictions. When you do this, the Type I error rate (which you set at .05) balloons, depending on the number of tests you want to conduct. There are 10 possible comparisons (no color vs. 25%, no color vs. 50%, no color vs. 75%, etc.), and the real Type I error rate is

$$1 - (1 - \alpha)^k,$$

where

- α is the Type I error rate (.05 in this example); and
- k is the number of comparisons.

So, instead of .05, the actual error rate that each comparison is being tested at is

$$1 - (1 - .05)^{10} = .40 \ (!!!!!)$$

This is a far cry from .05.

USING SPSS TO COMPUTE THE *F* RATIO

Computing the F Ratio Using SPSS

The *F* ratio is a tedious value to compute by hand—not difficult, just time-consuming. That's all there is to it. Using the computer is much easier and more accurate because it eliminates any computational errors. That said, you should be glad you have seen the value computed manually because it's an important skill to have and helps you understand the concepts behind the process. But also be glad that there are tools such as SPSS.

We'll use the data found in Chapter 13 Data Set 1 (the same data used in the above preschool example).

1. Enter the data in the Data Editor. Be sure that there is a column for group and that you have three groups represented in that column. In Figure 13.3, you can see that the cell entries are labeled Group and Language_Score.

| Figure 13.3 | Data From Chapter 13 Data Set 1 |

	Group	Language_Score
1	5 Hours	87
2	5 Hours	86
3	5 Hours	76
4	5 Hours	56
5	5 Hours	78
6	5 Hours	98
7	5 Hours	77
8	5 Hours	66
9	5 Hours	75
10	5 Hours	67
11	10 Hours	87
12	10 Hours	85
13	10 Hours	99
14	10 Hours	85
15	10 Hours	79
16	10 Hours	81
17	10 Hours	82

2. Click Analyze → Compare Means → One-Way ANOVA, and you will see the One-Way ANOVA dialog box shown in Figure 13.4.

| Figure 13.4 | One-Way ANOVA Dialog Box |

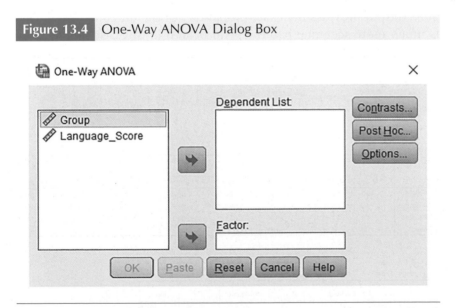

3. Click on the variable named Group and then move it to the Factor: box.

4. Click on the variable named Language_Score and then move it to the Dependent List: box.

5. Click Options, then Descriptives, and then Continue.

6. Click OK. SPSS will conduct the analysis and produce the output you see in Figure 13.5.

Understanding the SPSS Output

This SPSS output is straightforward and looks just like the table that we created earlier to show you how to compute the *F* ratio along with some descriptive statistics. Here's what we have:

1. The descriptive values are reported (sample size, means, etc.).

2. The source of the variance as between-groups, within-groups, and the total is identified.

3. The respective sum of squares is given for each source.

4. The degrees of freedom follows. This is followed by the mean square, which is the sum of squares divided by the degrees of freedom.

5. Finally, there's the obtained value and the associated level of significance.

Keep in mind that this hypothesis was tested at the .05 level. The SPSS output provides the exact probability of the outcome, .001. This is much more accurate than the *p*-value and, in this case, much more unlikely than .05.

Figure 13.5 SPSS Output for a One-Way Analysis of Variance

Descriptives

Language_Score

	N	Mean	Std. Deviation	Std. Error	95% Confidence Interval for Mean Lower Bound	Upper Bound	Minimum	Maximum
5 Hours	10	76.60	11.965	3.784	68.04	85.16	56	98
10 Hours	10	85.20	6.197	1.960	80.77	89.63	78	99
20 Hours	10	91.60	3.406	1.077	89.16	94.04	87	96
Total	30	84.47	9.951	1.817	80.75	88.18	56	99

ANOVA

Language_Score

	Sum of Squares	df	Mean Square	F	Sig.
Between Groups	1133.067	2	566.533	8.799	.001
Within Groups	1738.400	27	64.385		
Total	2871.467	29			

THE EFFECT SIZE FOR ONE-WAY ANOVA

In previous chapters, you saw how we used Cohen's d as a measure of effect size. Here, we change directions and use a value called eta squared or η^2. As with Cohen's d, η^2 has a scale of how large the effect size is:

- A small effect size is about .01.
- A medium effect size is about .06.
- A large effect size is about .14.

Now to the actual effect size.
The formula for η^2 is as follows:

$$\eta^2 = \frac{\text{Between-groups sums of squares}}{\text{Total sums of squares}}$$

and you get that information right from the source table that SPSS (or doing it by hand) generates.

In the example of the three groups we used earlier, the between-groups sums of squares equals 1,133.07, and the total sums of squares equals 2,871.27. The easy computation is . . .

$$\eta^2 = \frac{1,133.07}{2,871.27} = .39$$

and according to your small–medium–large guidelines for η^2, .39 is a large effect. The effect size is a terrific tool to aid in the evaluation of F ratios, as indeed it is with almost any test statistic.

OK, so you've run an ANOVA and you know that an overall difference exists among the means of three or four or more groups. But where does that difference lie?

You already know not to perform multiple t-tests. You need to perform what are called **post hoc**, or after-the-fact, comparisons. You'll compare each mean with every other mean and see where the difference lies, but what's most important is that the Type I error for each comparison is controlled at the same level as you set. There are a bunch of ways to do these comparisons, among them being the Bonferroni (your dear author's favorite statistical term). To complete this specific analysis using SPSS, click the Post Hoc option you see in the ANOVA dialog box (Figure 13.4), then click Bonferroni, then Continue,

(Continued)

(Continued)

and so on, and you'll see output something like that shown in Figure 13.6. Clearly, this analysis tells you that the significant pairwise differences between the groups contributing to the overall significant difference among all three groups lie between Groups 1 and 3; there is no pairwise difference between Groups 1 and 2 or between Groups 2 and 3. This pairwise stuff is very important because it allows you to understand the source of the difference among more than two groups.

Figure 13.6   Post Hoc Comparisons Following a Significant *F* Value

Post Hoc Tests

Multiple Comparisons

Dependent Variable: Language_Score

Bonferroni

(I) Group	(J) Group	Mean Difference (I-J)	Std. Error	Sig.	95% Confidence Interval Lower Bound	95% Confidence Interval Upper Bound
5 Hours	10 Hours	-8.600	3.588	.071	-17.76	.56
	20 Hours	-15.000*	3.588	.001	-24.16	-5.84
10 Hours	5 Hours	8.600	3.588	.071	-.56	17.76
	20 Hours	-6.400	3.588	.257	-15.56	2.76
20 Hours	5 Hours	15.000*	3.588	.001	5.84	24.16
	10 Hours	6.400	3.588	.257	-2.76	15.56

*. The mean difference is significant at the 0.05 level.

REAL-WORLD STATS

How interesting is it when researchers combine expertise from different disciplines to answer different questions? Just take a look at the journal in which this appeared: *Music and Medicine*! In this study, in which the authors examined anxiety among performing musicians, they tested five professional singers' and four flute players' arousal levels. In addition, they used a 5-point Likert-type scale to assess the subjects' nervousness. Every musician performed a relaxed and a strenuous piece with an audience (as if playing a concert) and without an audience (as though in rehearsal). Then the researchers used a one-way analysis of variance measuring heart rate (HR) and heart rate variability (HRV), which showed a significant difference across the four conditions (easy/rehearsal, strenuous/rehearsal, easy/concert, and strenuous/concert) within subjects. Moreover, there was no difference due to age, sex, or the instrument (song/flute) when those factors were examined.

Want to know more? Go online or to the library and find . . .

Harmat, L., & Theorell, T. (2010). Heart rate variability during singing and flute playing. *Music and Medicine, 2*, 10–17.

Real-World Stats

SUMMARY

Analysis of variance (ANOVA) is the most complex of all the inferential tests you will learn in *Statistics for People Who (Think They) Hate Statistics*. It takes a good deal of concentration to perform the manual calculations, and even when you use SPSS, you have to be on your toes to understand that this is an overall test—one part will not give you information about differences between pairs of treatments. If you choose to go on and do post hoc analysis, you're only then completing all the tasks that go along with the powerful tool. We'll learn about just one more test between averages, and that's a factorial ANOVA. This, the holy grail of ANOVAs, can involve two or more factors, but we'll stick with two and SPSS will show us the way.

Chapter Summary

TIME TO PRACTICE

1. When is analysis of variance a more appropriate statistical technique to use than a test between a pair of means?

2. What is the difference between a one-way analysis of variance and a factorial analysis of variance?

3. Using the following table, provide three examples of a simple one-way ANOVA, two examples of a two-factor ANOVA, and one example of a three-factor ANOVA. We show you some examples—you provide the others. Be sure to identify the grouping and the test variable as we have done here.

Design	Grouping Variable(s)	Test Variable
Simple ANOVA	Four levels of hours of training—2, 4, 6, and 8 hours	Typing accuracy
	Enter your example here.	Enter your example here.
	Enter your example here.	Enter your example here.
	Enter your example here.	Enter your example here.

(Continued)

(Continued)

Design	Grouping Variable(s)	Test Variable
Two-factor ANOVA	Two levels of training and gender (2 × 2 design)	Typing accuracy
	Enter your example here.	Enter your example here.
	Enter your example here.	Enter your example here.
Three-factor ANOVA	Two levels of training and two of gender and three of income	Voting attitudes
	Enter your example here.	Enter your example here.

Problem 4

4. Using the data in Chapter 13 Data Set 2 and SPSS, compute the F ratio for a comparison among the three levels representing the average amount of time that swimmers practice weekly (<15, 15–25, and >25 hours), with the outcome variable being their time for the 100-yard freestyle. Answer the question of whether practice time makes a difference. Don't forget to use the Options feature to get the means for the groups.

5. The data in Chapter 13 Data Set 3 were collected by a researcher who wants to know whether the amount of stress is different for three groups of employees. Group 1 employees work the morning/day shift, Group 2 employees work the day/evening shift, and Group 3 employees work the night shift. The null hypothesis is that there is no difference in the amount of stress between groups. Test this in SPSS and provide your conclusion.

Problem 6

6. The Noodle company wants to know what thickness of noodle consumers find most pleasing to their palate (on a scale of 1 to 5, with 1 being most pleasing), so the food manufacturer put it to the test. The data are found in Chapter 13 Data Set 4. Turns out that there is a significant difference ($F_{(2,57)} = 19.398$, $p < .001$), and thin noodles are most preferred. But what about differences among thin, medium, and thick noodles? Post hoc analysis to the rescue!

7. Why is it only appropriate to do a post hoc analysis if the F ratio is significant?

STUDENT STUDY SITE

⑤SAGE edge™

Get the tools you need to sharpen your study skills! Visit **edge.sagepub.com/salkind6e** to access practice quizzes, eFlashcards, original and curated videos, data sets, journal articles, and more!

14

Two Too Many Factors

Factorial Analysis of Variance—A Brief Introduction

Difficulty Scale ☺ (about as tough as it gets for the challenging ideas—but we're only touching on the main concepts here)

WHAT YOU'LL LEARN ABOUT IN THIS CHAPTER

✦ Using analysis of variance with more than one factor

✦ Understanding main and interaction effects

✦ Using SPSS to complete a factorial analysis of variance

✦ Computing the effect size for a factorial analysis of variance

INTRODUCTION TO FACTORIAL ANALYSIS OF VARIANCE

How people make decisions has fascinated psychologists for decades. The data resulting from their studies have been applied to such broad fields as advertising, business, planning, and even theology. Miltiades Proios and George Doganis investigated the effect of how the experience of being actively involved in the decision-making process (in a variety of settings) and age can have an impact on moral reasoning. The sample consisted of a total of 148 referees—56 who refereed soccer, 55 who refereed basketball, and 37 who refereed handball. Their ages ranged from 17 to 50 years, and gender was not considered an important variable. Within the entire sample, about 8% had not had any experience in social, political, or athletic settings where they fully participated in the decision-making process; about 53% were active but not fully participatory; and about 39% were active and did participate in the decisions made within that organization. A two-way (multivariate—see Chapter 18 for

Introduction to
Chapter 14

more about this) analysis of variance showed an interaction between experience and age on moral reasoning and goal orientation of referees.

Why a two-way analysis of variance? Easy—there were two independent factors, with the first being level of experience and the second being age. Here, just as with any analysis of variance procedure, there are

1. a test of the main effect for age,

2. a test of the main effect for experience, and

3. a test for the interaction between experience and age (which turned out to be significant).

The very cool thing about analysis of variance when more than one factor or independent variable is tested is that the researcher can look not only at the individual effects of each factor but also at the simultaneous effects of both, through what is called an interaction. We will talk more about this later in the chapter.

Want to know more? Go online or to the library and find . . .

Proios, M., & Doganis, G. (2003). Experiences from active membership and participation in decision-making processes and age in moral reasoning and goal orientation of referees. *Perceptual and Motor Skills, 96*(1), 113–126.

THE PATH TO WISDOM AND KNOWLEDGE

Here's how you can use the flowchart shown in Figure 14.1 to select ANOVA (but this time with more than one factor) as the appropriate test statistic. Follow along the highlighted sequence of steps.

1. We are testing for differences among scores of different groups, in this case, groups that differ on level of experience and age.

2. The participants are not being tested more than once.

3. We are dealing with more than two groups.

4. We are dealing with more than one factor or independent variable.

5. The appropriate test statistic is factorial analysis of variance.

Finding the Correct Test Statistic

As in Chapter 13, we have decided that ANOVA is the correct procedure (to examine differences among more than two groups or levels of the independent variable), but because we have more than one factor, factorial ANOVA is the right choice.

Figure 14.1 Determining That Factorial Analysis of Variance Is the Correct Test Statistic

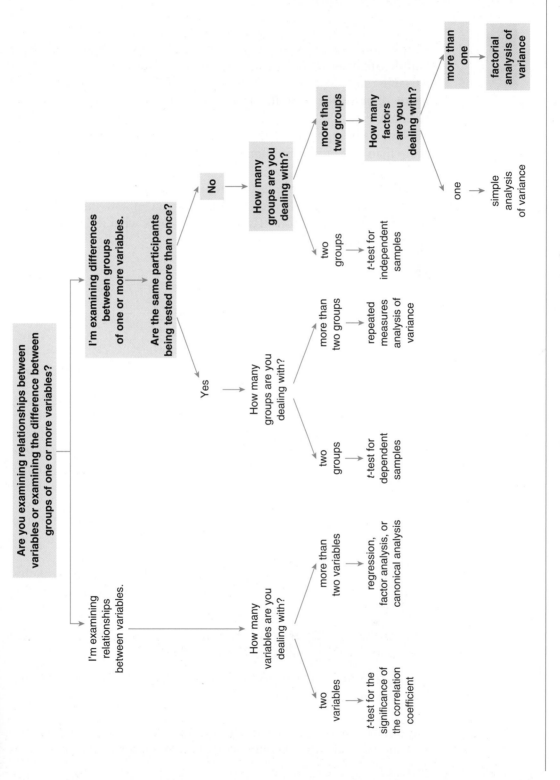

A NEW FLAVOR OF ANOVA

Factorial
ANOVA

You already know that ANOVA comes in at least one flavor, the simple analysis of variance we discussed in Chapter 13. With simple analysis of variance, there is one factor or one treatment variable (such as group membership) being explored, and there are more than two groups or levels within this factor or treatment variable.

Now, we bump up the entire technique a notch to so we can explore more than one factor simultaneously. This is called **factorial analysis of variance**.

Let's look at a simple example that includes two factors, gender (male or female) and treatment (high-impact or low-impact exercise program), and the outcome (weight loss). Here's what the experimental design would look like:

		Exercise Program	
		High Impact	*Low Impact*
Gender	*Male*		
	Female		

Then, we will look at what main effects and an interaction look like. We don't see a lot of data analysis here until a bit later in the chapter—mostly now it's just look and learn. Even though the exact probabilities of a Type I error are provided in the results (and we don't have to mess with statements like $p < .05$ and such), we'll use .05 as the criterion for rejection (or lack of acceptance) of the null.

There are three questions that you can ask and answer with this type of analysis:

1. Is there a difference between the effects on weight loss between two levels of exercise program, high impact and low impact?

2. Is there a difference between the effects on weight loss between two levels of gender, male and female?

3. Is the effect of being in the high- or low-impact program different for males or females?

Questions 1 and 2 deal with the presence of main effects, whereas Question 3 deals with the *interaction* between the two factors.

THE MAIN EVENT: MAIN EFFECTS IN FACTORIAL ANOVA

You might well remember that the primary task of analysis of variance is to test for the difference between two or more groups. When an analysis of the data reveals a difference between the levels of any factor, we talk about there being a **main effect**. In our example, there are 10 participants in each of the 4 groups, for a total of 40. And here's what the results of the analysis look like (and we used SPSS to compute this fancy-pants table). This is another form of the source table we introduced in Chapter 13.

Tests of Between-Subjects Effects

Dependent Variable: LOSS

Source	Type III Sum of Squares	df	Mean Square	F	Sig.
Corrected Model	3678.275	3	1226.092	8.605	.000
Intercept	232715.025	1	232715.025	1633.183	.000
TREATMEN	429.025	1	429.025	3.011	.091
GENDER	3222.025	1	3222.025	22.612	.000
TREATMEN *					
GENDER	27.225	1	27.225	.191	.665
Error	5129.700	36	142.492		
Total	241523.000	40			
Corrected Total	8807.975	39			

Pay attention to only the *Source* and the *Sig.* columns (which are highlighted). The conclusion we can reach is that there is a main effect for gender ($p = .000$), no main effect for treatment ($p = .091$), and no interaction between the two main factors ($p = .665$). So, as far as weight loss, it didn't matter whether one was in the high- or low-impact group, but it did matter if one were male or female. And because there was no interaction between the treatment factor and gender, there were no differential effects for treatment across gender.

If you plotted the means of these values, you would get something that looks like Figure 14.2.

You can see a big difference in distance on the loss axis between males and females (average score for all males is 85.25 and for females is 67.30), but for treatment (if you computed the averages), you would

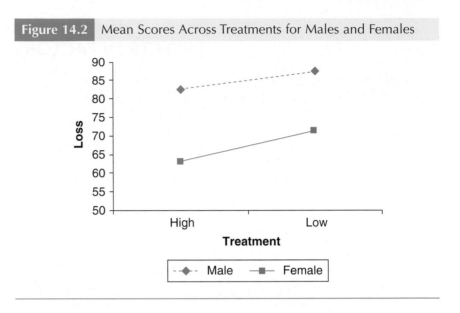

Figure 14.2   Mean Scores Across Treatments for Males and Females

find there to be little difference (with the average score across all highs being 73.00 and across all lows being 79.55). Now, of course, this is an analysis of variance, and the variability in the groups does matter, but in this example, you can see the differences between groups (such as males and females) within each factor (such as gender) and how they are reflected by the results of the analysis.

EVEN MORE INTERESTING: INTERACTION EFFECTS

Interaction Effects

Okay—now let's move to the interaction. Let's look at a different source table that indicates men and women are affected differentially across treatments, indicating the presence of an **interaction effect**. And, indeed, you will see some very cool outcomes. Once again, the most important parts are highlighted.

Here, there is no main effect for treatment or gender ($p = .127$ and .176, respectively), but yikes, there is one for the treatment by gender interaction ($p = .004$), which makes this a very interesting outcome. In effect, it does not matter whether you are in the high- or low-impact treatment group, or whether you are male or female, but it does matter a whole lot if you are in both conditions simultaneously such that the treatment does have a different impact on the weight loss of males than it does on females.

Figure 14.3 shows what a chart of the averages for each of the four groups looks like.

Tests of Between-Subjects Effects

Dependent Variable: LOSS

Source	Sum of Squares	df	Mean Square	F	Sig.
Corrected Model	1522.875	3	507.625	4.678	.007
Intercept	218892.025	1	218892.025	2017.386	.000
TREATMEN	265.225	1	265.225	2.444	.127
GENDER	207.025	1	207.025	1.908	.176
TREATMEN * GENDER	1050.625	1	1050.625	9.683	.004
Error	3906.100	36	108.503		
Total	224321.000	40			
Corrected Total	5428.975	39			

Figure 14.3   Averages Across Treatments for Males and Females

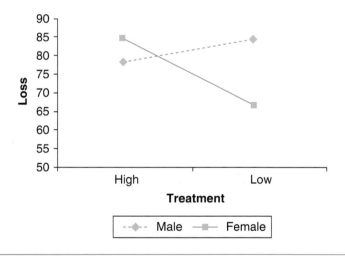

And here's what the actual means themselves look like (all compliments of SPSS):

	Male	Female
High Impact	73.70	79.40
Low Impact	78.80	64.00

What to make of this? Well, the interpretation here is pretty straightforward. Here's what we can say, and, being as smart as you are, you can recognize that these are the answers to the three questions we listed earlier:

1. There is no main effect for type of exercise.

2. There is no main effect for gender.

3. There is a clear interaction between treatment and gender, which means females lose more weight than males under the high-impact treatment condition and males lose more weight than females under the low-impact condition.

Indeed, one of the easiest ways to see the magnitude of an interaction (or the absence of one) is to plot the means as we did in Figure 14.3. The lines may not always cross as dramatically, and the appearance of the graph also depends upon what variable the x-axis explores, but charting the means is often the only way to get a sense of what's going on as far as main effects and interactions.

This is all pretty remarkable stuff. If you didn't know any better (and had never read this chapter), you would think that all you have to do is a simple *t*-test between the averages for males and females and then another simple *t*-test for the averages between those who participated in the high-impact and those who participated in the low-impact treatment—and you would have found nothing. But, using the idea of an interaction between main factors, you find out that there is a differential effect—an outcome that would have gone unnoticed otherwise. Indeed, if you can bear the cost of admission, interactions really are the most interesting outcomes in any factorial analysis of variance.

USING SPSS TO COMPUTE THE *F* RATIO

Computing the
F Ratio Using
SPSS

Here's a change for you. Throughout *Statistics for People Who (Think They) Hate Statistics,* we have provided you with examples of how to perform particular techniques the old-fashioned way (by hand, using a calculator) as well as by using a statistical analysis package such as SPSS. With the introduction of factorial ANOVA, we are illustrating the analysis using only SPSS. It is not any more of an intellectual challenge to complete a factorial ANOVA using a calculator, but it certainly is more laborious. For that reason, we are not going to cover the computation by hand but go right to the computation of the important values and spend more time on the interpretation.

We'll use the data that show there is a significant interaction, as seen here.

Treatment →	High Impact	High Impact	Low Impact	Low Impact
Gender →	Male	Female	Male	Female
	76	65	88	65
	78	90	76	67
	76	65	76	67
	76	90	76	87
	76	65	56	78
	74	90	76	56
	74	90	76	54
	76	79	98	56
	76	70	88	54
	55	90	78	56

Here are the steps and the computation of the *F*-test statistic. The reason you don't see the "famous eight steps" (so OK, there are 10) is that this is the first (and only) time throughout the book that we don't do the computations by hand but use only the computer. The analysis (as we said before) is just too labor-intensive for a course at this level.

1. State the null and research hypotheses.

There are actually three null hypotheses, shown here (Formulas 14.1a, 14.1b, and 14.1c) that state that there is no difference between the means for the two factors and that there is no interaction. Here we go.

First for the treatment . . .

$$H_0: \mu_{high} = \mu_{low} \qquad (14.1a)$$

And now for gender . . .

$$H_0: \mu_{male} = \mu_{female} \qquad (14.1b)$$

And now for the interaction between treatment and gender . . .

$$H_0: \mu_{high\cdot male} = \mu_{high\cdot female} = \mu_{low\cdot male} = \mu_{low\cdot female} \qquad (14.1c)$$

The research hypotheses, shown in Formulas 14.2a, 14.2b, and 14.2c, state that there is a difference between the means of the groups and that there is an interaction. Here they are.

First for the treatment . . .

$$H_1: \overline{X}_{high} \neq \overline{X}_{low} \qquad (14.2a)$$

And now for gender . . .

$$H_1: \overline{X}_{male} \neq \overline{X}_{female} \qquad (14.2b)$$

And now for the interaction between treatment and gender . . .

$$H_1: \overline{X}_{high\text{-}male} \neq \overline{X}_{high\text{-}female} \neq \overline{X}_{low\text{-}male} \neq \overline{X}_{low\text{-}female} \qquad (14.2c)$$

2. Set the level of risk (or the level of significance or Type I error) associated with the null hypothesis.

The level of risk or Type I error or level of significance is .05 here. Once again, the SPSS analysis will provide the exact p value, but the level of significance used is totally at the discretion of the researcher.

3. Select the appropriate test statistic.

Using the flowchart shown in Figure 14.1, we determined that the appropriate test is a factorial ANOVA.

4. Compute the test statistic value (called the obtained value).

We'll use SPSS for this, and here are the steps. We'll use the above data, which are available on the website for the book and listed in Appendix C and available directly from me at njs@ku.edu.

Enter the data in the Data Editor or open the file. Be sure that there is a column for each factor, including treatment, gender, and loss, as you see in Figure 14.4.

5. Click Analyze → General Linear Model → Univariate, and you will see the Factorial ANOVA dialog box shown in Figure 14.5.

6. Click on the variable named Loss and then move it to the Dependent Variable: box.

7. Click on the variable named Treatment and then move it to the Fixed Factor(s): box.

8. Click on the variable named Gender and then move it to the Random Factor(s): box.

9. Click Options and then Descriptive Statistics and then click Continue.

10. Click OK. SPSS will conduct the analysis and produce the output you see in Figure 14.6 (as you saw earlier in the chapter).

Figure 14.4   Data From Chapter 14 Data Set 1

	Treatment	Gender	Loss
1	High Impact	Male	76
2	High Impact	Male	78
3	High Impact	Male	76
4	High Impact	Male	76
5	High Impact	Male	76
6	High Impact	Male	74
7	High Impact	Male	74
8	High Impact	Male	76
9	High Impact	Male	76
10	High Impact	Male	55

Figure 14.5   Factorial ANOVA Dialog Box

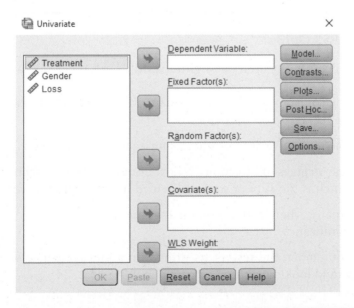

Figure 14.6 SPSS Output for a Factorial Analysis of Variance

Tests of Between-Subjects Effects

Dependent Variable: Loss

Source		Type III Sum of Squares	df	Mean Square	F	Sig.
Intercept	Hypothesis	218892.025	1	218892.025	1057.322	.020
	Error	207.025	1	207.025[a]		
Treatment	Hypothesis	265.225	1	265.225	.252	.704
	Error	1050.625	1	1050.625[b]		
Gender	Hypothesis	207.025	1	207.025	.197	.734
	Error	1050.625	1	1050.625[b]		
Treatment * Gender	Hypothesis	1050.625	1	1050.625	9.683	.004
	Error	3906.100	36	108.503[c]		

a. MS(Gender)

b. MS(Treatment * Gender)

c. MS(Error)

 Wondering why the SPSS output is labeled *univariate* analysis of variance? We knew you were. Well, in SPSS talk, this is an analysis that looks at only one dependent or outcome variable—in this case, weight loss. If we had more than one variable as part of the research question (such as attitude toward eating), then it would be a *multivariate* analysis of variance, which not only looks at group differences and more than one dependent variable but also controls for the relationship between the dependent variables. More about this in Chapter 18.

Understanding the SPSS Output

This SPSS output is pretty straightforward. Here's what we have:

1. The source of the variance—main effects and the interaction are identified.

2. The respective sum of squares for each source is given.

3. The degrees of freedom follows, and then the mean square, which is the sum of squares divided by the degrees of freedom.

4. Finally, there's the obtained value and the exact level of significance.

5. For gender, the results, as printed in a journal article or report, would look something like $F_{(1,36)} = .197$, $p = .734$.

6. For the treatment, the results, as printed in a journal article or report, would look something like $F_{(1,36)} = .252$, $p = .704$.

7. And for the interaction, the results, as printed in a journal article or report, would look something like $F_{(1,36)} = 9.683$, $p = .004$.

We're done! With the only significant outcome being the interaction between gender and treatment.

COMPUTING THE EFFECT SIZE FOR FACTORIAL ANOVA

A different formula is used to compute the effect size for a factorial ANOVA, but the idea is the same. We are still making a judgment about the magnitude of a difference that we observe.

For a factorial ANOVA, the statistic is called omega squared represented by ω^2, and here's the formula:

$$\omega^2 = \frac{SS_{\text{between groups}} - (df_{\text{between groups}})(MS_{\text{within groups}})}{MS_{\text{within groups}} + SS_{\text{total}}},$$

where

ω^2 is the value of the effect size;

SS_{between} is the sum of squares between treatments;

df_{between} is the total degrees of freedom;

MS_{within} is the mean sum of squares within treatments; and

SS_{total} is the corrected total sums of squares.

Don't be thrown by this formula; the terms are ones that you have seen in Chapter 13, and the interpretation of the final ω^2 value is the same as the interpretation of the other effect sizes we discussed earlier as well.

For the effect size for the factor named Gender in the example on page 272, we then have the following formula . . .

$$\omega^2 = \frac{3222.025 - (1)(142.492)}{142.492 + 8807.975} = .34$$

with an effect size of .34 for gender. According to our guidelines, this effect size of .34, reflecting the strength of the association, is large.

REAL-WORLD STATS

You've seen the news: Certain disorders primarily occurring in childhood, such as autism spectrum disorders and attention-deficit/hyperactivity disorder (ADHD), have been increasing in frequency. This may be because of changing diagnostic criteria, or it may reflect an actual increase in their occurrence. In any case, certain behaviors and outcomes need to be accurately measured as a first step in considering such diagnoses. That's what this study is all about. Measures of behavioral inhibition were used to investigate the factorial validity, ecological validity, and reliability of five performance-based measures of behavioral inhibition in a sample of seventy 3–, 4–, and 5-year-old children. A 2 × 3 factorial analysis of variance using sex and age-group (two factors) as independent variables showed significant main effects for sex and age and a nonsignificant interaction. Some of the measures correlated with teacher rating, and some did not, leading the researchers to conclude that some of the measures of behavioral inhibition might be useful, but others need further exploration.

Want to know more? Go online or to the library and find . . .

Real-World
Stats

Floyd, R. G., & Kirby, E. A. (2001). Psychometric properties of measures of behavioral inhibition with preschool-age children: Implications for assessment of children at risk for ADHD. *Journal of Attention Disorders, 5,* 79–91.

SUMMARY

Chapter
Summary

Now that we are done, done, done with testing differences between means, we'll move on to examining the significance of correlations, or the relationship between two variables.

TIME TO PRACTICE

1. When would you use a factorial ANOVA rather than a simple ANOVA to test the significance of the difference between the average of two or more groups?

2. Create a drawing or plan for a 2 × 3 experimental design that would lend itself to a factorial ANOVA. Be sure to identify the independent and dependent variables.

Problem 3

3. Using the data in Chapter 14 Data Set 2 and SPSS, complete the analysis and interpret the results for level of severity and type of treatment for

pain relief. It is a 2 × 3 experiment, like the kind you created in the answer to Question 2.

4. Use the data from Chapter 14 Data Set 3 and answer the following questions about an analysis of whether level of stress and gender have an impact on caffeine consumption (measured as cups of coffee per day). Problem 4

    a. Is there a difference among the levels of caffeine consumption for the high-stress, low-stress, and no-stress groups?

    b. Is there a difference between males and females, regardless of stress group?

    c. Are there any interactions?

5. This is the extra-credit question: For the data set named Chapter 14 Data Set 4, are girls better (as evaluated by a test of skill ranging from 1 to 10, with 10 being best) than boys and, if so, under what conditions?

STUDENT STUDY SITE

⑤SAGE edge™

Get the tools you need to sharpen your study skills! Visit **edge.sagepub.com/ salkind6e** to access practice quizzes, eFlashcards, original and curated videos, data sets, journal articles, and more!

15

Cousins or Just Good Friends?

Testing Relationships Using the Correlation Coefficient

Difficulty Scale ☺☺☺☺ (easy— you don't even have to figure anything out!)

WHAT YOU'LL LEARN ABOUT IN THIS CHAPTER

✦ Testing the significance of the correlation coefficient

✦ Interpreting the correlation coefficient

✦ Distinguishing between significance and meaningfulness (again!)

✦ Using SPSS to analyze correlational data and how to understand the results of the analysis

INTRODUCTION TO TESTING THE CORRELATION COEFFICIENT

In his research article on the relationship between the quality of a marriage and the quality of the relationship between the parent and the child, Daniel Shek told us that there are at least two possibilities. First, a poor marriage might enhance parent–child relationships. This is because parents who are dissatisfied with their marriage might substitute their relationship with their children for the emotional gratification lacking in their marriage. Or, according to the spillover hypothesis, a poor marriage might damage the parent–child relationship by setting the stage for increased difficulty in parenting children.

Introduction to Chapter 15

Shek examined the link between marital quality and parent–child relationships in 378 Chinese married couples over a 2-year period. He found that higher levels of marital quality were related to higher levels of quality of parent–child relationships; this was found for concurrent

Testing the
Correlation
Coefficient

measures (at the present time) as well as longitudinal measures (over time). He also found that the strength of the relationship between parents and children was the same for both mothers and fathers. This is an obvious example of how the use of the correlation coefficient gives us the information we need about whether sets of variables are related to one another. Shek computed a whole bunch of different correlations across mothers and fathers as well at time 1 and time 2, but all with the same purpose: to see if there was a significant correlation between the variables. Remember that this does not say anything about the causal nature of the relationship, only that the variables are associated with one another.

Want to know more? Go online or to the library and find . . .

Shek, D. T. L. (1998). Linkage between marital quality and parent-child relationship: A longitudinal study in the Chinese culture. *Journal of Family Issues, 19*, 687–704.

THE PATH TO WISDOM AND KNOWLEDGE

Selecting the
Appropriate
Test Statistic

Here's how you can use the flowchart to select the appropriate test statistic, the test for the correlation coefficient. Follow along the highlighted sequence of steps in Figure 15.1.

1. The relationship between variables, and not the difference between groups, is being examined.
2. Only two variables are being used.
3. The appropriate test statistic to use is the *t*-test for the correlation coefficient.

COMPUTING THE TEST STATISTIC

Determining
the Critical
Value

Here's something you'll probably be pleased to read: The correlation coefficient can act as its own test statistic. This makes things much easier because you don't have to compute any test statistics, and examining the significance is very easy indeed.

Let's use, as an example, the following data that examine the relationship between two variables, the quality of marriage and the quality of parent–child relationships. These hypothetical (made-up) scores can range from 0 to 100, and the higher the score, the happier the marriage and the better the parenting.

Figure 15.1 Determining That a *t*-Test for the Correlation Coefficient Is the Correct Test Statistic

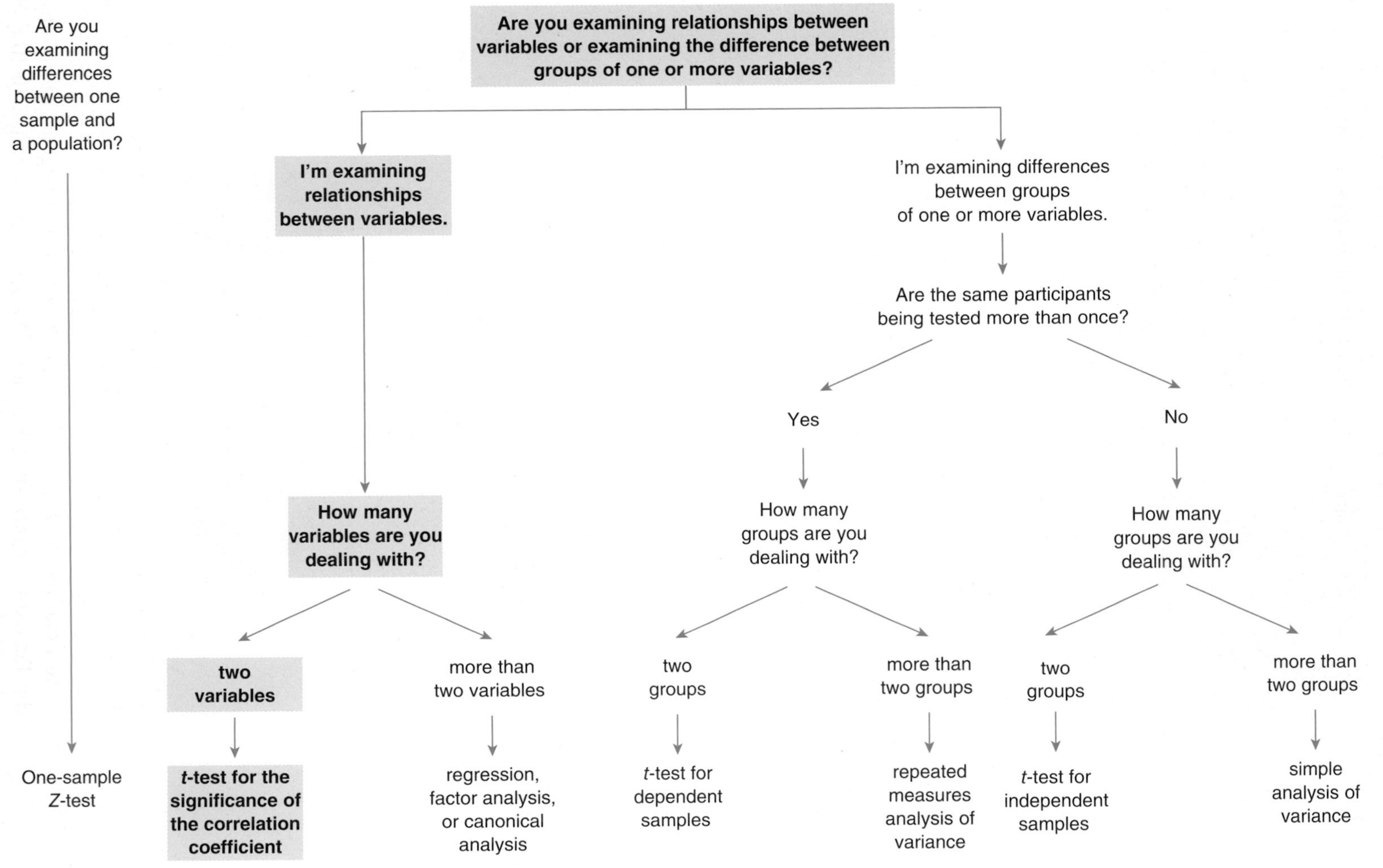

Are you examining differences between one sample and a population?

Are you examining relationships between variables or examining the difference between groups of one or more variables?

I'm examining differences between groups of one or more variables.

I'm examining relationships between variables.

Are the same participants being tested more than once?

How many variables are you dealing with?

No

Yes

How many groups are you dealing with?

How many groups are you dealing with?

more than two variables

two variables

more than two groups

two groups

two groups

more than two groups

regression, factor analysis, or canonical analysis

t-test for the significance of the correlation coefficient

simple analysis of variance

t-test for independent samples

t-test for dependent samples

repeated measures analysis of variance

One-sample Z-test

279

Quality of Marriage	Quality of the Parent–Child Relationship
76	43
81	33
78	23
76	34
76	31
78	51
76	56
78	43
98	44
88	45
76	32
66	33
44	28
67	39
65	31
59	38
87	21
77	27
79	43
85	46
68	41
76	41
77	48
98	56
98	56
99	55
98	45
87	68
67	54
78	33

You can use Formula 5.1 from Chapter 5 (on page 85) to compute the Pearson correlation coefficient. When you do, you will find that $r = .437$. Now let's go through the steps of actually testing the value for significance and deciding what the value means.

If you need a quick review of the basics of correlations, take a look back at Chapter 5.

Here are the famous eight steps and the computation of the *t*-test statistic.

1. State the null and research hypotheses.

   The null hypothesis states that there is no relationship between the quality of the marriage and the quality of the relationship between parents and children. The research hypothesis is a two-tailed, nondirectional research hypothesis because it posits that a relationship exists between the two variables but not the direction of that relationship. Remember that correlations can be positive (or direct) or negative (or indirect), and the most important characteristic of a correlation coefficient is its absolute value or size and not its sign (positive or negative).

   The null hypothesis is shown in Formula 15.1:

   $$H_0: \rho_{xy} = 0 \tag{15.1}$$

   The Greek letter ρ, or rho, represents the population estimate of the correlation coefficient.

   The research hypothesis (shown in Formula 15.2) states that there is a relationship between the two values and that the relationship differs from a value of zero:

   $$H_1: r_{xy} \neq 0 \tag{15.2}$$

2. Set the level of risk (or the level of significance or Type I error) associated with the null hypothesis.

   The level of risk or Type I error or level of significance is .05.

3. and 4. Select the appropriate test statistic.

   Using the flowchart shown in Figure 15.1, we determined that the appropriate test is for the correlation coefficient. In this instance, we do not need to compute a test statistic because the sample r value (r_{xy} = .437) is, for our purposes, the test statistic.

5. Determine the value needed for rejection of the null hypothesis using the appropriate table of critical values for the particular statistic.

   Table B.4 lists the critical values for the correlation coefficient.

Our first task is to determine the degrees of freedom (df), which approximates the sample size. For this particular test statistic, the degrees of freedom is $n - 2$, or $30 - 2 = 28$, where n is equal to the number of pairs used to compute the correlation coefficient. This is the degrees of freedom only for this test statistic and not necessarily for any other.

Using this number (28), the level of risk you are willing to take (.05), and a two-tailed test (because there is no direction to the research hypothesis), the critical value is .381 (using $df = 25$ because that's more conservative than using 30). So, at the .05 level, with 28 degrees of freedom for a two-tailed test, the value needed for rejection of the null hypothesis is .381.

6. Compare the obtained value and the critical value.

The obtained value is .437, and the critical value for rejection of the null hypothesis that the two variables are not related is .381.

7. and 8. Make a decision.

Now comes our decision. If the obtained value (or the value of the test statistic) is more extreme than the critical value (or the tabled value), the null hypothesis cannot be accepted. If the obtained value does not exceed the critical value, the null hypothesis is the most attractive explanation.

In this case, the obtained value (.437) does exceed the critical value (.381)—it is extreme enough for us to say that the relationship between the two variables (quality of marriage and quality of parent–child relationships) did occur due to something other than chance.

One tail or two? It's pretty easy to conceptualize what a one-tailed versus a two-tailed test is when it comes to differences between means. And it may even be easy for you to understand a two-tailed test of the correlation coefficient (where any difference from zero is what's tested). But what about a one-tailed test? It's really just as easy. A directional test of the research hypothesis that there is a relationship posits that relationship as being either direct (positive) or indirect (negative). So, for example, if you think that there is a positive correlation between two variables, then the test is one tailed. Similarly, if you hypothesize that there is a negative correlation between two variables, the test is one tailed as well. It's only when you don't predict the direction of the relationship that the test is two tailed. Got it?

Okay, we cheated a little. Actually, you can compute a *t* value (just like for the test for the difference between means) for the significance of the correlation coefficient. The formula is not any more difficult than any other you have dealt with up to now, but you won't see it here. The point is that some smart statisticians have computed the critical *r* value for different sample sizes (and, likewise, degrees of freedom) for one- and two-tailed tests at different levels of risk (.01, .05), as you see in Table B.4. But if you are reading along in an academic journal and see that a correlation was tested using a *t* value, you'll now know why.

So How Do I Interpret $r_{(28)} = .437$, p < .05?

- *r* represents the test statistic that was used.
- 28 is the number of degrees of freedom.
- .437 is the obtained value using the formula we showed you in Chapter 5.
- $p < .05$ (the really important part of this little phrase) indicates that the probability is less than 5% on any one test of the null hypothesis that the relationship between the two variables is due to chance alone. Because we defined .05 as our criterion for the research hypothesis to be more attractive than the null hypothesis, our conclusion is that a significant relationship exists between the two variables. This means that as the level of marital quality increases, so does the level of quality of the parent–child relationship. Similarly, as the level of marital quality decreases, so does the level of quality of the parent–child relationship.

Causes and Associations (Again!)

You're probably thinking that you've heard enough of this already, but this is so important that we really can't emphasize it enough. So, we'll emphasize it again. Just because two variables are related to one another (as in the previous example), that has no bearing on whether one causes the other. In other words, having a terrific marriage of the highest quality in no way ensures that the parent–child relationship will be of a high quality as well. These two variables may be correlated because some of the same traits that might make a person a good spouse also make that person a good parent (patience, understanding, willingness to sacrifice), but it's certainly possible to see how someone could be a good husband or wife and have a terrible relationship with his or her children.

Remember the crimes and ice cream example from Chapter 5? It's the same here. Just because things are related and share something in

The Correlation Game!

Basics of Correlations

common with one another has no bearing on whether there is a causal relationship between the two.

Correlation coefficients are used for lots of different purposes, and you're likely to read journal articles in which they've been used to estimate the reliability of a test. But you already know all this because you read Chapter 6 and mastered it! In Chapter 6, you may remember that we talked about such reliability coefficients as test–retest (the correlation of scores at two points in time), parallel forms (the correlation between scores on different forms), and internal consistency (the intercorrelation between items). Correlation coefficients also are the standard units used in more advanced statistical techniques, such as those we discuss in Chapter 18.

Significance Versus Meaningfulness (Again, Again!)

In Chapter 5, we reviewed the importance of the use of the coefficient of determination for understanding the meaningfulness of the correlation coefficient. You may remember that you square the correlation coefficient to determine the amount of variance in one variable that is accounted for by another variable. In Chapter 9, we also went over the general issue of significance versus meaningfulness.

But we should mention and discuss this topic again. Even if a correlation coefficient is significant (as was the case in the example in this chapter), it does not mean that the amount of variance accounted for is meaningful. For example, in this case, the coefficient of determination for a simple Pearson correlation value of .437 is equal to .190, indicating that 19% of the variance is accounted for and a whopping 81% of the variance is not. This leaves lots of room for doubt, doesn't it?

So, even though we know that there is a positive relationship between the quality of a marriage and the quality of a parent–child relationship and they tend to "go together," the relatively small correlation of .437 indicates that lots of other things are going on in that relationship that may be important as well. So, if ever you wanted to apply a popular saying to statistics, here goes: "What you see is not always what you get."

USING SPSS TO COMPUTE A CORRELATION COEFFICIENT (AGAIN)

Using SPSS

Here, we are using Chapter 15 Data Set 1, which has two measures of quality—one of marriage (time spent together in one of three categories, with the value of 1 indicating a lower quality of marriage than a value of 3) and one of parent–child relationships (strength of affection, with the higher score being more affectionate).

1. Enter the data in the Data Editor (or just open the file). Be sure you have two columns, each for a different variable. In Figure 15.2, the columns were labeled Qual_Marriage (for Quality of Marriage) and Qual_PC (Quality of Parent–Child Relationship).

Figure 15.2 Chapter 15 Data Set 1 Calculator

	Qual_Marriage	Qual_PC
1	1	58.7
2	1	55.3
3	1	61.8
4	1	49.5
5	1	64.5
6	1	61.0
7	1	65.7
8	1	51.4
9	1	53.6
10	1	59.0
11	2	64.4
12	2	55.8

2. Click Analyze → Correlate → Bivariate, and you will see the Bivariate Correlations dialog box, as shown in Figure 15.3.

Figure 15.3 Bivariate Correlations Dialog Box

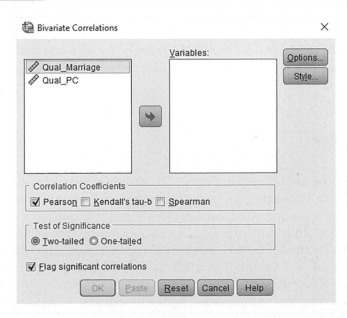

3. Double-click on the variable named Qual_Marriage to move it to the Variables: box, and then double-click on the variable named Qual_PC to move it to the Variables: box.

4. Click Pearson and click Two-tailed.

5. Click OK. The SPSS output is shown in Figure 15.4.

Figure 15.4	SPSS Output for Testing the Significance of the Correlation Coefficient

Correlations

		Qual_PC	Qual_Marriage
Qual_PC	Pearson Correlation	1	.081
	Sig. (2-tailed)		.637
	N	36	36
Qual_Marriage	Pearson Correlation	.081	1
	Sig. (2-tailed)	.637	
	N	36	36

Understanding the SPSS Output

This SPSS output is simple and straightforward.

The correlation between the two variables of interest is .081, which is significant at the .637 level. But to be (much) more precise, the probability of committing a Type I error is .637. That means that the likelihood of rejecting the null when true (that the two variables are not related) is about 63.7%—spookily high and very bad odds! Seems like quality of parenting and quality of marriage, at least for this set of data, are unrelated.

REAL-WORLD STATS

With all the attention given to social media (in the social media, of course), it's interesting to look at how researchers evaluate "social" well-being in college students who study and live outside the United States. This study investigated the relationship between the five personality domains and social well-being among 236 students enrolled at the University of Tehran. Correlations showed that some of the relationship between personality factors and dimensions of social well-being, such as neuroticism, was negatively related to social acceptance, social contribution, and social coherence. Conscientiousness was positively related to social contribution. Openness was positively related to social contribution, and social coherence and agreeableness were related to social acceptance and social contribution. Male students scored significantly higher than female students on social well-being. The researchers discussed these findings in light of cultural differences between countries and the general relationship between personality traits and well-being. Interesting stuff!

Want to know more? Go online or to the library and find . . .

Joshanloo, M., Rastegar, P., & Bakhshi, A. (2012). The Big Five personality domains as predictors of social wellbeing in Iranian university students. *Journal of Social and Personal Relationships, 29*, 639–660.

Real-World Stats

SUMMARY

Correlations are powerful tools that point out the direction of a relationship and help us to understand what two different outcomes share with one another. Remember that correlations work only when you are talking about associations and never when you are talking about causal effects.

Chapter Summary

TIME TO PRACTICE

1. Given the following information, use Table B.4 in Appendix B to determine whether the correlations are significant and how you would interpret the results.

   a. The correlation between speed and strength for 20 women is .567. Test these results at the .01 level using a one-tailed test.

   b. The correlation between the number correct on a math test and the time it takes to complete the test is −.45. Test whether this correlation is significant for 80 children at the .05 level of significance. Choose either a one- or a two-tailed test and justify your choice.

   c. The correlation between number of friends and grade point average (GPA) for 50 adolescents is .37. Is this significant at the .05 level for a two-tailed test?

2. Use the data in Chapter 15 Data Set 2 to answer the questions below. Do the analysis manually or by using SPSS.

   a. Compute the correlation between motivation and GPA.

   Problem 2

   b. Test for the significance of the correlation coefficient at the .05 level using a two-tailed test.

    c. True or false? The more highly you are motivated, the more you will study. Which answer did you select and why?

3. Use the data in Chapter 15 Data Set 3 to answer the questions below. Do the calculations manually or use SPSS.

    a. Compute the correlation between income and level of education.

    b. Test for the significance of the correlation.

    c. What argument can you make to support the conclusion that "lower levels of education cause low income"?

Problem 4

4. Use the data in Chapter 15 Data Set 4 to answer the questions below. Compute the correlation coefficient manually.

    a. Test for the significance of the correlation coefficient at the .05 level using a two-tailed test between hours of studying and grade.

    b. Interpret this correlation. What do you conclude about the relationship between the number of hours spent studying and the grade received on a test?

    c. How much variance is shared between the two variables?

    d. How do you interpret the results?

5. A study was completed that examined the relationship between coffee consumption and level of stress for a group of 50 undergraduates. The correlation was .373, and it was a two-tailed test done at the .01 level of significance. First, is the correlation significant? Second, what's wrong with the following statement: "As a result of the data collected in this study and our rigorous analyses, we have concluded that if you drink less coffee, you will experience less stress"?

6. Use the data below to answer the questions that follow. Do these calculations manually.

Age in Months	Number of Words Known
12	6
15	8
9	4
7	5
18	14
24	20
15	7
16	6
21	18
15	17

   a. Compute the correlation between age in months and number of words known.

   b. Test for the significance of the correlation at the .05 level of significance.

   c. Go way back and recall what you learned in Chapter 5 about correlation coefficients and interpret this correlation.

7. Discuss the general idea that just because two things are correlated, one does not necessarily cause the other. Provide an example (other than ice cream and crime!).

8. Chapter 15 Data Set 5 contains four variables:

   • Age (in years)
   • Shoe_Size (small, medium, and large as 1, 2, and 3)
   • Intelligence (as measured by a standardized test)
   • Level_of_Education (in years)

Which variables are significantly correlated, and, more important, what correlations are meaningful?

STUDENT STUDY SITE

⑤SAGE edge™

Get the tools you need to sharpen your study skills! Visit **edge.sagepub.com/ salkind6e** to access practice quizzes, eFlashcards, original and curated videos, data sets, journal articles, and more!

16

Predicting Who'll Win the Super Bowl

Using Linear Regression

Difficulty Scale ☺ (as hard as they get!)

WHAT YOU'LL LEARN ABOUT IN THIS CHAPTER

+ Understanding how prediction works and how it can be used in the social and behavioral sciences

+ Understanding how and why linear regression works when predicting one variable on the basis of another

+ Judging the accuracy of predictions

+ Understanding how multiple regression works and why it is useful

INTRODUCTION TO LINEAR REGRESSION

You've seen it all over the news—concern about obesity and how it affects work and daily life. A set of researchers in Sweden were interested in looking at how well mobility disability and/or obesity predicted job strain and whether social support at work can modify this association. The study included more than 35,000 participants, and differences in job strain mean scores were estimated using linear regression, the exact focus of what we are discussing in this chapter. The results found that level of mobile disability did predict job strain and that social support at work significantly modified the association among job strain, mobile disability, and obesity.

Introduction to Chapter 16

Want to know more? Go to the library or go online . . .

Norrback, M., De Munter, J., Tynelius, P., Ahlstrom, G., and Rasmussen, F. (2016). The association of mobility disability, weight status and job strain: A cross-sectional study. *Scandinavian Journal of Public Health, 44,* 311–319.

WHAT IS PREDICTION ALL ABOUT?

Here's the scoop. Not only can you compute the degree to which two variables are related to one another (by computing a correlation coefficient as we did in Chapter 5), but you can also use these correlations to predict the value of one variable based on the value of another. This is a very special case of how correlations can be used, and it is a very powerful tool for social and behavioral sciences researchers.

The basic idea is to use a set of previously collected data (such as data on variables X and Y), calculate how correlated these variables are with one another, and then use that correlation and the knowledge of X to predict Y. Sound difficult? It's not really, especially once you see it illustrated.

For example, a researcher collects data on total high school grade point average (GPA) and first-year college GPA for 400 students in their freshman year at the state university. He computes the correlation between the two variables. Then, he uses the techniques you'll learn about later in this chapter to take a *new* set of high school GPAs and (knowing the relationship between high school GPA and first-year college GPA from the previous set of students) predict what first-year GPA should be for a new sample of 400 students. Pretty nifty, huh?

Here's another example. A group of teachers is interested in finding out how well retention works. That is, do children who are retained in kindergarten (and not passed on to first grade) do better in first grade when they get there? Once again, these teachers know the correlation between being retained and first-grade performance from *prior* years; they can apply it to a new set of students and predict first-grade performance based on kindergarten performance.

How does regression work? Data are collected on past events (such as the existing relationship between two variables) and then applied to a future event given knowledge of only one variable. It's easier than you think.

 The higher the absolute value of the correlation coefficient, regardless of whether it is direct or indirect (positive or negative), the more accurate the prediction is of one variable from the other based on that correlation. That's because the more two variables share in common, the more you know about the second variable based on your knowledge of the first variable. And you may already surmise that when the correlation is perfect (+1.0 or −1.0), then the prediction is perfect as well. If $r_{xy} = -1.0$ or +1.0 and if you know the value of X, then you also know the exact value of Y. Likewise, if $r_{xy} = -1.0$ or +1.0 and you know the value of Y, then you also know the exact value of X. Either way works just fine.

What we'll do in this chapter is go through the process of using linear regression to predict a Y score from an X score. We'll begin by discussing the general logic that underlies prediction, then review some simple line-drawing skills, and, finally, discuss the prediction process using specific examples.

Why the prediction of Y from X and not the other way around? Convention. Seems like a good idea to have a consistent way to identify variables, so the Y variable becomes the dependent variable or the one being predicted and the X variable becomes the independent variable and is the variable used to predict the value of Y. And when predicted, the Y value is represented as Y' (read as **Y prime**)—the predicted value of Y.

THE LOGIC OF PREDICTION

Before we begin with the actual calculations and show you how correlations are used for prediction, let's understand the argument for why and how prediction works. We will continue with the example of predicting college GPA from high school GPA.

Prediction is the computation of future outcomes based on a knowledge of present ones. When we want to predict one variable from another, we need to first compute the correlation between the two variables. Table 16.1 shows the data we will be using in this example. Figure 16.1 shows the scatterplot (see Chapter 5) of the two variables that are being computed.

To predict college GPA from high school GPA, we have to create a **regression equation** and use that to plot what is called a **regression line**. A regression line reflects our best guess as to what score on the Y variable (college GPA) would be predicted by a score on the X variable (high school GPA). For all the data you see in Table 16.1, the regression line is drawn so that it minimizes the distance between

Correlations
and Causality

Correlations
and Causality
Again!

Table 16.1	Total High School GPA and First-Year College GPA

High School GPA	First-Year College GPA
3.50	3.30
2.50	2.20
4.00	3.50
3.80	2.70
2.80	3.50
1.90	2.00
3.20	3.10
3.70	3.40
2.70	1.90
3.30	3.70

itself and each of the points on the predicted (Y') variable. You'll learn shortly how to draw that line, shown in Figure 16.2.

What does the regression line you see in Figure 16.2 represent?

First, it's the regression of the Y variable on the X variable. In other words, Y (college GPA) is being predicted from X (high school GPA). This regression line is also called the **line of best fit**. The line fits these data because it minimizes the distance between each individual point and the regression line. For example, if you take all of these points and try to find the line that best fits them all at once, the line you see in Figure 16.2 is the one you would use.

Second, it's the line that allows us our best guess (at estimating what college GPA would be, given each high school GPA). For example, if high school GPA is 3.0, then college GPA should be around (remember, this is only an eyeball prediction) 2.8. Take a look at Figure 16.3 to see how we did this. We located the predictor value (3.0) on the x-axis, drew a perpendicular line from the x-axis to the regression line, then drew a horizontal line to the y-axis, and finally *estimated* what the *predicted* value of Y would be.

Third, the distance between each individual data point and the regression line is the **error in prediction**—a direct reflection of the correlation between the two variables. For example, if you look at data point (3.3, 3.7), marked in Figure 16.4, you can see that this (X, Y) data point is above the regression line. The distance between that point and the line is the error in prediction, as marked in Figure 16.4, because if the prediction were perfect, then all the predicted points would fall where? Right on the regression or prediction line.

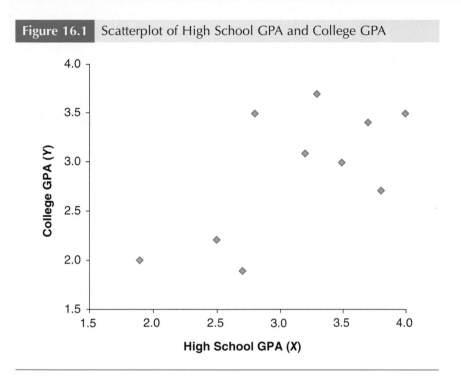

Figure 16.1 Scatterplot of High School GPA and College GPA

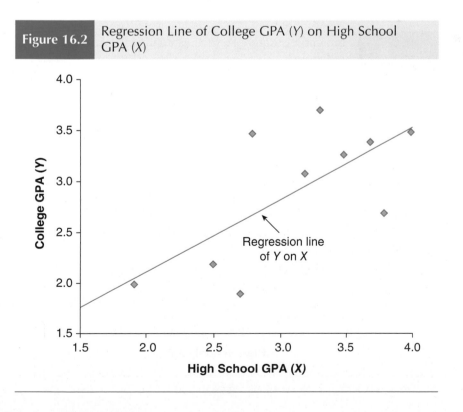

Figure 16.2 Regression Line of College GPA (Y) on High School GPA (X)

| Figure 16.3 | Estimating College GPA Given High School GPA |

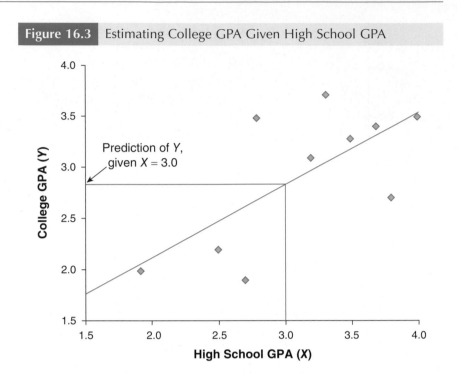

| Figure 16.4 | Prediction Is Rarely Perfect: Estimating the Error in Prediction |

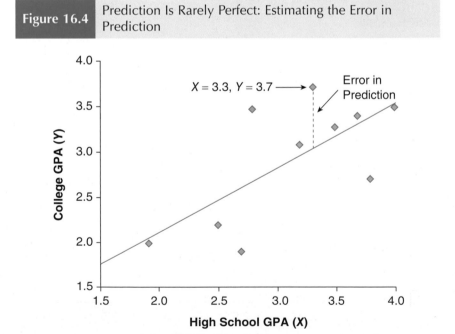

Fourth, if the correlation were perfect, all the data points would align themselves along a 45° angle, and the regression line would pass through each point (just as we said earlier in the third point).

Given the regression line, we can use it to precisely predict any future score. That's what we'll do right now—create the line and then do some prediction work.

DRAWING THE WORLD'S BEST LINE (FOR YOUR DATA)

The simplest way to think of prediction is that you are determining the score on one variable (which we'll call Y—the **criterion** or **dependent variable**) based on the value of another score (which we'll call X—the **predictor** or **independent variable**).

Drawing a Regression Line

The way that we find out how well X can predict Y is through the creation of the regression line we mentioned earlier in this chapter. This line is created from data that have already been collected. The equations are then used to predict scores using a new value for X, the predictor variable.

Formula 16.1 shows the general formula for the regression line, which may look familiar because you probably used something very similar in your high school and college math courses. It's the formula for any straight line.

$$Y' = bX + a, \tag{16.1}$$

where

- Y' is the predicted score of Y based on a known value of X;
- b is the slope, or direction, of the line;
- X is the score being used as the predictor; and
- a is the point at which the line crosses the y-axis.

Let's use the same data shown earlier in Table 16.1, along with a few more calculations that we will need thrown in.

X	Y	X²	Y²	XY
3.5	3.3	12.25	10.89	11.55
2.5	2.2	6.25	4.84	5.50
4.0	3.5	16.00	12.25	14.00

(Continued)

(Continued)

X	Y	X²	Y²	XY
3.8	2.7	14.44	7.29	10.26
2.8	3.5	7.84	12.25	9.80
1.9	2.0	3.61	4.00	3.80
3.2	3.1	10.24	9.61	9.92
3.7	3.4	13.69	11.56	12.58
2.7	1.9	7.29	3.61	5.13
3.3	3.7	10.89	13.69	12.21
Total	31.4	29.3	102.50	89.99	94.75

From this table, we see that . . .

- ΣX, or the sum of all the X values, is 31.4.
- ΣY, or the sum of all the Y values, is 29.3.
- ΣX^2, or the sum of each X value squared, is 102.5.
- ΣY^2, or the sum of each Y value squared, is 89.99.
- ΣXY, or the sum of the products of X and Y, is 94.75.

Formula 16.2 is used to compute the slope of the regression line (b in the equation for a straight line):

$$b = \frac{\Sigma XY - (\Sigma X \Sigma Y / n)}{\Sigma X^2 - [(\Sigma X)^2 / n]} \qquad (16.2)$$

In Formula 16.3, you can see the computed value for b, the slope of the line:

$$b = \frac{94.75 - [(31.4 \times 29.3)/10]}{102.5 - [(31.4)^2 / 10]}$$

$$b = \frac{2.749}{3.904} = 0.704 \qquad (16.3)$$

Formula 16.4 is used to compute the point at which the line crosses the y-axis (a in the equation for a straight line):

$$a = \frac{\Sigma Y - b \Sigma X}{n} \qquad (16.4)$$

In Formula 16.5, you can see the computed value for a, the intercept of the line:

$$a = \frac{29.3 - (0.704 \times 31.4)}{10}$$

$$a = \frac{7.19}{10} = 0.719$$

(16.5)

Now, if we go back and substitute b and a into the equation for a straight line $(Y = bX + a)$, we come up with the final regression line:

$$Y' = 0.704X + 0.719$$

Why the Y' and not just a plain Y? Remember, we are using X to predict Y, so we use Y' to mean the predicted and not the actual value of Y.

So, now that we have this equation, what can we do with it? Predict Y, of course.

For example, let's say that high school GPA equals 2.8 (or $X = 2.8$). If we substitute the value of 2.8 into the equation, we get the following formula:

$$Y' = 0.704(2.8) + 0.719 = 2.69$$

So, 2.69 is the predicted value of Y (or Y') given X is equal to 2.8. Now, for any X score, we can easily and quickly compute a predicted Y score.

Linear Regression

You can use this formula and the known values to compute predicted values. That's most of what we just talked about. But you can also plot a regression line to show how well the scores (what you are trying to predict) actually fit the data from which you are predicting. Take another look at Figure 16.2, the plot of the high school–college GPA data. It includes a regression line, which is also called a **trend line**. How did we get this line? Easy. We used the same charting skills you learned in Chapter 5 to create a scatterplot; then we selected Add Fit Line in the SPSS Chart Editor. Poof! Done!

You can see the trend is positive (in that the line has a positive slope) and that the correlation is .6835—very positive. And you can see that the data points do not align directly on the line, but they are pretty close, which indicates that there is a relatively small amount of error.

Not all lines that fit best between a bunch of data points are straight. Rather, they could be curvilinear, just as you can have a curvilinear relationship between your variables, as we discussed in Chapter 5. For example, the relationship between anxiety and performance is such that when people are not at all anxious or very anxious, they don't perform very well. But if they're moderately anxious, then performance can be enhanced. The relationship between these two variables is curvilinear, and the prediction of Y from X takes that into account.

HOW GOOD IS YOUR PREDICTION?

How can we measure how good a job we have done predicting one outcome from another? We know that the higher the absolute magnitude of the correlation between two variables, the better the prediction. In theory, that's great. But being practical, we can also look at the difference between the predicted value (Y') and the actual value (Y) when we first compute the formula of the regression line.

For example, if the formula for the regression line is $Y' = 0.704X + 0.719$, the predicted Y (or Y') for an X value of 2.8 is $0.704(2.8) + 0.719$, or 2.69. We know that the actual Y value that corresponds to an X value is 3.5 (from the data set shown in Table 16.1). The difference between 3.5 and 2.69, or 0.81, is known as an error in prediction.

Another measure of error that you could use is the coefficient of determination (see Chapter 5), which is the percent of error that is *reduced* in the relationship between variables. For example, if the correlation between two variables is .4 and the coefficient of determination is 16% or $.4^2$, the reduction in error is 16% (since initially we suspect the relationship between the two variables starts at 0 or 100% error (no predictive value at all).

If we take all of these differences, we can compute the average amount that each data point differs from the predicted data point, or the **standard error of estimate**. This is a kind of standard deviation that reflects variability about the line of regression. The value tells us how much imprecision there is in our estimate. As you might expect, the higher the correlation between the two values (and the better the prediction), the lower this standard error of estimate will be. In fact, if the correlation between the two variables is perfect (either $+1$ or -1), then the standard error of estimate is zero. Why? Because if prediction is perfect, all of the actual data points fall on the regression line, and there's no error in estimating Y from X.

 The predicted Y', or dependent variable, need not always be a continuous one, such as height, test score, or problem-solving skills. It can be a categorical variable, such as admit/don't admit, Level A/Level B, or Social Class 1/Social Class 2. The score that's used in the prediction is "dummy coded" to be a 1 or a 2 and then used in the same equation.

USING SPSS TO COMPUTE THE REGRESSION LINE

Let's use SPSS to compute the regression line that predicts Y' from X. The data set we are using is Chapter 16 Data Set 1. We will be using the number of hours of training to predict how severe injuries will be if someone is injured playing football.

There are two variables in this data set:

Variable	Definition
Training (X)	Number of hours per week of strength training
Injuries (Y)	Severity of injuries on a scale from 1 to 10

Computing a Regression Line

Here are the steps to compute the regression line that we discussed in this chapter. Follow along and do it yourself.

1. Open the file named Chapter 16 Data Set 1.

2. Click Analyze → Regression → Linear. You'll see the Linear Regression dialog box shown in Figure 16.5.

3. Click on the variable named Injuries and then move it to the Dependent: variable box. It's the *dependent* variable because its value *depends* on the value of number of hours of training. In other words, it's the variable being predicted.

4. Click on the variable named Training and then move it to the Independent(s): variable box.

5. Click OK, and you will see the partial results of the analysis, as shown in Figure 16.6.

We'll get to the interpretation of this output in a moment. First, let's have SPSS overlay a regression line on the scatterplot for these data like the one you saw earlier in Figure 16.2.

6. Click Graphs → Legacy Dialogs → Scatter/Dot.

7. Click Simple Scatter and then click Define. You'll see the simple Scatterplot dialog box.

8. Click Injuries and move it to the variable label to the Y Axis: box. Remember, the predicted variable is represented by the y-axis.

9. Click Training and move it to the variable label to the X Axis: box.

10. Click OK, and you will see the scatterplot as shown in Figure 16.7.

    Now let's draw the regression line.

11. If you are not in the chart editor, double-click on the chart to select it for editing.

12. Click on the Add Fit Line at Total button that looks like this: ⬈.

13. Close the Properties box that opened when you selected the Add Fit Line at Total button and then close the chart editor window. The completed scatterplot, with the regression line, is shown in Figure 16.8 along with the multiple regression value R^2, which equals 0.21. As you will read more about shortly, the multiple regression correlation coefficient is the regression of all the X values on the predicated value.

Figure 16.5   Linear Regression Dialog Box

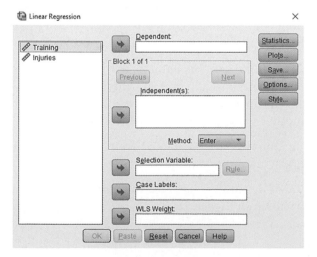

| Figure 16.6 | Results of the SPSS Analysis |

Coefficients[a]

Model		Unstandardized Coefficients		Standardized Coefficients	t	Sig.
		B	Std. Error	Beta		
1	(Constant)	6.847	1.004		6.818	.000
	Training	-.125	.046	-.458	-2.727	.011

a. Dependent Variable: Injuries

| Figure 16.7 | A Scatterplot Generated Using SPSS |

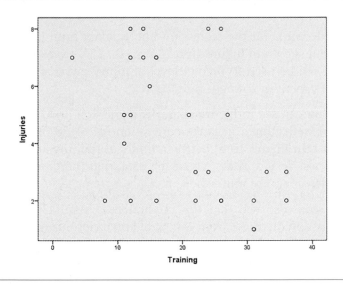

| Figure 16.8 | SPSS Scatterplot With Regression Line |

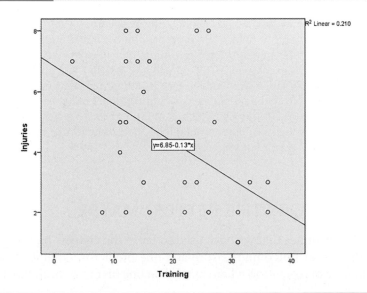

When you have the properties dialog box open for drawing the regression line, notice that there is a set of Confidence Intervals options. When clicked, these show you a boundary within which there is a specific probability as to how good the prediction is. For example, if you click Mean and specify 95%, the graph will show you the boundaries surrounding the regression line, within which there is a 95% chance of the predicted scores occurring.

Understanding the SPSS Output

The SPSS output tells us several things.

1. The formula for the regression line is taken from the first set of output shown in Figure 16.6 as $Y' = -0.125X + 6.847$. This equation can be used to predict level of injury given any number of hours spent in strength training.

2. As you can see in Figure 16.8, the regression line has a negative slope, reflecting a negative correlation (of –.458, which is what Beta is in Figure 16.6) between hours of training and severity of injuries. So it appears, given the data, that the more one trains, the fewer severe injuries occur.

3. You can also see that the prediction is significant—in other words, predicting Y from X is based on a significant relationship between the two variables such that the test of significance for both the constant (Training) and the predicted variable (Injuries) is significantly different from zero (which it would be if there was no predictive value for X predicting Y.

So just d how good is the prediction? Well, the SPSS output (which we did not show you) also indicates that the standard error of estimate for Injuries (the predicted variable) is 2.182, which means that there is a 95% chance (remember it is 1 standard deviation removed from the mean) the prediction will fall between the mean of all injuries (which is 4.33) and ±2.182. So, based on the correlation coefficient, the prediction is reasonably good, but not great.

THE MORE PREDICTORS THE BETTER? MAYBE

All of the examples that we have used so far in the chapter have been for one criterion or outcome measure and one predictor variable. There is also the case of regression where more than one predictor or independent

variable is used to predict a particular outcome. If one variable can predict an outcome with some degree of accuracy, then why couldn't two do a better job? Maybe so, but there's a big caveat—read on.

For example, if high school GPA is a pretty good indicator of college GPA, then how about high school GPA plus number of hours of extracurricular activities? So, instead of . . .

$$Y' = bX + a$$

the model for the regression equation becomes . . .

$$Y' = bX_1 + bX_2 + a,$$

where

- X_1 is the value of the first independent variable;
- X_2 is the value of the second independent variable;
- b is the regression weight for that particular variable; and
- a is the intercept of the regression line, or where the regression line crosses the y-axis.

As you may have guessed, this model is called **multiple regression** (multiple independent variables, right?). So, in theory anyway, you are predicting an outcome from two independent variables rather than one. But you want to add additional independent variables only under certain conditions. Read on . . .

Any variable you add has to make a unique contribution to understanding the dependent variable. Otherwise, why use it? What do we mean by *unique*? The additional variable needs to explain differences in the predicted variable that the first predictor does not. That is, the two variables in combination should predict Y better than any one of the variables would do alone.

In our example, level of participation in extracurricular activities could make a unique contribution. But should we add a variable such as the number of hours each student studied in high school as a third independent variable or predictor? Because number of hours of study is probably highly related to high school GPA (another of our predictor variables, remember?), study time probably would not add very much to the overall prediction of college GPA. We might be better off looking for another variable (such as ratings on letters of recommendation) rather than collecting the data on study time.

Take a look at Figure 16.9, which is the result of a multiple regression analysis that adds the number of extracurricular activity hours to the data you saw in Table 16.1. You can see how both high school GPA and number of hours of extracurricular activity are significant contributors to first-year college GPA. This is a powerful way of examining what and how *more than one* independent variable contribute to prediction of another variable.

| Figure 16.9 | A Multiple Regression Analysis |

Coefficients[a]

Model		Unstandardized Coefficients		Standardized Coefficients	t	Sig.
		B	Std. Error	Beta		
1	(Constant)	.411	.485		.847	.425
	High_School_GAP	.494	.158	.480	3.133	.017
	Etxra_Curr_Act	.070	.016	.653	4.261	.004

a. Dependent Variable: First_Year_College_GPA

The Big Rule(s) When It Comes to Using Multiple Predictor Variables

If you are using more than one predictor variable, try to keep the following two important guidelines in mind:

1. When selecting an independent variable to predict an outcome, select a predictor variable (X) that is related to the predicted variable (Y). That way, the two share something in common (remember, they should be correlated).

2. When selecting more than one independent variable (such as X_1 and X_2), try to select variables that are independent or uncorrelated with one another but are both related to the outcome or predicted (Y) variable.

In effect, you want only independent or predictor variables that are related to the dependent variable and are unrelated to each other. That way, each one makes as distinct a contribution as possible to predicting the dependent or predicted variable.

How many predictor variables are too many? Well, if one variable predicts some outcome, and two are even more accurate, then why not three, four, or five predictor variables? In practical terms, every time you add a variable, an expense is incurred. Someone has to go collect the data, it takes time (which is $$$ when it comes to research budgets), and so on. From a theoretical sense, there is a fixed limit on how many variables can contribute to an understanding of what we are trying to predict. Remember that it is best when the predictor or independent variables are independent or unrelated to each other. The problem is that once you get to three or four variables, fewer things can remain unrelated. Better to be accurate and conservative than to include too many variables and waste money and the power of prediction.

REAL-WORLD STATS

How children feel about what they do is often very closely related to how well they do what they do. The aim of this study was to analyze the consequences of emotion during a writing exercise. In the model this research follows, motivation and affect (the experience of emotion) play an important role during the writing process. Fourth and fifth graders were instructed to write autobiographical narratives with no emotional content, positive emotional content, and negative emotional content. The results showed no effect regarding these instructions on the proportion of spelling errors, but the results did reveal an effect on the length of narrative the children wrote. A simple regression analysis (just like the ones we did and discussed in this chapter) showed a correlation and some predictive value between working memory capacity and the number of spelling errors in the neutral condition only. Since the model on which the researchers based much of their preliminary thought about this topic states that emotions can increase the cognitive load or the amount of "work" necessary in writing, that becomes the focus of the discussion in this research article.

Want to know more? Go online or to the library and find . . .

Fartoukh, M., Chanquoy, L., & Piolat, A. (2012). Effects of emotion on writing processes in children. *Written Communication, 29,* 391–411.

Real-World
Stats

SUMMARY

Prediction is a special case of simple correlation, and it is a very powerful tool for examining complex relationships. This chapter might have been a little more difficult than others, but you'll be well served by what you have learned, especially if you can apply it to the research reports and journal articles that you have to read. We are now at the end of lots of chapters on inference, and we're about to move on in the next part of this book to using statistics when the sample size is very small or when the assumption that the scores are distributed in a normal way is violated.

Chapter
Summary

TIME TO PRACTICE

1. How does linear regression differ from analysis of variance?

2. Chapter 16 Data Set 2 contains the data for a group of participants who took a timed test. The data are the average amount of time the participants took on each item (Time) and the number of guesses it took to get each item correct (Correct).

Problem 2

a. What is the regression equation for predicting response time from number correct?

  b. What is the predicted response time if the number correct is 8?

  c. What is the difference between the predicted and the actual number correct for each of the predicted response times?

3. Betsy is interested in predicting how many 75-year-olds will develop Alzheimer's disease and is using as predictors level of education and general physical health graded on a scale from 1 to 10. But she is interested in using other predictor variables as well. Answer the following questions:

  a. What criteria should she use in the selection of other predictors? Why?

  b. Name two other predictors that you think might be related to the development of Alzheimer's disease.

  c. With the four predictor variables—level of education and general physical health and the two new ones that you named in (b)—draw out what the model of the regression equation would look like.

4. Go to the library and locate three different research studies in your area of interest that use linear regression. It's okay if the studies contain more than one predictor variable. Answer the following questions for each study:

  a. What is one independent variable? What is the dependent variable?

  b. If there is more than one independent variable, what argument does the researcher make that these variables are independent of one another?

  c. Which of the three studies seems to present the least convincing evidence that the dependent variable is predicted by the independent variable, and why?

5. Here's where you can apply the information in one of this chapter's tips and get a chance to predict a Super Bowl winner! Joe Coach was curious to know whether the average number of games won in a year predicts Super Bowl performance (win or lose). The X variable was the average number of games won during the past 10 seasons. The Y variable was whether the team ever won the Super Bowl during the past 10 seasons. Here are the data:

Team	Average Number of Wins Over 10 Years	Bowl? (1 = yes and 0 = no)
Savannah Sharks	12	1
Pittsburgh Pelicans	11	0
Williamstown Warriors	15	0

(Continued)

Team	Average Number of Wins Over 10 Years	Bowl? (1 = yes and 0 = no)
Bennington Bruisers	12	1
Atlanta Angels	13	1
Trenton Terrors	16	0
Virginia Vipers	15	1
Charleston Crooners	9	0
Harrisburg Heathens	8	0
Eaton Energizers	12	1

   a. How would you assess the usefulness of the average number of wins as a predictor of whether a team ever won a Super Bowl?

   b. What's the advantage of being able to use a categorical variable (such as 1 or 0) as a dependent variable?

   c. What other variables might you use to predict the dependent variable, and why would you choose them?

6. Check your calculation of the correlation coefficient of the relationship between coffee consumption and stress done in Chapter 15, Question 5. If you want to know whether coffee consumption predicts group membership . . .

   a. What is the predictor or the independent variable?

   b. What is the criterion or the dependent variable?

   c. Do you have an idea what R^2 will be?

7. Time to try out multiple predictor variables. Take a look at the data shown here where the outcome is becoming a great chef. We suspect that variables such as number of years of experience cooking, level of formal culinary education, and number of different positions (sous chef, pasta station, etc.) all contribute to rankings or scores on the Great Chef Test.

By this time, you should be pretty much used to creating equations from data like these, so let's get to the real questions:

Years of Experience	Level of Education	# Positions	Score on Great Chef Test
5	1	5	88
6	2	4	78

(Continued)

Years of Experience	Level of Education	# Positions	Score on Great Chef Test
12	3	9	56
21	3	8	88
7	2	5	97
9	1	8	90
13	2	8	79
16	2	9	85
21	2	9	60
11	1	4	89
15	2	7	88
15	3	7	76
1	3	3	78
17	2	6	98
26	2	8	91
11	2	6	88
18	3	7	90
31	3	12	98
27	2	16	88

    a. Which are the best predictors of a chef's score?

    b. What score can you expect from a person with 12 years of experience and a Level 2 education who has held five positions?

8. Take a look at Chapter 16 Data Set 3, where number of home sales (Number_Homes_Sold) is being predicted by years in the business (Years_In_Business) and level of education in years (Level_Of_Education). Why is level of education such a poor contributor to the overall prediction (using both years in the business and level of education combined) of number of homes sold? What's the best predictor and how do you know? (Hint: These are sort of trick questions. Before you go ahead and analyze the data, look at the raw data in the file for the characteristics you know are important for one variable to be correlated with another.)

9. For any combinations of predicted and predictor variables, what should be the nature of the relationship between them?

STUDENT STUDY SITE

⑤SAGE edge™

Get the tools you need to sharpen your study skills! Visit **edge.sagepub.com/ salkind6e** to access practice quizzes, eFlashcards, original and curated videos, data sets, journal articles, and more!

PART V

More Statistics!
More Tools! More Fun!

Dan shows off his data collection.

W ow—the bulk of the book (and probably the course) is behind you and you're ready to be introduced to some extra new ideas in this part of *Statistics for People Who (Think They) Hate Statistics*. Our goal in Part V of the book is to provide you with an overview of different tools that are used for a variety of purposes.

In Chapter 17, we introduce the flip side of inferential statistics—the always interesting and useful nonparametrics (in particular the chi-square statistic) used when the distribution of scores you are working with is not a normal distribution or violates some other important assumptions.

From there, Chapter 18 reviews some more advanced statistical procedures such as factor analysis and structural equation modeling. You are sure to hear about these, and others, in your studies. Although in-depth coverage is a bit beyond the level of an introductory course, we talk briefly about each technique and provide examples. This chapter will give you a peek at what some of the more advanced tools can do and how they work.

Chapter 19, new to this edition, introduces an approach to data that is quickly becoming very popular—data mining. As our amount of data grows and our technological ability to deal with it all keeps pace, researchers are finding more and more uses for data mining. Here, you'll learn how you can use SPSS to look for patterns in very large data sets. What's fun about this chapter is that to get some practice, we use a database of more than 52,000 names of babies born in New York State since 2007, and you can download it and follow along.

Finally, the last chapter in this part of the book reviews some software that can be used for statistical analysis in lieu of SPSS. Some of it is free, some of it costs, but all of it offers sets of tools you might find valuable.

Okay, this is Part V of six parts, and you are well on your way to being fully equipped as an expert in introductory statistics.

17

What to Do When You're Not Normal

Chi-Square and Some Other Nonparametric Tests

Difficulty Scale ☺ ☺ ☺ ☺ (easy)

WHAT YOU'LL LEARN ABOUT IN THIS CHAPTER

✦ Understanding what nonparametrics are and how they are used

✦ Analyzing data using a chi-square test for goodness of fit

✦ Analyzing data using a chi-square test of independence

✦ When and how nonparametric statistics should be used

INTRODUCTION TO NONPARAMETRIC STATISTICS

Almost every statistical test that we've covered so far in *Statistics for People Who (Think They) Hate Statistics* assumes that the data set with which you are working has certain characteristics. For example, one assumption underlying a *t*-test between means (be the means independent or dependent) is that the variances of each group are homogeneous, or similar. And this assumption can be tested. Another assumption of many **parametric statistics** is that the sample is large enough to represent the population. Statisticians have found that it takes a sample size of about 30 to fulfill this assumption. Many of the statistical tests we've covered so far are also robust, or powerful, enough that even if one of these assumptions is violated, the test is still valid.

But what do you do when these assumptions may be violated? The original research questions are certainly still worth asking and answering. That's when we use **nonparametric statistics** (also called distribution-free statistics). These tests don't follow the same "rules"

Introduction to Chapter 17

315

(meaning they don't require the same, possibly restrictive assumptions as the parametric tests we've reviewed), but the nonparametrics are just as valuable. The use of nonparametric tests also allows us to analyze data that come as frequencies, such as the number of children in different grades or the percentage of people receiving Social Security. And very often, these variables are measured at the nominal or categorical level. (See Chapter 2 for more information about nominal scales of measurement.)

For example, if we wanted to know whether the characteristics of people who voted for school vouchers in the most recent election were what we would expect by chance, or if there was really a pattern of preference according to one or more variables, we would then use a nonparametric technique called *chi-square*.

In this chapter, we will cover chi-square, one of the most commonly used nonparametric tests, and provide a brief review of some others just so you can become familiar with some of the tests that are available.

INTRODUCTION TO THE
GOODNESS OF FIT (ONE-SAMPLE) CHI-SQUARE

Chi-square is an interesting nonparametric test that allows you to determine if what you observe in a distribution of frequencies is what you would expect to occur by chance. A one-sample chi-square includes only one dimension, variable, or factor, such as in the example you'll see here, and is usually referred to as a **goodness of fit test** (just how well do the data you collect fit the pattern you expected?). A two-sample chi-square includes two dimensions, variables, or factors and is usually referred to as a **test of independence**. For example, it might be used to test whether preference for school vouchers is *independent* of political party affiliation and gender.

We'll look at both of these types of chi-square procedures.

For a goodness of fit example, here are data from a sample selected at random from the 1990 US Census data collected in Sonoma County, California. As you can see, the table organizes information about level of education (the only factor).

Level of Education		
No College	**Some College**	**College Degree**
25	42	17

The question of interest here is whether the number of respondents is equally distributed across all three levels of education. To answer this question, the chi-square value was computed and then tested for significance. In this example, the chi-square value is equal to 11.643, which is significant beyond the .05 level. The conclusion is that the numbers of respondents at the various levels of education for this sample are not equally distributed. In other words, the values are not what we would expect by chance.

The rationale behind the one-sample chi-square or goodness of fit test is that for any set of occurrences, you can easily compute what you would expect by chance. You do this by dividing the total number of occurrences by the number of classes or categories. In this example using data from the Census, the observed total number of occurrences is 84. We would expect that, by chance, 84/3 (84, which is the total of all frequencies, divided by 3, which is the total number of categories), or 28, respondents would fall into each of the three categories of level of education.

Then we look at how different what we would expect by chance is from what we observe. If there is no difference between what we expect and what we observe, the chi-square value is equal to zero.

Let's look more closely at how the chi-square value is computed.

COMPUTING THE GOODNESS OF FIT CHI-SQUARE TEST STATISTIC

The one-sample goodness of fit chi-square test compares what is observed with what would be expected by chance. Formula 17.1 shows how to compute the chi-square value for a one-sample chi-square test.

Chi Square

$$\chi^2 = \Sigma \frac{(O - E)^2}{E}, \qquad (17.1)$$

where

Determining the Critical Value

- χ^2 is the chi-square value;
- Σ is the summation sign;
- O is the observed frequency; and
- E is the expected frequency.

Here are some sample data we'll use to compute the chi-square value for a one-dimensional analysis.

Preference for School Voucher			
For	Maybe	Against	Total
23	17	50	90

And here are the famous eight steps to test this statistic.

1. State the null and research hypotheses.

The null hypothesis shown in Formula 17.2 states that there is no difference in the frequency or the proportion of occurrences in each category.

$$H_0: P_1 = P_2 = P_3 \qquad\qquad (17.2)$$

The P in the null hypothesis represents the percentage of occurrences in any one category. This null hypothesis states that the percentages of cases in Category 1 (For), Category 2 (Maybe), and Category 3 (Against) are equal. We are using only three categories, but the number could be extended to fit the situation as long as each of the categories is *mutually exclusive*, meaning that any one observation cannot be in more than one category. For example, you can't be in both the male and female categories (the Census only lets you check one). Or you can't be both for and against the school voucher plan at the same time.

The research hypothesis shown in Formula 17.3 states that there is a difference in the frequency or proportion of occurrences in each category.

$$H_1: P_1 \neq P_2 \neq P_3 \qquad\qquad (17.3)$$

2. Set the level of risk (or the level of significance or Type I error) associated with the null hypothesis.

The Type I error rate is set at .05. How did we decide on this value and not some other, like .01 or .001? As we have emphasized in past chapters, we made the (somewhat) arbitrary decision to take this amount of risk.

3. Select the appropriate test statistic.

Any test between frequencies or proportions of mutually exclusive categories (such as For, Maybe, and Against) requires the use of chi-square. The flowchart we used all along at the beginning of Chapters 10 through 16 to select the type of statistical test to use is not applicable to nonparametric procedures.

4. Compute the test statistic value (called the obtained value).

Let's go back to our voucher data and construct a worksheet that will help us compute the chi-square value.

Category	O (observed frequency)	E (expected frequency)	D (difference)	$(O - E)^2$	$(O - E)^2/E$
For	23	30	7	49	1.63
Maybe	17	30	13	169	5.63
Against	50	30	20	400	13.33
Total	90	90			

Here are the steps we took to prepare the above information.

1. Enter the categories (Category) of For, Maybe, and Against. Remember that these three categories are mutually exclusive. Any data point can be in only one category at a time.

2. Enter the observed frequency (O), which reflects the data that were collected.

3. Enter the expected frequency (E), which is the total of the observed frequencies (90) divided by the number of categories (3), or 90/3 = 30.

4. For each cell, subtract the expected frequency from the observed frequency (D), or vice versa. It does not matter which is subtracted from the other because these differences are squared in the next step.

5. Square the difference between the observed and expected value. You can see these values in the column named $(O - E)^2$.

6. Divide the squared difference between the observed and the expected frequencies by the expected frequency. You can see these values in the column marked $(O - E)^2/E$.

7. Sum up this last column, and you have the total chi-square value of 20.6.

5. Determine the value needed for rejection of the null hypothesis using the appropriate table of critical values for the particular statistic.

Here's where we go to Appendix B, Table B.5 for the list of critical values for the chi-square test.

Our first task is to determine the degrees of freedom (*df*), which approximates the number of categories in which data have been organized. For this particular test statistic, the degrees of freedom is *r* − 1, where *r* equals rows, or 3 − 1 = 2.

Using this number (2) and the level of risk you are willing to take (defined earlier as .05), you can use the chi-square table to look up the critical value. It is 5.99. So, at the .05 level, with 2 degrees of freedom, the value needed for rejection of the null hypothesis is 5.99.

6. Compare the obtained value with the critical value.

The obtained value is 20.6, and the critical value for rejection of the null hypothesis that the frequency of occurrences in Groups 1, 2, and 3 are equal is 5.99.

7. and 8. Decision time!

Now comes our decision. If the obtained value is more extreme than the critical value, the null hypothesis cannot be accepted. If the obtained value does not exceed the critical value, the null hypothesis is the most attractive explanation.

In this case, the obtained value exceeds the critical value—it is extreme enough for us to say that the distribution of respondents across the three groups is not equal. Indeed, there is a difference in the frequency of people voting for, maybe, or against school vouchers.

Why "goodness of fit"? This name suggests that the statistic addresses the question of how well or "good" a set of data "fits" an existing set. The set of data is, of course, what you observe. The "fit" part suggests that there is another set of data to which the observed set can be matched. This standard is the set of expected frequencies that are calculated in the course of computing the χ^2 value. If the observed data fit, the χ^2 value is just too close to what you would expect by chance and does not differ significantly. If the observed data do not fit, then what you observed is different from what you would expect.

So How Do I Interpret $\chi^2_{(2)} = 20.6$, p < .05?

Using an
Applet to
Compute
Chi-Square
Probabilities

- χ^2 represents the test statistic.
- 2 is the number of degrees of freedom.
- 20.6 is the value obtained by using the formula we showed you earlier in the chapter.
- $p < .05$ (the really important part of this little phrase) indicates that the probability is less than 5% on any one test of the null hypothesis that the frequency of votes is equally distributed across all categories by chance alone. Because we defined .05 as our criterion to deem the research hypothesis more attractive than the null hypothesis, our conclusion is that there is a significant difference among the three sets of scores.

INTRODUCTION TO THE TEST OF INDEPENDENCE CHI-SQUARE

We've been mostly talking about the goodness of fit type of chi-square test, but another one—that's just a bit more advanced—is well worth exploring: the chi-square test of independence (also called a chi-square test for association).

With this test, two different dimensions, almost always at the nominal level of measurement, are examined to see whether they are related. Take, for example, the following fourfold table that includes men and women, those voting and those not.

		Voter Participation	
		Vote	Don't Vote
Gender	Men	50	20
	Women	40	10

What you need to keep in mind is that (1) there are two dimensions (gender and voting participation) and (2) the question asked by the chi-square test of independence is whether, in this case, gender and voting participation are independent of each other.

As with a one-dimensional chi-square, the closer the observed and expected values are to one another, the less likely the dimensions are to be independent of one another. The less similar the observed and expected values are, the more likely the two dimensions are independent of one another.

The test statistic is computed in the same manner as we did for a goodness of fit test (what a nice surprise); expected values are computed and used along with observed values to compute a chi-square value, which is then tested for significance. However, unlike with the one-dimensional test, the expected values for a test of independence are computed in a different fashion, which we will review below.

COMPUTING THE TEST OF INDEPENDENCE CHI-SQUARE TEST STATISTIC

Here are the steps we took to perform a test of independence given the voting–gender data you saw earlier. We followed the same steps as those you saw for the goodness of fit test and used the same formula. Here's our summary chart . . .

	Men	Women	Total
Vote	37	32	69
Don't Vote	20	31	51
Total	57	63	120

But note that while the formula is the same, the expected values are computed differently. In particular, the expected value for any cell is the product of the row total multiplied by the column total divided by the total sum of all observations.

So, for example, the expected value for Men who Vote is as follows . . .

$$\frac{69 \times 57}{120} = 32.775$$

1. Enter the categories (Category) of Men Vote, Men Don't Vote, Women Vote, Women Don't Vote. Remember that all these categories are mutually exclusive; any data point can be in only one category at a time.

2. Enter the observed frequency (O), which reflects the data that were collected.

3. Enter the expected frequency (E), which is the row total times the column total divided by the total number of observations.

4. For each cell, subtract the expected frequency from the observed frequency (D), or vice versa. It does not matter which is subtracted from the other because these differences are squared in the next step.

5. Square the difference between the observed and expected value. You can see these values in the column named $(O - E)^2$.

6. Divide the squared difference between the observed and the expected frequencies by the expected frequency. You can see these values in the column marked $(O - E)^2/E$.

7. Sum up this last column, and you have the total chi-square value of 2.441.

   Here is a summary table of the observed values (O), expected value (E), and the square of the observed minus the expected value for each cell $(O - E)^2$ calculations . . .

Computing the Chi-Square Value

Observed (O)			
	Men	**Women**	**Total**
Vote	37	32	69
Don't Vote	20	31	51
Total	57	63	120

Expected (E)			
	Men	**Women**	
Vote	32.78	36.23	
Don't Vote	24.23	26.78	

$(O - E)^2/E$			
	Men	**Women**	
Vote	0.54	0.49	
Don't Vote	0.74	0.67	

The obtained chi-square value is .54 + .49 + .74 + .67 = 2.44.

Here's where we go to Appendix B, Table B.5 for the list of critical values for the chi-square test.

Our first task is to determine the degrees of freedom (df), which approximates the number of categories in which data have been organized. For a chi-square test of independence statistic, the degrees of freedom are $(r - 1)(c - 1)$, where r equals rows $(2 - 1 = 1)$ and c equals columns $(2 - 1 = 1)$. So $(r - 1)(c - 1)$ equals $(1)(1)$ or 1.

Using this number (1) and the level of risk you are willing to take (defined earlier as .05), you can use the chi-square table to look up the critical value. It is 3.84. So, at the .05 level, with 2 degrees of freedom, the value needed for rejection of the null hypothesis is 3.84.

The obtained value is 2.44, and the critical value for rejection of the null hypothesis that voting participation and gender are independent is 3.84.

8. Now comes our decision. If the obtained value is more extreme than the critical value, the null hypothesis cannot be accepted. If the obtained value does not exceed the critical value, the null hypothesis is the most attractive explanation. In this case, the obtained value does not exceed the critical value. Indeed, gender and voting participation are independent.

USING SPSS TO PERFORM CHI-SQUARE TESTS

Goodness of Fit and SPSS

Here's how to perform a simple, one-sample chi-square test using SPSS. We are using the data set named Chapter 17 Data Set 1, which was used in the school vouchers example above.

1. Open the data set. For a one-sample chi-square test, you need only enter the number of occurrences into each column, using a different value for each possible outcome. In this example, there would be a total of 90 data points in column 1: 23 would be entered as 1s (or For), 17 would be entered as 2s (or Maybe), and 50 would be entered as 3s (or Against).

2. Click Analyze → Nonparametric Tests → Legacy Dialogs → Chi-Square, and you will see the dialog box shown in Figure 17.1.

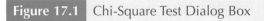

Figure 17.1   Chi-Square Test Dialog Box

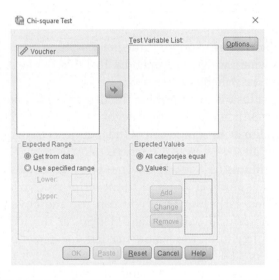

3. Double-click on the variable named Voucher.

4. Click OK. SPSS will conduct the analysis and produce the output you see in Figure 17.2.

Figure 17.2   SPSS Output for a Chi-Square Analysis

Voucher

	Observed N	Expected N	Residual
For	23	30.0	-7.0
Maybe	17	30.0	-13.0
Against	50	30.0	20.0
Total	90		

Test Statistics

	Voucher
Chi-Square	20.600[a]
df	2
Asymp. Sig.	.000

Understanding the SPSS Output

The SPSS output for the chi-square test shows you exactly what you read about earlier.

1. The categories For (coded as 1), Maybe (coded as 2), and Against (coded as 3) are listed along with their Observed N.

2. This is followed by the Expected N, which in this case is 90/3, or 30.

3. The chi-square value of 20.600 and degrees of freedom appear under the Test Statistics section of the output.

The exact level of significance (in the figure, it's named Asymp. Sig.) is so small (less than .0001, for example) that SPSS computes it only as .000. A very unlikely outcome! So, it is highly likely that these three categories are not equal in frequency.

Test of Independence and SPSS

Here's how to perform a two-dimensional chi-square test of independence using SPSS. We are using the data set named Chapter 17 Data Set 2, which was used in the gender/voting participation example above. As you can see in Figure 17.3, showing cases 63 through 75, there is one variable named Gender (which can assume values of Male or Female) and one variable named Vote (which can take on the value of Yes or No).

Tests of
Independence

Figure 17.3   Chapter 17 Data Set 2 Showing Two Variables

63	Women	Yes
64	Women	Yes
65	Women	Yes
66	Women	Yes
67	Women	Yes
68	Women	Yes
69	Women	Yes
70	Men	No
71	Men	No
72	Men	No
73	Men	No
74	Men	No
75	Men	No

1. For a test of independence chi-square test, you need to enter the number of occurrences into each column, using a different value for each possible outcome. In this example, there would be a total of 120 data points with a 1 or 2 in the column representing Gender and a 1 or 2 in the column representing Voting Participation. We used the Values option in the Variable View to define the 1s as Male and Yes, respectively, and the 2s as Female and No, respectively so that we could see labels in the columns rather than numbers.

2. Click Analyze → Descriptive Statistics → Cross Tabs, and you will see the dialog box shown in Figure 17.4.

3. Double-click on the variable named Gender to move it to the Rows(s): box.

4. Double-click on the variable named Vote to move it to the Column(s): box.

5. Click Statistics and in Crosstabs: Statistics, click Chi-square.

6. Click OK. SPSS will conduct the analysis and produce the output you see in Figure 17.5.

Figure 17.4   The Crosstabs Dialog Box

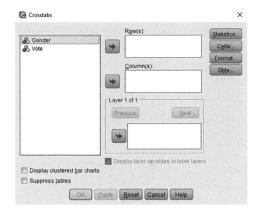

Figure 17.5   SPSS Output for a Chi-Square Analysis

Gender * Vote Crosstabulation

Count

		Vote		Total
		Yes	No	
Gender	Men	37	20	57
	Women	32	31	63
Total		69	51	120

Chi-Square Tests

	Value	df	Asymptotic Significance (2-sided)	Exact Sig. (2-sided)	Exact Sig. (1-sided)
Pearson Chi-Square	2.441[a]	1	.118		
Continuity Correction[b]	1.897	1	.168		
Likelihood Ratio	2.454	1	.117		
Fisher's Exact Test				.141	.084
Linear-by-Linear Association	2.421	1	.120		
N of Valid Cases	120				

A very cool chi-square value generator is available for both types of chi-square tests at http://www.quantpsy.org/chisq/chisq.htm, brought to us by Dr. Kristopher J. Preacher of Vanderbilt University. Once you do calculations by hand so you really understand how chi-square values are created, use this tool along with SPSS to make your stats life even easier!

Understanding the SPSS Output

The SPSS output for the chi-square test of independence shows the following.

1. The frequencies for the various categories are listed such as Men–Yes (37), Men–No (20), Women–Yes (32), Women–No (31).

2. The Pearson chi-square value of 2.441 and degrees of freedom (1) appear under the Chi-Square Tests section of the output. And the results show the same value as was computed manually earlier in the chapter.

3. The exact level of significance (in the figure, it's named Asymptotic Significance) is .118, and so the results are not significant at the .05 level. In other words, gender and voting participation are not independent of one another.

   Our interpretation? Voting participation is not a function of gender, nor is gender related to voting participation.

OTHER NONPARAMETRIC TESTS YOU SHOULD KNOW ABOUT

You may never need a nonparametric test to answer any of the research questions that you propose. On the other hand, you may very well find yourself dealing with samples that are very small (or at least fewer than 30 cases) or data that violate some of the important assumptions underlying parametric tests.

Actually, a primary reason why you may want to use nonparametric statistics is a function of the measurement level of the variable you are assessing. We'll talk more about that in the next chapter, but for now, most data that

(Continued)

(Continued)

are categorical and are placed in categories (such as the Sharks and Jets) or that are ordinal and are ranked (1st, 2nd, and 3rd place) call for nonparametric tests of the kind you see in Table 17.1.

If that's the case, try nonparametrics on for size. Table 17.1 provides all you need to know about some other nonparametric tests, including their name, what they are used for, and a research question that illustrates how each might be used. Keep in mind that the table represents only a few of the many different tests that are available.

| **Table 17.1** | Nonparametric Tests to Analyze Data in Categories and by Ranks |

Test Name	When the Test Is Used	A Sample Research Question
McNemar test for significance of changes	To examine "before and after" changes	How effective is a phone call to undecided voters in influencing them to vote a certain way on an issue?
Fisher's exact test	To compute the exact probability of outcomes in a 2 × 2 table	What is the exact likelihood of getting six heads on a toss of six coins?
Chi-square one-sample test (the one we focused on earlier in this chapter)	To determine whether the number of occurrences across categories is random	Did brands Fruities, Whammies, and Zippies each sell an equal number of units during the recent sale?
Kolmogorov–Smirnov test	To see whether scores from a sample came from a specified population	How representative is a set of judgments about certain children at an elementary school of the entire student body of that school?
The sign test, or median test	To compare the medians from two samples	Is the median income of people who voted for Candidate A greater than the median income of people who voted for Candidate B?
Mann–Whitney U test	To compare two independent samples	Did the transfer of learning, measured by number correct, occur faster for Group A than for Group B?
Wilcoxon rank test	To compare the magnitude as well as the direction of differences between two groups	Is preschool twice as effective as no preschool experience for helping develop children's language skills?

Test Name	When the Test Is Used	A Sample Research Question
Kruskal–Wallis one-way analysis of variance	To compare the overall difference between two or more independent samples	How do rankings of supervisors differ among four regional offices?
Friedman two-way analysis of variance	To compare the overall difference between two or more independent samples on more than one dimension	How do rankings of supervisors differ as a function of regional office and gender?
Spearman rank correlation coefficient	To compute the correlation between ranks	What is the correlation between class rank at the end of the senior year of high school and class rank at the end of the freshman year of college?

REAL-WORLD STATS

Using a modern tool to explore the content of the Bible is really exciting (and clever) and a great illustration of how chi-square is used in the real world. Professor Houk conducted what is called a syllable–word frequency pattern analysis, using the chi-square test, to identify significant differences across different chapters of Genesis (yes, that Genesis). The thesis? That the various passages were written by the same person. The analysis of the various passages argued against the thesis that the stories were developed separately, given that there is no significant difference in the frequency of the occurrence of certain words and such.

Want to know more? Go online or to the library and find . . .

Houk, C. B. (2002). Statistical analysis of Genesis sources. *Journal for the Study of the Old Testament, 27,* 75–105.

Real-World Stats

SUMMARY

Chi-square is one of many different types of nonparametric statistics that help you answer questions based on data that violate the basic assumptions of the normal distribution or when a data set is just too small for other statistical methods to work. These nonparametric tests are a very valuable tool, and even as limited an introduction as this will help you understand their use in studies you may read and begin your own exploration of their possibilities.

Chapter Summary

TIME TO PRACTICE

1. When does the obtained chi-square value equal zero? Provide an example of how this might happen.

2. Using the following data, test the question of whether an equal number of Democrats, Republicans, and Independents voted during the most recent election. Test the hypothesis at the .05 level of significance. Do your calculations by hand.

Political Affiliation		
Republican	**Democrat**	**Independent**
800	700	900

Problem 3

3. Using the following data, test the question of whether an equal number of boys (code = 1) and girls (code = 2) participate in soccer at the elementary level at the .01 level of significance. (The data are available as Chapter 17 Data Set 3.) Use SPSS or some other statistical program to compute the exact probability of the chi-square value. What's your conclusion?

Gender	
Boys = 1	**Girls = 2**
45	55

4. School enrollment officials observed a change in the distribution of the number of students across grades and were not sure whether the new distribution was what they should have expected. Test the following data for goodness of fit at the .05 level.

Grade	1	2	3	4	5	6
Number of Students	309	432	346	432	369	329

Problem 5

5. Half the marketing staff of a candy company argue that all these candy bars taste the same and are barely different from one another. The other half disagrees. Who is right? The data are ready for you to analyze as Chapter 17 Data Set 4.

Candy Bar	# of People Out of 100 Who Prefer
Nuts & Grits	9
Bacon Surprise	27
Dimples	16
Froggy	17
Chocolate Delight	31

6.   Here are the results of a survey that examines preference for peanut or plain M&Ms and exercise level. Are these variables related? Do the calculations by hand or using SPSS.

		Exercise Level		
		High	**Medium**	**Low**
M&M Preference	Plain	160	400	175
	Peanut	150	500	250

7.   In Chapter 17 Data Set 5, you will find entries for two variables: age category (young, middle-aged, and old) and strength following weight training (weak, moderate, strong). Are these two factors independent of one another?

		Strength		
		Weak	**Moderate**	**Strong**
Age	Young	12	18	22
	Middle	20	22	20
	Old	9	10	5

STUDENT STUDY SITE

⑤SAGE edge™

Get the tools you need to sharpen your study skills! Visit **edge.sagepub.com/ salkind6e** to access practice quizzes, eFlashcards, original and curated videos, data sets, journal articles, and more!

18

Some Other (Important) Statistical Procedures You Should Know About

Difficulty Scale ☺ ☺ ☺ ☺ (moderately easy—just some reading and an extension of what you already know)

WHAT YOU'LL LEARN ABOUT IN THIS CHAPTER

✦ More advanced statistical procedures and when and how they are used

In *Statistics for People Who (Think They) Hate Statistics,* we have covered only a small part of the whole body of statistics. We don't have room in this textbook for everything—there's a lot of it! More important, when you are starting out, it's important to keep things simple and direct.

Introduction to Chapter 18

However, that does not mean that in a research article you read or in a class discussion, you won't come across other analytical techniques that might be important for you to know about. So, for your edification, here are nine of those techniques with descriptions of what they do and examples of studies that used the technique to answer a question.

MULTIVARIATE ANALYSIS OF VARIANCE

You won't be surprised to learn that there are many different renditions of analysis of variance (ANOVA), each one designed to fit a particular "the averages of more than two groups being compared" situation. One of these, multivariate analysis of variance (MANOVA), is used when there is more than one dependent variable. So, instead of looking at just one outcome or dependent variable, the analysis can use

Multivariate
Analysis of
Variance

more than one. If the dependent or outcome variables are related to one another (which they usually are—see the Tech Talk note in Chapter 13 about multiple *t*-tests), it would be hard to determine clearly the effect of the treatment variable on any one outcome. Hence, MANOVA to the rescue.

For example, Jonathan Plucker from Indiana University examined the impact of gender, race, and grade level on how gifted adolescents dealt with pressures at school. The MANOVA analysis that he used was a 2 (gender: male and female) × 4 (race: Caucasian, African American, Asian American, and Hispanic) × 5 (grade: 8th through 12th) MANOVA. The *multivariate* part of the analysis was the five subscales of the Adolescent Coping Scale. Using a multivariate technique, the effects of the independent variables (gender, race, and grade) can be estimated for each of the five scales, independently of one another.

Want to know more? Go online or to the library and find . . .

Plucker, J. A. (1998). Gender, race, and grade differences in gifted adolescents' coping strategies. *Journal for the Education of the Gifted, 21,* 423–436.

REPEATED MEASURES ANALYSIS OF VARIANCE

Repeated
Measures
Analysis of
Variance

Here's another kind of analysis of variance. Repeated measures analysis of variance is very similar to any other analysis of variance, in which you'll recall that the means of two or more groups are tested for differences (see Chapter 13 if you need a review). In a repeated measures ANOVA, participants are tested on one factor more than once. That's why it's called *repeated*, because you repeat the process at more than one point in time on the same factor.

For example, Brenda Lundy, Tiffany Field, Cami McBride, Tory Field, and Shay Largie examined same-sex and opposite-sex interaction with best friends among juniors and seniors in high school. One of the researchers' main analyses was ANOVA with three factors: gender (male or female), friendship (same-sex or opposite-sex), and year in high school (junior or senior year). The repeated measure was year in high school, because the measurement was repeated across the same subjects.

Want to know more? Go online or to the library and find . . .

Lundy, B., Field, T., McBride, C., Field, T., & Largie, S. (1998). Same-sex and opposite-sex best friend interactions among high school juniors and seniors. *Adolescence, 33,* 279–289.

ANALYSIS OF COVARIANCE

Here's our last rendition of ANOVA. Analysis of covariance (ANCOVA) is particularly interesting because it basically allows you to equalize initial differences between groups. Let's say you are sponsoring a program to increase running speed and want to compare how fast two groups of athletes can run a 100-yard dash. Because strength is often related to speed, you have to make some correction so that strength does not account for any differences in the results at the end of the program. Rather, you want to see the effects of training with strength removed. You would measure participants' strength before you started the training program and then use ANCOVA to adjust final speed based on initial strength.

Analysis of Covariance

Michaela Hynie, John Lydon, and Ali Taradash from McGill University used ANCOVA in their investigation of the influence of intimacy and commitment on the acceptability of premarital sex and contraceptive use. They used ANCOVA with social acceptability of sexual activity as the dependent variable (in which they were looking for group differences) and ratings of a particular scenario as the covariate. ANCOVA ensured that differences in social acceptability were corrected using ratings, so this difference was controlled.

Want to know more? Go online or to the library and find . . .

Hynie, M., Lydon, J., & Taradash, A. (1997). Commitment, intimacy, and women's perceptions of premarital sex and contraceptive readiness. *Psychology of Women Quarterly, 21,* 447–464.

MULTIPLE REGRESSION

You learned in Chapter 16 how the value of one variable can be used to predict the value of another, an analysis often of interest to social and behavioral sciences researchers. For example, it's fairly well established that parents' literacy behaviors (such as having books in the home) are related to how much and how well their children read. So, it would seem quite interesting to look at such variables as parents' age, education level, literacy activities, and shared reading with children to see what each contributes to early language skills and interest in books. Paula Lyytinen, Marja-Leena Laakso, and Anna-Maija Poikkeus did exactly that, using stepwise regression analysis to examine the contribution of parental background variables to children's literacy. They found that mothers' literacy activities and mothers' level of education contributed significantly to children's language skills, whereas mothers' age and shared reading did not.

Multiple Regression

Want to know more? Go online or to the library and find . . .

Lyytinen, P., Laakso, M. L., & Poikkeus, A.-M. (1998). Parental contributions to child's early language and interest in books. *European Journal of Psychology of Education, 13,* 297–308.

META-ANALYSIS

Unless you've been on spring break for the entire semester and took a sabbatical from reading even the local newspaper, you know that these are the days of BIG data. You have "quants" in the business world crunching data to gain the slightest competitive advantage. And there are full-fledged stats guys such as Nate Silver who, using his own probability models, very accurately predicted the outcome of the 2012 presidential election.

One of the tools that these number crunchers—and many social and behavioral scientists—use is meta-analysis. Researchers combine data from several studies to examine patterns and trends. Given how much data are collected today, meta-analysis is a powerful tool to help organize disparate information and guide policy decision making.

For example, this methodology is useful when it comes to studying gender differences. There are so many studies showing so many differences (and lack thereof) that a tool like meta-analysis could certainly help clarify the meaning of all those studies' outcomes. Regarding job performance, there are many different views of gender differences. Philip Roth, Kristen Purvis, and Philip Bobko suggested that males generally are evaluated higher than females across a variety of situations, which include job performance. To clarify the nature of the differences that have been found in job performance between genders, these authors conducted a meta-analysis of job performance measures from field studies. They found that while job performance ratings favored females, ratings of promotion potential were higher for males. This is pretty interesting stuff! And it's a good example of how the technique of meta-analysis gives an accurate, 30,000-foot view of findings across many different studies.

Want to know more? Go online or to the library and find . . .

Roth, P. L., Purvis, K. L., & Bobko, P. (2012). A meta-analysis of gender group differences for measures of job performance in field studies. *Journal of Management, 38,* 719–739.

DISCRIMINANT ANALYSIS

Since group membership is so often a variable of interest, discriminant analysis is an especially good technique to use when one wants to see what sets of variables discriminate between two sets of individuals. For example, it could answer such questions as what sets of variables differentiate those patients who get Treatment A from those who get Treatment B. It gives us an idea of how various assessments might distinguish one group from another.

Researchers from Youngstown State University studied students enrolled in undergraduate economic statistics classes. Specifically, they were interested in what differentiates users and nonusers of web-based instruction (WBI) as a supplement to traditional classroom lecture and problem-solving approaches. They used discriminant analysis to compare the perceptions of users and nonusers. The users concluded that distance learning using the Internet was not only a good method of obtaining general information but also a useful tool in improving their academic performance in a quantitative economics class. Nonusers thought the university should provide financial assistance for going online and that WBI should not be required for graduation. And now you know why you should be a fan of web-based learning!

Want to know more? Go online or to the library and find . . .

Usip, E. E., & Bee, R. H. (1998). Economics: A discriminant analysis of students' perceptions of web-based learning. *Social Science Computer Review, 16*(1), 16–29.

FACTOR ANALYSIS

Factor analysis is a technique based on how well various items are related to one another and form clusters or factors. Each factor represents several variables, and factors turn out to be more efficient than individual variables at representing outcomes in certain studies. In using this technique, the goal is to represent those things that are related to one another by a more general name—this is the *factor*. And the names you assign to factors are not a willy-nilly choice; the names reflect the content and the ideas underlying how the variables might be related.

For example, David Wolfe and his colleagues at the University of Western Ontario attempted to understand how experiences of maltreatment occurring before children were 12 years old affected peer and dating relationships during adolescence. To do this, the researchers collected data on many variables and then looked at the relationships among all of them. Those that seemed to contain items that were related (and belonged to a group that made theoretical sense) were deemed factors, such as the factor named Abuse/Blame in this study. Another factor, named Positive Communication, was made up of 10 items, all of which were related to each other.

Want to know more? Go online or to the library and find . . .

Wolfe, D. A., Wekerle, C., Reitzel-Jaffe, D., & Lefebvre, L. (1968). Factors associated with abusive relationships among maltreated and non-maltreated youth. *Developmental Psychopathology, 10,* 61–85.

PATH ANALYSIS

Here's another statistical technique that examines correlations but allows researchers to suggest direction, or causality, in the relationship between factors. Path analysis basically examines the direction of relationships through first a postulation of some theoretical relationship between variables and then a test to see if the direction of these relationships is substantiated by the data.

For example, Anastasia Efklides, Maria Papadaki, Georgia Papantoniou, and Gregoris Kiosseoglou examined individual experiences of having difficulty in learning mathematics. To do this, they administered several types of tests (such as those in the area of cognitive ability) and found that feelings of difficulty are mainly influenced by cognitive (problem solving) rather than affective (emotional) factors. One of the most interesting uses of path analysis is that a technique called structural equation modeling is used to present the results in a graphical representation of the relationships among all of the different factors under consideration. In this way, you can actually see what relates to what and with what degree of strength. Then, you can judge how well the data fit the model that was previously suggested. Cool.

Want to know more? Go online or to the library and find . . .

Efklides, A., Papadaki, M., Papantoniou, G., & Kiosseoglou, G. (1998). Individual differences in feelings of difficulty: The case of school mathematics. *European Journal of Psychology of Education, 13,* 207–226.

STRUCTURAL EQUATION MODELING

Structural
Equation
Modeling

Structural equation modeling (SEM) is a relatively new technique that has become increasingly popular since it was introduced in the early 1960s. Some researchers feel that it is an umbrella term for techniques such as regression, factor analysis, and path analysis. Others believe that it stands on its own as an entirely separate approach. It's based on relationships among variables (as are the previous three techniques we described).

The major difference between SEM and other advanced techniques, such as factor analysis, is that SEM is *confirmatory* rather than *exploratory*. In other words, the researcher is more likely to use SEM to confirm whether a certain model that has been proposed works (meaning the data fit that model). In contrast, exploratory techniques set out to discover a particular relationship with less (but not no) model building beforehand.

For example, Heather Gotham, Kenneth Sher, and Phillip Wood examined the relationships among young adult alcohol use disorders, preadulthood variables (gender, family history of alcoholism, childhood stressors, high school class rank, religious involvement, neuroticism, extraversion, psychoticism), and young adult developmental tasks (baccalaureate degree completion, full-time employment, marriage). Using structural equation modeling techniques, they found that preadulthood variables were more salient predictors of developmental task achievement than was a diagnosis of having a young adult alcohol use disorder.

Want to know more? Go online or to the library and find . . .

Gotham, H. J., Sher, K. J., & Wood, P. K. (2003). Alcohol involvement and developmental task completion during young adulthood. *Journal of Studies on Alcohol, 64*, 32–42.

Real-World
Stats

SUMMARY

Even though you probably will not be using these more advanced procedures anytime soon, that's all the more reason to know at least something about them, because you will certainly see them mentioned in various research publications and may even hear them mentioned in another class you are taking. With a basic understanding of these combined with your grasp of the fundamentals (from all the chapters in the book up to this one), you can be confident of having mastered a good deal of important information about introductory (and even some intermediate) statistics.

Chapter
Summary

STUDENT STUDY SITE

⑤SAGE edge™

Get the tools you need to sharpen your study skills! Visit **edge.sagepub.com/ salkind6e** to access practice quizzes, eFlashcards, original and curated videos, data sets, journal articles, and more!

19 Data Mining

An Introduction to Getting the Most Out of Your BIG Data

WHAT YOU'LL LEARN ABOUT IN THIS CHAPTER

◆ Understanding what data mining is.

◆ Understanding how data mining can help make sense out of very large sets of data.

◆ Using SPSS to apply some basic data mining tools

◆ Applying pivot tables to the analysis of large data sets

By the time you are reading this sixth edition of *Statistics for People Who (Think They) Hate Statistics*, the amount of data possessed by scientists, politicians, sports columnists, health care professionals, businesspeople, and just about everyone who deals with data will already be astronomical, and it will still be growing and growing and growing. It's not called BIG data for nothing!

What is Data Mining?

How big? Well, these days, information is measured in terms of **exabytes** or about 1,152,921,504,606,846,976 bytes (with each byte representing a 1 or a 0) or about 1 quintillion bytes. And there are about 1,000 exabytes of information out there now (and that amount is growing rapidly). That's a lot of stuff.

Why so much data, and what can we do with it?

Technological advances allow for more interconnectedness among institutions and people and machines (see the Internet of Things soon to be playing on your local desktop), and this same technology allows for the collection of every single piece of data that is available, if privacy and other concerns can be addressed (and the jury is surely still out on those issues). Think *The Internet of Everything*.

You'd be surprised at how much data is "Big" and how much of it is all around us, including . . .

- Health care records
- Social media interactions
- Detailed analysis of sporting events
- What you purchased online, when you bought it, what it cost, and what you might be interested in purchasing next
- Your daily physical activity
- Human genome mapping
- Weather data analysis and prediction
- Traffic patterns
- Even such seemingly innocuous things as how many Twinkies are sold each year (okay, okay . . . 500 *million*)

And business, business, business—it loves big data because the people looking to sell products and services believe (somewhat rightly so) that these reams of data can be analyzed to understand patterns (which is what analyzing big data is all about) of consumer behavior at the most micro of levels. In turn, the results of the analysis of this information can be used to predict, and understand, and influence consumer buying habits. Ever wonder how Amazon knows what you've shopped for recently and then suggests alternative products? *Analytics*, that's how—just another term for data mining.

All of these data are invaluable, but there is so much information that it is very difficult to make sense out of it. What's needed is a set of tools that can look for patterns, and that's where data mining (and SPSS) come in.

 Okay, so big data it is, and there are tons of jobs for those who have the quantitative skills to dig into it (these folks are often referred to as "quants," as in the movie *The Big Short*). But do be aware: *Big* data is no substitute for *good* data, and we have a tendency to think that bigger is better (this *is* true only on Thanksgiving). So, while big data will surely come your way in one form or another, the same questions need to be asked about it as about all data sets, including where the data came from, whether they can help answer the question being asked, whether they are valid and reliable, who collected them and with what intention or purpose, and so on. We suspect that in the years to come, the analysis of big data will yield remarkable findings in many areas, but we also suspect that the possible pitfalls of big data, and its uses, will need to be addressed as well.

This chapter deals with the simple question of how to make sense of data sets that are very large. Our goal is to make these data sets more manageable through the use of tools that allow us to easily extract the information that we want. To that end, this chapter provides a very general introduction to using SPSS for **data mining** (which is looking for patterns in large data sets) and discusses different features of SPSS that can help you deal with large data sets. You'll also get a brief introduction to pivot tables and cross-tabulation tables, which are tables that can be rearranged with a click of a mouse button to view information in unique ways.

Not to anyone's surprise, there is an entire industry devoted to data mining, and these companies, firms, etc., tend to operate according to a CRoss-Industry Standard Process for DataMining (CRISP-DM). These step-by-step guidelines help ensure consistency across data-mining efforts.

Now here are the big caveats for this chapter.

First, SPSS offers an advanced set of tools for data and text mining named SPSS Modeler. This is an add-on to the usual SPSS package and is very unlikely to be readily available through your school (ditto for home installations). So, in hopes of making this chapter as relevant as possible to as many users of this book (and others) as possible, we'll focus on SPSS features that do not rely on the modeler.

Second, this chapter is an introduction to how to work with large data sets using SPSS. Whatever you read here can be used with small data sets as well, but we want to show you how these tools can help make sense out of very large data sets. Smaller data sets can almost always be examined visually to look for trends or patterns.

Third, and more about this later, the data set we will work with in this chapter is one that was procured from a website, and it changes at least every month or so with the acquisition of new data. So, what you see on your screen may not be what you see in the figures in this chapter. However, while the content may be a bit different, the format and instructions will remain the same.

OUR SAMPLE DATA SET—WHO DOESN'T LOVE BABIES?

Basically, BIG data is a very large collection of either cases or variables, but usually both. Conceptually, big data represents a data set that is too large to just "eyeball" and get a sense of what trends might be present, what outliers might be in the set, or what important patterns may be less than obvious but are there.

For illustrative purposes, we have selected a database titled *Baby Names: Beginning 2007* from https://health.data.ny.gov/Health/Baby-Names-Beginning-2007/jxy9-yhdk/. This is a database from New York

State containing more than 52,000 (big enough?) baby names. Baby names are displayed by the year of birth, the baby's name, the name of the county or borough where the mother resided according to the child's birth certificate, the child's gender, and the count (or frequency) with which the name appears in the data set. The variables are these:

- Year of birth
- First name
- County of birth
- Count (or number of times that name appears)
- Gender

For example, in Figure 19.1 you can see the first few sample records.

Figure 19.1 The First Few Records in the Baby Names Database

	Year_of_Birth	First_Name	County	Count	Gender
1	2014	DAVID	Kings	245	M
2	2014	MOSHE	Kings	236	M
3	2014	ETHAN	Queens	223	M
4	2014	JACOB	Kings	221	M
5	2014	ETHAN	Kings	213	M
6	2014	JAYDEN	Kings	210	M
7	2014	SOPHIA	Queens	206	F

The Baby Names database changes over time because more information is added as it becomes available. This is the case with many online databases, especially when they are sponsored by government agencies. So, while what you see as examples in this chapter reflect the data available when this chapter was written, the results you get might be different in content. However, the format will be the same. In other words, there may be a difference in the numbers you see but not how they are displayed.

Thousands of databases, like the Baby Names database, are available to the public, and there are plenty of places you can look for them. Some databases allow for a search (only), such as the Grand Comics Database (http://comics.org), and other databases contain data that you can download, usually presented in a variety of formats.

In this case, *Baby_Names* was available for downloading as a *.csv* file from Data.gov, where you can find more than 100,000 data sets in categories such as health, weather, and consumer and scientific research, among many others. SPSS can easily read, and convert, a *.csv* file into a *.sav* file.

It's fun to browse through the collection of databases at Data.gov. You might also want to look at DataUSA (http://datausa.io), an MIT project that makes much of the data available at Data.gov transparent and easily understood and offers a very cool visual interface to help with understanding it.

There is a whole other side to SPSS that is called SPSS Syntax. This is where you can use commands (such as COMPUTE and SUM) to perform SPSS operations with more control and specificity than when using menu commands, as we have done throughout most of this book. Using menu commands is more convenient and faster but offers less control. SPSS Syntax is in a sense SPSS's programming language. It's beyond our scope to get into SPSS Syntax in this chapter, but you should know that it exists and feel welcome to try it. Just open a new Syntax window from the File → New menu and apply different functions to any data set. See what happens!

COUNTING OUTCOMES

There are many "counting" commands within SPSS that are very useful for aggregating or understanding the nature of data in a large data set. Here are just a few of these commands with examples of how they are used.

Counting With Frequencies

You have this large data set of baby names and you want to count the number of occurrences by year. Here's how . . .

Counting with Frequencies

1. Select Analyze → Descriptive Statistics → Frequencies and you will see the Descriptives dialog box as shown in Figure 19.2.

2. Click on Year_of_Birth and move it to the Variables(s): box.

3. Click OK and you will see a count of baby names by year as shown in Figure 19.3.

Figure 19.2   The Frequencies Dialog Box

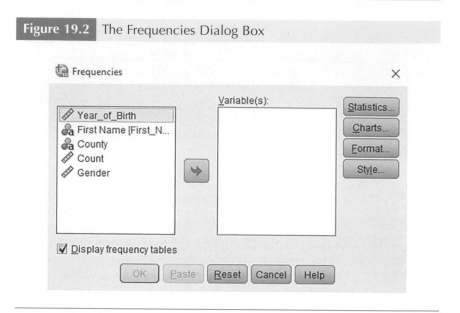

Figure 19.3   Results of a Simple Frequencies Analysis

Year_of_Birth

		Frequency	Percent	Valid Percent	Cumulative Percent
Valid	2007	6367	12.2	12.2	12.2
	2008	6481	12.4	12.4	24.6
	2009	6312	12.1	12.1	36.7
	2010	6192	11.9	11.9	48.5
	2011	6216	11.9	11.9	60.4
	2012	6164	11.8	11.8	72.2
	2013	6158	11.8	11.8	84.0
	2014	8362	16.0	16.0	100.0
	Total	52252	100.0	100.0	

Should you find it necessary, you can perform this same operation on any variable, including text or string variables such as First_Name, the results of which you can see in Figure 19.4. This shows, for example, that among the 52,000+ names (actually 52,252), Abigail appeared 235 times (accounting for 0.4% of all names) and Aahil appeared once. It would be highly unlikely that you could ascertain the frequency of any baby name by eyeballing the data set of 52,000+ names, but using a tool like SPSS pulls the frequencies right up.

| Figure 19.4 | Results of Using Frequencies to Count Text or String Values |

First Name

		Frequency	Percent	Valid Percent	Cumulative Percent
Valid	AADEN	3	.0	.0	.0
	AAHIL	1	.0	.0	.0
	AALIYAH	99	.2	.2	.2
	AARAV	2	.0	.0	.2
	AARIZ	1	.0	.0	.2
	AARON	113	.2	.2	.4
	AARYA	1	.0	.0	.4
	AAYAN	3	.0	.0	.4
	ABBA	1	.0	.0	.4
	ABBY	10	.0	.0	.4
	ABDIEL	1	.0	.0	.4
	ABDOUL	1	.0	.0	.5
	ABDOULAYE	3	.0	.0	.5
	ABDUL	6	.0	.0	.5
	ABDULLAH	14	.0	.0	.5
	ABDULRAHMAN	1	.0	.0	.5
	ABE	1	.0	.0	.5
	ABEL	11	.0	.0	.5
	ABIGAIL	235	.4	.4	1.0

Another very effective SPSS tool is the Crosstabs option on the Analyze menu, which allows you to cross-tabulate more than one variable. **Cross-tabulation tables**, or crosstabs (for short, and that's exactly what you see in Figure 19.4), summarize categorical data (such as year or sex in the Baby Names database we have been working with) to create a table with row and column totals as well as cell counts.

For example, if we wanted to know how many males and females were born in each of the years contained in the data set, we would follow these steps:

1. Select Analyze → Descriptive Statistics → Crosstabs and you will see the Crosstabs dialog box in Figure 19.5.
2. Move the variable named Year_of_Birth to the Row(s): box.
3. Move the variable named Gender to the Column(s): box.
4. Click OK and you will see the results of the analysis in Figure 19.6.

Figure 19.5 The Crosstabs Dialog Box

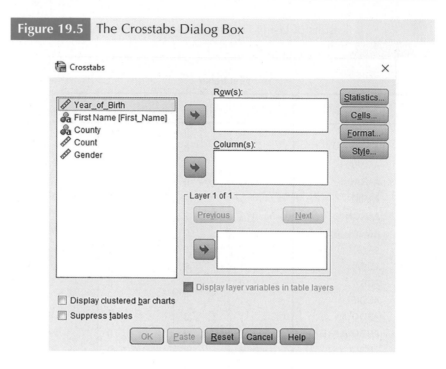

Figure 19.6 Results of a Crosstabs Analysis

Year_of_Birth * Gender Crosstabulation

Count

		Gender		Total
		F	M	
Year_of_Birth	2007	3002	3365	6367
	2008	3039	3442	6481
	2009	2917	3395	6312
	2010	2925	3267	6192
	2011	2918	3298	6216
	2012	2872	3292	6164
	2013	2836	3322	6158
	2014	4121	4241	8362
Total		24630	27622	52252

For example, 3,365 male names were assigned to new babies in 2007 and 4,121 female names in 2014. Once again, such observations are only reasonable to pursue once you can grab and analyze all the data in a large data set, which you can easily do with a tool like SPSS.

PIVOT TABLES AND CROSS-TABULATION: FINDING HIDDEN PATTERNS

Once you have the "raw" information organized, as you have seen throughout this chapter, it's time to think about how you might further "mine" that information, looking for patterns that can help you make important decisions.

To this end, a very useful tool is a **pivot table**, a special kind of table that allows you to easily visualize and manipulate rows and columns as well as the contents of cells.

There are many different ways to create, use, and modify pivot tables in SPSS. We'll show you examples of the ways we think are the most direct and easiest to use and understand.

Creating a Pivot Table

Creating a Pivot Table

When pivot tables first became a feature of statistical analysis programs, creating and using them was tedious and time-consuming. Fortunately, with newer versions of SPSS, a pivot table is only a few clicks away. After that, manipulating the table is as simple as dragging the title of a row or a column, or the contents of cells, to a new location.

All pivot tables need something to "pivot" on, so SPSS needs some type of output on which to operate. In other words, you have to first create a table of results in order to create and use a pivot table. Note that all tables in SPSS can be pivoted—it's just a matter of knowing what you want and how you want it displayed.

Let's use the baby names example and create a pivot table to look at the number of female and male names that appeared from 2007 to 2014 (which is year of birth). If you want to follow along, you should have on your screen the crosstabs analysis that is shown in Figure 19.6.

A pivot table can be created from almost any SPSS output, and the use of pivot tables isn't restricted to crosstabs tables, descriptive output, or frequency output. Whatever output you can generate (almost), you can transform into a pivot table.

1. Double-click on the table of data from which you want to create the pivot table. When you do this, you will see a dotted line outlining the table on the right and bottom border, as you see in Figure 19.7.

2. Right-click in the selected table and select the Pivoting Trays option, and you will see the Pivoting Trays as shown in Figure 19.8.

Figure 19.7   Selecting a Table to Create a Pivot Table

Year_of_Birth * Gender Crosstabulation				
Statistics **Count**				
		Gender		
		F	M	Total
Year_of_Birth	2007	3002	3365	6367
	2008	3039	3442	6481
	2009	2917	3395	6312
	2010	2925	3267	6192
	2011	2918	3298	6216
	2012	2872	3292	6164
	2013	2836	3322	6158
	2014	3239	2870	6109
Total		23748	26251	49999

Figure 19.8   The Pivoting Trays

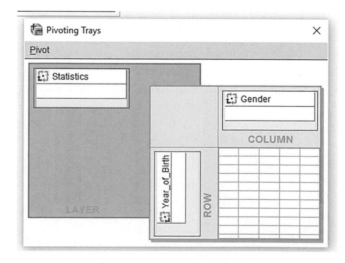

That was easy and fast. Here you can see that SPSS is creating a pivot table with Gender in COLUMN and Year_of_Birth in ROW.

And now, by dragging and arranging rows and columns as we see fit, we can organize the data any way that we want, given the original information available and the fields in which the original data were organized.

For example, if you wan t to see a table that reveals year of birth within gender, drag the field named Year_of_Birth to the COLUMNS area within Gender. You can see the result in Figure 19.9. Same info, different appearance.

Figure 19.9 A Pivot Table Showing Year of Birth by Gender

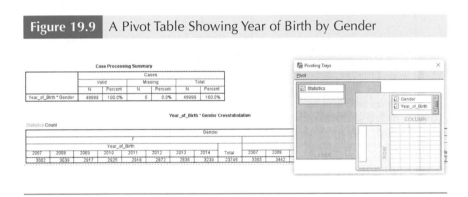

Remember that if you want to create a table with rows or columns encompassing more than one variable, just drag those variables to columns or rows as the question you're asking dictates.

Your ability to change the appearance of variables or fields included in a pivot table is only limited by what variables or fields you first define with row and column headings when you enter the original data. So, as a general rule, be as granular (a fancy word for *detailed*) as possible, since you never know what you'll want to explore in your table until the time comes for exploration.

Modifying a Pivot Table

Once you have the information as shown in Figure 19.9, the modification and reimagining of a pivot table is a cinch.

In Figure 19.10, you see a gender-by-county crosstabs table, created as easily as the other crosstabs tables shown in this chapter. Only this time, we clicked the Display clustered bar charts check box (and who knows why these are suddenly referred to as charts here—SPSS calls them graphs elsewhere). In Figure 19.11, you can also see a chart illustrating the results.

Figure 19.10	A Pivot Table

County * Gender Crosstabulation

Count

		Gender		
		F	M	Total
County	Albany	44	71	115
	ALBANY	319	495	814
	Allegany	0	3	3
	ALLEGANY	9	25	34
	Bronx	377	347	724
	BRONX	1120	1273	2393
	Broome	33	57	90
	BROOME	201	371	572
	Cattaraugus	7	12	19

Figure 19.11	A Bar Chart Illustrating the Results of a Crosstabs Analysis

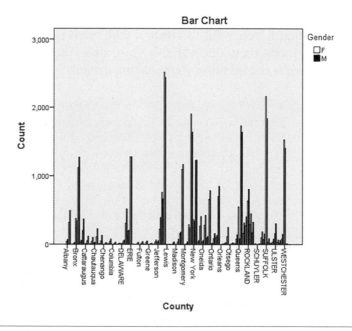

You can also create a table and then use drop-down menus to access the information that you want.

For example, we created a pivot table that shows count by year of birth and then nested each of these variables under Statistics, as you see in Figure 19.12. Now, we can use the drop-down menus to examine any combination of values from that table. For example, if we want to know how many names appeared exactly 9 times in 2009, we'll just select

Figure 19.12 Using Drop-down Menus to Select Different Values in a Pivot Table

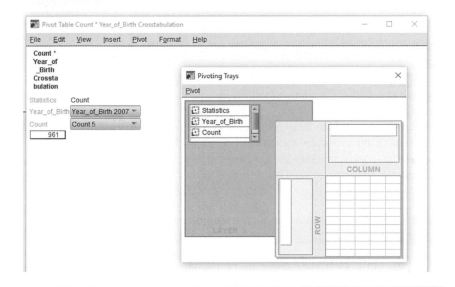

Figure 19.13 Showing Exact Results From a Pivot Table

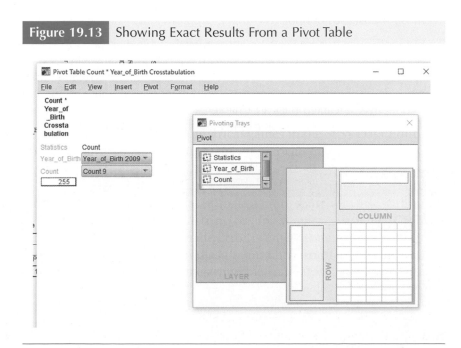

those values from the drop-down menu. This gives us Figure 19.13, where we can see that there were 255 different names that appeared with a total frequency of 9 times.

 Want to fancy up your crosstabs tables? Just right-click and select TableLooks and then select the color and design combinations you want to use.

SUMMARY

It would be very difficult to cover even an introduction to data mining in an introductory book, but in this chapter and of *Statistics for People . . .* , we wanted you to see how SPSS can help you search for patterns in big data sets using crosstabs tables and pivot tables. Hopefully, you will be inspired to search data sets such as those you'll find in Chapter 21, "The 10 (or More) Best Internet Sites for Statistics Stuff."

TIME TO PRACTICE

1. Using the data from the file named Chapter 19 Data Set 1, create a crosstabs table for category (ranging from 1 to 10, with 10 being most industrious) as a function of proficiency scores (ranging from 1 to 5, with 5 being most proficient). How many participants are really (the most) industrious but not very (the least) proficient?

2. Using Chapter 19 Data Set 2, create a crosstabs table that shows the number of level-4 athletes who are female among 1,000 students. In the data set for the variable Gender, 1 = male and 2 = female. Pivot table time! Now graph the results and produce something visually interesting and informative. (Hearken back to Chapter 4, "A Picture Really Is Worth a Thousand Words," for tips on making graphs people will love.)

Problem 2

STUDENT STUDY SITE

⑤SAGE edge™

Get the tools you need to sharpen your study skills! Visit **edge.sagepub.com/ salkind6e** to access practice quizzes, eFlashcards, original and curated videos, data sets, journal articles, and more!

20 A Statistical Software Sampler

Difficulty Scale ☺ ☺ ☺ ☺ ☺ (a cinch!)

WHAT YOU'LL LEARN ABOUT IN THIS CHAPTER

✦ All about other types of software that allow you to analyze, chart, and better understand your data

You need not be a nerd or anything of the sort to appreciate and enjoy what the various computer programs can do for you in your efforts to learn and use basic statistics. The purpose of this chapter is to give you an overview of some of the more commonly used programs and some of their features, and a quick look at how they work. But before we go into these descriptions, here are some words of advice.

You can find a mega listing of software programs and links to the home pages of the companies that have created these programs at http://en.wikipedia.org/wiki/List_of_statistical_packages. We're only reviewing a few of the ones that we really like in this chapter, so if you are looking for a package (free or paid), take a look at this Wiki site and spend some time poking around.

SELECTING THE PERFECT STATISTICS SOFTWARE

Here are some tried-and-true suggestions for making sure that you get what you want from a stats program.

Software for
Statistics

1. Whether the software program is expensive (like SPSS) or less so (like Statistica), be sure you try it out before you buy it. Almost every stats program listed offers a demo (usually on its website) that you can download, and in some cases (but increasingly rarely), you can even ask the company to send you a demo version on CD. These versions may even be fully featured and last for up to 30 days, giving you plenty of time to try before you buy.

2. While we're mentioning price, buying directly from the manufacturer might be the *most* expensive way to go, especially if you buy outright without inquiring about discounts for students and faculty (what they sometimes call an educational discount). Your school bookstore may offer a discount, and a mail-order company might have even a better deal (again, ask about an educational discount). You can find these sellers' toll-free phone numbers listed in any popular computer magazine. You may also find that some stats books come with a limited/student version of the software, and in some cases, it is fully functioning and ready to go. Best of all, many schools and other institutions license the software from the manufacturer, allowing you to download it at home as well as use it on a campus computer. Sweet.

3. Many of the vendors who produce statistical analysis software offer two flavors. The first is the commercial version, and the second is the academic version. They are usually the same in content (but may be limited in very specific ways, such as the number of variables you can test) and always differ, sometimes dramatically, in price. If you are going for the academic version, be sure that it is the same as the fully featured commercial version, and if not, then ask yourself if you can live with the differences. Why is the academic version so much cheaper? The company hopes that if you are a student, when you graduate, you'll move into some fat-cat job and buy the full version and upgrade forever. Also, often academic versions are only available as rentals so their usefulness expires after 6 months or a year . . . read the fine print. Companies such as Student Discounts (http://studentdiscounts.com) and OnTheHub (http://onthehub.com) offer SPSS, for example, at very reasonable rental rates.

4. It's hard to know exactly what you'll need before you get started, but some packages come in modules, and you often don't have to buy all of them to get the tools you need for the job you have to do. Read the company's brochures and website and call and ask questions.

5. Shareware and freeware are other options, and there are plenty of such programs available. *Shareware* is a method of distributing software so you pay for it only if you like it. Sounds like the honor system,

doesn't it? Well, it is. The suggested donations are almost always very reasonable; the shareware is often better than many commercial products; and, if you do pay, you help ensure that the clever author will continue his or her efforts at delivering new versions that are even better than the one you have. *Freeware* is exactly that: free.

More on Selecting a Statistical Software Package

6. Don't buy any software that does not offer telephone technical support or, at the least, some type of email contact or live online chat. To test this, call the tech support number (before you buy!) and see how long it takes for someone to pick up the phone. If you're on hold for 20 minutes, that may indicate that the company doesn't take tech support seriously enough to get to users' questions quickly. Or, if you email them and never hear back, or if their online chat service never seems to be working, look for another product.

7. Almost all the big stats packages do the same things—the difference is in the way that they do them. For example, SPSS, Minitab, and JMP all do a nice job of analyzing data and are acceptable. But it's the little things that might make a difference. For example, perhaps you want to import data from another software application. Some programs may allow this and others not. If any one specific feature is important to you, be sure before you buy that the program has it.

8. Make sure you have the hardware to run the program you want to use. For example, most software is not limited by the number of cases and variables you want to analyze (unless you are using a trial version). The only limit is storage availability, which is becoming less of an issue given that many programs now interact with cloud-based servers and such. And if you have a "slow" machine and less than 1 gigabyte of RAM (random access memory), then you're likely to be waiting around and watching that hourglass while your CPU does its thing. Be sure you have the hardware you need to run a program before you download the demo.

9. Operating systems are always changing, but sometimes software does not keep up. For example, some of these packages work only with a Windows operating system, and at that, some do not work with the newest version of Windows, Windows 10. If in doubt, call and be sure that your system has the stuff it takes to make the program work. Some stats software works with both Windows and the Mac OS, and many packages now are Linux compatible as well.

10. Finally, some companies only offer downloads over the Internet, so you would not get an actual program disk. For the most part, this is fine, but for those of you who really, really, really like to cross your *t*'s and dot your *i*'s, it might make you a bit crazy not to have a CD that you can call your own. Once again, check ahead.

WHAT'S OUT THERE

There are more statistical analysis programs available than you would ever need. Here's a listing of some of the most popular and their outstanding features. Remember that many of these do the same thing. If at all possible, as emphasized in the preceding section, try before you buy. Explore, explore, explore!

 The Wikipedia List of statistical packages (at http://en.wikipedia.org/wiki/List_of_statistical_packages) lists (with links to many) more than 50 open source programs (open to user changes), 3 programs in the public domain, 6 freeware programs, and more proprietary software (which means you pay) than you will ever need to look at. This is a great place to start in any hunt for what's available that will fit your specific or general needs.

First, the Free Stuff

Don't do anything until you've looked at the list of free statistical software available at the Wikipedia site we mentioned above and at http://freestatistics.altervista.org/?p=stat. There are loads of packages here that you can download and use and that perform many of the procedures that you have learned about in *Statistics for People* . . . . We can't possibly review all of them here, but spend some time tooling around and see what fits your needs.

OpenStat

My all-time favorite? OpenStat (http://openstat.en.softonic.com/), written by Dr. Bill Miller from Iowa State University. What's great about it? Well, first, it is entirely free—no "free for 124 days" stuff or anything like that. Next, it is very similar to SPSS and, in some ways, even easier to use (see Figure 20.1). It also features loads of options (far more than many commercial, for-fee programs), including a bunch of nonparametric tests, measurement tools, and even tools for financial analysis (as in, how much can I save if I pay off my student loan early?). And finally, for those of you who like to tinker around, the "Open" in OpenStat means that it is open source. The program is written in C++, and if you know just a bit about that language, you can tweak the program as you see fit. It's a great all-around deal. And, on the site, you can learn about a new program that is compatible with Linux and the Mac OSX operating system.

Figure 20.1	The OpenStat Data Entry Window

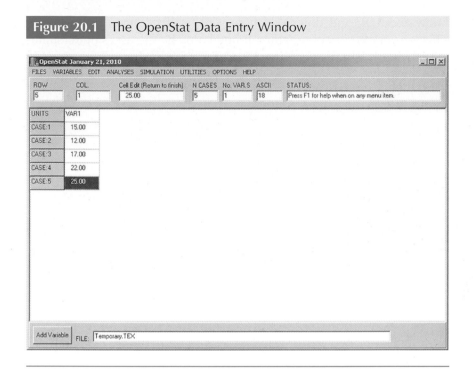

PSPP—(Almost) SPSS

If you just can't shake that SPSS thing and really don't have hundreds of bucks to spare, take a look at PSPP (creative name, huh?) at http://www.gnu.org/software/pspp/, where there is an open source version of a program that mimics SPSS very closely. You can really do a lot here, at no charge, and get a very fair approximation of what SPSS looks like and does. And, if you like to tinker, the open source nature of this program (and this entire movement) welcomes you to actually change the code and make suggestions as to how to improve the program.

R

And if you are way ready for big-time computing, then turn to the open source program named R (probably after the two authors, Robert Gentleman and Ross Ihaka). There are some commercial versions of R that offer formal support, but there is a *huge* community of R users that can be of help as well. And R is also open source (like PSPP) and has its own journal (*The R Journal*). It works on Linux, Unix, Windows, and Mac platforms. It does take some getting used to. However, because it is command line controlled (which means you don't point and click but rather enter commands on a line), it is uniquely flexible in what you can tell it to do and how. Find everything you wanted to know about R, and download it as well, at http://www.r-project.org. This is a very powerful program, limited only by the user's understanding of what it can do. See Figure 20.2 for a screenshot.

Figure 20.2 A Screenshot From R

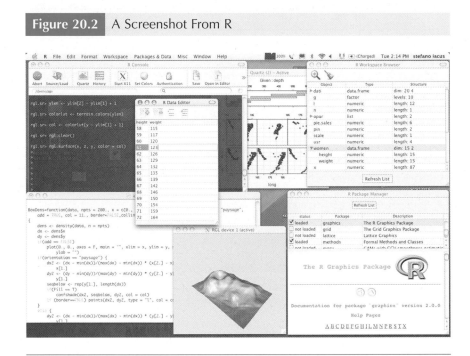

Time to Pay

JMP

JMP (now in Version 12 and part of SAS) is billed as the "statistical discovery" software. It operates on Mac, Windows, and Linux platforms and "links statistics with graphics to interactively explore, understand, and visualize data." One of JMP's features is to present a graph accompanying every statistic so you can always see the results of the analysis both as statistical text and as a graphic. And all this is done automatically without you requesting it.

Want more information? Try http://www.jmp.com.

Cost: $1,620 for the single-user license version and $2,995 for a 3-year academic license. The JMP Student Edition appears in many textbooks but is not available as an independent product.

Minitab

This is one of the first stats programs that was made available for the personal computer, and it is now in Version 17 (it's been around a while!), which means that it's seen its share of changes over the years in response to users' needs. Here are some of the more outstanding features of this version:

- Quality Trainer™, an online course to help train the user (which you can also purchase individually)
- ReportPad™ for generating reports
- Project Manager, which organizes analysis
- Smart Dialog Boxes™, which remember recent settings

In Figure 20.3, you can see a sample of what Minitab output looks like for a simple regression analysis.

Want more information? Try http://www.minitab.com.

Cost: $1,495 for the full-fledged single-user license version, but there are various rental options for different prices for different time periods, including a 6-month rental license for $30. This is the product to choose if you're going to spend money and want a powerful set of tools.

STATISTICA

StatSoft (owned by Dell) offers a family of STATISTICA (version 13.1) products for Windows (up through the latest, Windows 10) but, sorry, no Mac version. Some of the features that are particularly nice about this powerful program are the self-prompting dialog boxes (you click

Figure 20.3 Minitab Output Showing a Simple Regression Analysis

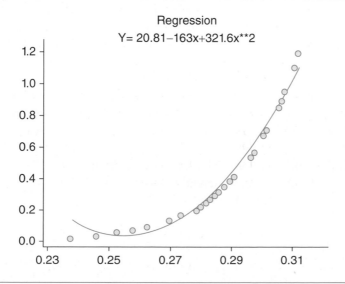

OK, and STATISTICA tells you what to enter); the customization of the interface; easy integration with other programs; STATISTICA Visual Basic, which allows you to access more than 10,000 functions and use this development environment to design special applications; and the ability to use macros to automate tasks. A nice bonus at the website is an Electronic Statistics Textbook, which you can use to access information about many different topics.

Want more information? Try http://www.statsoft.com/.

Cost: $795 for the base license (and there are lots and lots of modules you can add) and there's a free Basic Academic Bundle (a 12-month rental). This is another good deal.

SPSS

SPSS is one of the most popular big-time statistical packages in use, and that's why we have a whole appendix devoted to it in the very book you are holding in your hands! And also why we go through the computer exercises in these pages. It comes with a variety of modules that cover all aspects of statistical analysis, including both basic and advanced statistics, and a version exists for almost every platform. It was recently purchased by IBM (after a brief renaming as PWAS, and thank heavens the new owners got rid of that).

Want more information? Try http://www.spss.com.

Cost: Lots! $2,610 a year (!!!) for the standard version (but $1,170 per year for the base version) and many different plans and time frames for the academic market.

STATISTIX for Windows

Version 10 of STATISTIX offers a menu-driven interface that makes it particularly easy to learn to use, and it is almost as powerful as the other programs mentioned here (but all only for Windows folks). The company offers free technical support, and—ready for this?—a real, 450-page paper manual. And when you call technical support, you talk with the actual programmers, who know what they're talking about (my question was answered in 10 seconds!). Figure 20.4 shows you some STATISTIX output from a paired-sample *t* test. All around, a good deal.

Want more information? Try http://www.statistix.com.

Cost: $495 for the commercial version and $395 for the academic version, and both are Windows only.

Figure 20.4 STATISTIX Output for a Paired *t*-Test

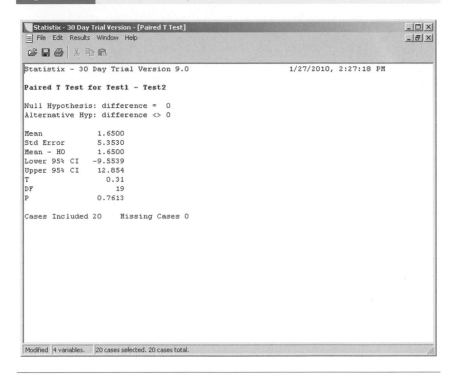

SUMMARY

That's the end of Part V and just about the end of *Statistics for People Who (Think They) Hate Statistics*. But read on! The next chapter includes the ten best Internet sites in the universe for information about statistics, followed by Chapter 22, the ten commandments of data collection. Have fun with both of these.

PART VI

Ten Things (Times Two) You'll Want to Know and Remember

"Everyone turn to chapter 12,432..."

21

The Ten (or More) Best (and Most Fun) Internet Sites for Statistics Stuff

O f course you use (and love) the Internet. It's probably your first stop when you want to find out about something new. Well, just like recipes, music-sharing apps, and the latest NCAA basketball scores, a ton of material on statistics is available for those who are new to (and those who are not so new to) its study and application. If you're not yet using the Internet as a specific tool in your learning and research activities, you are missing out on an extraordinary resource.

What's available on the Internet will not make up for a lack of studying or motivation—nothing will do that—but you can certainly find a great deal of information that will enhance your whole college experience. And this doesn't even begin to include all the fun you can have!

So, now that you're a certified novice statistician, here are some Internet sites that you might find very useful should you want to learn more about statistics.

 Although the locations of websites on the Internet are more stable than ever, they still can change frequently. The URL (the uniform resource locator—what we all call the "web address") that worked today might not work tomorrow. If you don't get to where you want to go, use Google or some other search engine and search on the name of the website—perhaps there's a new URL or another web address that works.

HOW ABOUT STUDYING STATISTICS IN STOCKHOLM?

The World Wide Web Virtual Library: Statistics is the name of the page, and indeed it is a virtual worldwide library. The site (from the good people at the University of Florida at http://www.stat.ufl.edu/vlib/statistics.html) includes information on just about every facet of the topic, including data sources, job announcements, departments, divisions and schools of statistics (descriptions of programs all over the world), statistical research groups, institutes and associations, statistical services, statistical archives and resources, statistical software vendors and software, statistical journals, mailing list archives, and related fields. Tons of great information is available here. Make it a stop along the way.

WHO'S WHO AND WHAT'S HAPPENED?

The History of Statistics page located at http://anselm.edu/homepage/jpitocch/biostats/biostatshist.html contains portraits and bibliographies of famous statisticians and a time line of important contributions to the field of statistics. Do names like Bernoulli, Galton, Fisher, and Spearman pique your curiosity? How about the development of the first test between two averages during the early 20th century? It might seem a bit boring until you have a chance to read about the people who make up this field and their ideas—in sum, pretty cool ideas and pretty cool people.

And of course, Wikipedia at http://en.wikipedia.org/wiki/History_of_statistics does a great job of introducing the history of this topic, as does eMathZone at http://www.emathzone.com/tutorials/basic-statistics/history-of-statistics.html.

IT'S ALL HERE

SurfStat Australia (https://surfstat.anu.edu.au/surfstat-home/contents.html) is the online component of a basic stats course taught at the University of Newcastle, Australia, but it has grown far beyond just the notes originally written by Annette Dobson in 1987 and updated over several years' use by Anne Young, Bob Gibberd, and others. Among other things, SurfStat (and you can see part of the detailed table of contents for the course in Figure 21.1) contains a complete interactive statistics text. Besides the text, there are exercises, a list of other statistics sites on the Internet, and a collection of Java applets (cool little programs you can use to work with different statistical procedures).

HYPERSTAT

This online tutorial with 18 lessons, at http://davidmlane.com/hyperstat/, offers nicely designed and user-friendly coverage of the important basic topics. What we really like about the site is the glossary, which uses

Figure 21.1	A Partial Table of Contents for the SurfStat Course

STATISTICAL INFERENCE

- POPULATIONS, SAMPLES, ESTIMATES AND REPEATED SAMPLING
  - 4-1-1.html
    - Definitions
    - Bias
      Example - Health survey conducted in the Hunter Region
      Example - Heights of women, 25 - 29

- POINT ESTIMATION AND INTERVAL ESTIMATION
  - 4-1-2.html
    - Sample Mean
    - Confidence Interval

- RESULTS FROM PROBABILITY THEORY
  - 4-1-3.html
    - Law of Averages
    - Law of Large Numbers
    - Central Limit Theorem

- ONE CONTINUOUS VARIABLE
  - 4-1-4.html
    - Sampling Distribution of the Sample Mean
      Properties of the Sampling Distribution

  - 4-1-5.html
    - Interpretation of Confidence Intervals
    - Comments
    - Hypothesis Testing
      Example - Bolt production
    - The First Approach
    - Remarks
    - The Second Approach
    - Steps for hypothesis testing

hypertext to connect different concepts to one another. For example, in Figure 21.2, you can see topics that are hyperlinked (the live underlined words) to others. Click on any of those and zap! You're there (or at least on your way).

Figure 21.2	Sample HyperStat Screen

RVLS home > HyperStat Online > Glossary

Glossary
A-C D-F G-J J-L M-O P-R S-U V-Z

alternative hypothesis
Analysis of variance
Bar graph
Between-subjects variable
Bias
Biased sample
Bimodal distribution
Binomial distribution
Box plot
Carryover effects
Categorical variable
Central tendency
Chi square distribution
Chi square independence test
Coefficient
Conditional probability
Confidence interval
Confounding
Consistency
Contingency table
Continuous variable
Correction for continuity
Correlation
Counterbalancing
Critical value
Criterion variable
Cumulative distribution

A-C D-F G-J J-L M-O P-R S-U V-Z

Descriptive Statistics

Next Section: Inferential Statistics

One important use of descriptive statistics is to summarize a collection of data in a clear and understandable way. For example, assume a psychologist gave a personality test measuring shyness to all 2500 students attending a small college. How might these measurements be summarized? There are two basic methods: numerical and graphical. Using the numerical approach one might compute statistics such as the mean and standard deviation. These statistics convey information about the average degree of shyness and the degree to which people differ in shyness. Using the graphical approach one might create a stem and leaf display and a box plot. These plots contain detailed information about the distribution of shyness scores.

Graphical methods are better suited than numerical methods for identifying patterns in the data. Numerical approaches are more precise and objective.

Since the numerical and graphical approaches compliment each other, it is wise to use both.

Next Section: Inferential Statistics

DATA? YOU WANT DATA?

Data from the US Government

There are data all over the place, ripe for the picking. Here are just a few. What to do with these? Download them to be used as examples in your work or as examples of analysis that you might want to do. You can use these as models:

- Statistical Reference Datasets at http://itl.nist.gov/div898/strd/
- United States Census Bureau (a huge collection and a goldmine of data courtesy of American FactFinder) at http://factfinder.census.gov/faces/nav/jsf/pages/index.xhtml
- The Data and Story Library, with great annotations of the data, at http://lib.stat.cmu.edu/DASL/ (look for the "stories" link)
- Tons of economic data sets provided by Growth Data Sets at http://www.bris.ac.uk/Depts/Economics/Growth/datasets.htm

Even more data sets are available through the federal government (besides the Census Bureau and others highlighted above). Your tax money supports these resources, so why not use them? For example, there's FedStats (at http://fedstats.sites.usa.gov), where more than 70 agencies in the US government publish statistics of interest to the public. The Federal Interagency Council on Statistical Policy maintains this site to provide easy access to the full range of statistics and information produced by these agencies for public use. Here you can find country profiles contributed by the CIA; public school student, staff, and faculty data (from the National Center for Education Statistics); and the Atlas of United States Mortality (from the National Center for Health Statistics). What a ton of data!

And most states have their data available for the clicking. We got the data we used in Chapter 17 for demonstration purposes from the Data.gov website, specifically from http://catalog.data.gov/dataset?q=baby+names&sort=score+desc%2C+name+asc. We clicked on the CSV option (which allows us to open the data set in Excel) and went from there.

MORE AND MORE RESOURCES

The University of Michigan's Research Guides site (http://guides.lib.umich.edu) has hundreds and hundreds of resource links, including to information on banking; book publishing; the elderly; and, for those of you with allergies, pollen count. Browse to see what you find or search for exactly what you need—either way, you are guaranteed to find something interesting.

ONLINE STATISTICAL TEACHING MATERIALS

You're so good at this statistics stuff that you might as well start helping your neighbor and colleague in class. If that's the case, turn to Teaching Statistics: A Bag of Tricks at http://poldw.com/teaching-statistics-a-bag-of-tricks-columbia-university.html, bought to us by Andrew Gelman and Deborah Nolan (you'll need to register). You can also find more about teaching statistics from Vanderbilt University, https://cft.vanderbilt .edu/guides-sub-pages/teaching-statistics/.

AND, OF COURSE, YOUTUBE . . .

Yes, you can now find stats stuff on YouTube in the form of Statz Rappers (http://www.youtube.com/watch?v=JS9GmU5hr5w), a group of talented young men and women who seem to be having a great time making just a bit of fun of their stats course—a very fitting stop on this path of what the Internet holds for those interested in exploring statistics. But there's also much more serious information. See, for example, http://www.youtube.com/watch?v=HvDqbzu0i0E from KhanAcademy .org—another cool place that is full of thousands—yes, thousands—of video tutorials on everything from algebra to economics to investing to of course (you guessed it) statistics!

AND, FINALLY . . .

At http://animatedsoftware.com/statglos/statglos.htm#index, you'll find the mother and father lode of definitions of statistics terms, bought to us by Dr. Howard S. Hoffman. Informative and very much fun.

22

The Ten Commandments of Data Collection

Now that you know how to analyze data, you would be well served to read something about collecting them. The data collection process can be long and rigorous. Even if it involves only a simple, one-page questionnaire given to a group of students, parents, patients, or voters, data collection may very well be the most time-consuming part of your project. But as many researchers realize, good data collection is essential to good research outcomes.

Data Collection and Analysis

Here they are: the ten commandments for making sure your data get collected in such a way that they are usable. Unlike the original Ten Commandments, these should not be carved in stone (because they can certainly change), but if you follow them, you can avoid lots of aggravation.

Commandment 1. As you begin thinking about a research question, also begin thinking about the type of data you will have to collect to answer that question. Interview? Questionnaire? Paper and pencil? Computer? Find out how other people have done it in the past by reading the relevant journals in your area of interest. Then consider doing what they did. At least one of the lessons here is to not repeat others' mistakes. If something didn't work for them, it's most likely not going to work for you.

Commandment 2. As you think about the type of data you will be collecting, think about where you will be getting the data. If you are using the library for historical data or accessing files of data that have already been collected, such as Census data (available through the US Census Bureau at http://www.census.gov and other locations online), you will have few logistical problems. But what if you want to assess the interaction between newborns and their parents? The attitude of teachers toward unionizing? The age at which people over 50 think they are old? All of these questions require people to provide the answers, and finding people can be tough. Start now.

Commandment 3. Make sure that the data collection forms you use are clear and easy to use. Practice on a set of pilot data so you can make sure it is easy to transfer data from the original scoring sheets or the data collection form to a digital format. And then have some colleagues complete the form to make sure it works.

Commandment 4. Always make a duplicate copy of the data file and the data collection sheets and keep them in a separate location. Keep in mind that there are two types of people: those who have lost their data and those who will. In fact, make two backups of your electronic data! These days, you can use online data backup services such as Carbonite (http://www.carbonite.com), Mozy (http://www.mozy.com), or CrashPlan (http://www.code42.com/crashplan/) in addition to your own physical backup.

Commandment 5. Do not rely on other people to collect or transfer your data unless you have personally trained them and are confident that they understand the data collection process as well as you do. It is great to have people help you, and it helps keep morale up during those long data collection sessions. But unless your helpers are competent beyond question, you could easily sabotage all your hard work and planning.

Commandment 6. Plan a detailed schedule of when and where you will be collecting your data. If you need to visit 3 schools and each of 50 children needs to be tested for a total of 10 minutes at each school, that is 25 hours of testing. That does not mean you can allot a mere 25 hours from your schedule for this activity. What about travel from one school to another? What about the child who is in the bathroom when it is his turn and you have to wait 10 minutes until he comes back to the classroom? What about the day you show up and Cowboy Bob is the featured guest . . . and on and on. Be prepared for anything, and allocate 25% to 50% more time in your schedule for unforeseen events.

Commandment 7. As soon as possible, cultivate possible sources for your subject pool. Because you already have some knowledge in your own discipline, you probably also know of people who work with the population you are interested in or who might be able to help you gain access to these samples. If you are in a university community, it is likely that hundreds of other people are competing for the same subject sample that you need. Instead of competing, why not try a more out-of-the-way (maybe 30 minutes away) school district or social group or civic organization or hospital, where you might be able to obtain a sample with less competition?

Commandment 8. Try to follow up on subjects who missed their testing session or interview. Call them back and try to reschedule. Once you get in the habit of skipping possible participants, it becomes too easy to cut the sample down to too small a size. And you can never tell—the people who drop out might be dropping out for reasons related to what you are studying. This can mean that your final sample of people is qualitatively different from the sample with which you started.

Commandment 9. Never discard the original data, such as the test booklets, interview notes, and so forth. Other researchers might want to use the same data, or you may have to return to the original materials to glean further information from them.

And Commandment 10? Follow the previous nine commandments. No kidding!

Appendix A

SPSS Statistics in Less Than 30 Minutes

This appendix will teach you enough about IBM SPSS to complete the exercises in *Statistics for People Who (Think They) Hate Statistics*. Learning SPSS is not rocket science—just take your time, work as slowly as you need to, and ask a fellow student or your instructor for help if necessary.

You are probably familiar with other Windows applications, and you will find that many SPSS features operate exactly the same way. We assume you know about dragging, clicking, double-clicking, and working with Windows or the Mac (the two versions of SPSS for these operating systems are very similar). If you do not, you can refer to one of the many popular trade computer books for help. SPSS works with Microsoft Windows XP, Vista, and versions 7 and 8, and the latest version of the operating system, 10. For the Mac, it works on Mountain Lion 10.8 and later. This appendix focuses almost exclusively on the Windows version because it is much more popular than the Mac version, but if you are a Mac user, you should have no difficulty at all following the instructions and examples.

This appendix is an introduction to SPSS (Version 23) and shows you just some of the things it can do. Almost all of the information in this appendix also can be applied to earlier versions of SPSS, from versions 11 through 22.

Throughout the examples in this appendix, we will use the sample data set shown in Appendix C named *Sample Data Set.sav*. You are welcome to enter those data manually or download the set from the SAGE website at **edge.sagepub.com/salkind6e** or contact the author at njs@ku.edu.

STARTING SPSS

Like other Windows-based applications, SPSS is organized as a group and is available on the Start menu. This group was created when you first installed SPSS. To start SPSS, follow these steps:

1. Click Start and then point to Programs.

2. Find and click the SPSS icon. When you do this, you will see the SPSS opening screen, as shown in Figure A.1. You should note that some computers are set up differently, and your SPSS icon might be located on the desktop. In that case, to open SPSS, just double-click on the icon.

The SPSS Opening Window

As you can see in Figure A.1, the opening window presents a series of options that allow you to select from an introduction to using the data editor to working with output and more. Should you not want to see this screen each time you open SPSS, then click on the "Don't show this dialog in the future" box in the lower left corner of the window.

For our purposes, you will click Cancel, and once you do this, the Data View window (also called the Data Editor) you see in Figure A.2 becomes active. This is where you enter data you want to use with SPSS once those data have been defined. If you think the Data Editor is similar

Figure A.1   The Opening SPSS Screen

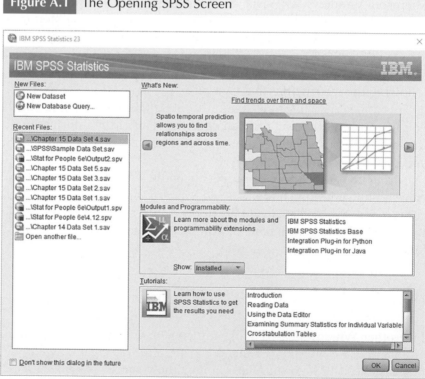

to a spreadsheet in form and function, you are right. In form, it certainly is, because the Data Editor consists of rows and columns just like an Excel spreadsheet. Values can be entered and then manipulated. In function as well, the Data Editor is much like a spreadsheet. Values that are entered can be transformed, sorted, rearranged, and more.

Although you cannot see it when SPSS first opens, there is another open (but not active) window as well. This is the Variable View, where variables are defined and the parameters for those variables are set.

The Viewer displays statistical results and charts that you create. An example of the Viewer section of the Output window is shown in Figure A.3. A data set is created using the Data Editor, and once the set is analyzed or graphed, you examine the results of the analysis in the Viewer.

Figure A.2   The Data View Window

| Untitled1 [DataSet0] - IBM SPSS Statistics Data Editor | — ☐ × |

File   Edit   View   Data   Transform   Analyze   Graphs   Custom   Utilities   Add-ons   Window   Help

Visible: 0 of 0 Variables

	var	var	var	var	var	var	var	var
1								
2								
3								
4								
5								
6								
7								
8								
9								
10								
11								
12								
13								
14								
15								
16								

Data View   Variable View

IBM SPSS Statistics Processor is ready   Unicode:OFF

Figure A.3   The Viewer

| *Output1 [Document1] - IBM SF

File   Edit   View   Data   Tra

☐+ Output
 ├ Log
 ├ Descriptives
 │ ├ Title
 │ ├ Notes
 │ ├ Active Dataset
 │ └ Descriptive Statist

THE SPSS TOOLBAR AND STATUS BAR

The use of the Toolbar—the set of icons below the menus—can greatly facilitate your SPSS activities. If you want to know what an icon on the Toolbar does, just roll the mouse pointer over it, and you will see a tip telling you what the tool does. Some of the buttons on the Toolbar are dimmed, meaning they are not active.

The Status Bar, located at the bottom of the SPSS window, is another useful on-screen tool. Here, you can see a one-line report telling you which activity SPSS is currently involved in. The message *IBM SPSS for Windows Processor is Ready* tells you that SPSS is ready for your directions or input of data. As another example, *Running Means . . .* tells you that SPSS is in the middle of the procedure named Means.

USING SPSS HELP

SPSS Help is only a few mouse clicks away, and it is especially useful when you are in the middle of a data file and need information about an SPSS feature. SPSS Help is so comprehensive that even if you are a new SPSS user, it can show you the way.

You can get help in SPSS by using the Help menu you see in Figure A.4.

There are 12 options on the Help menu (which is greatly expanded from earlier versions of SPSS), and 6 of these are directly relevant to helping you.

- *Topics* gives you a list of topics for which you can get help.
- *Tutorial* offers you a short tutorial on all aspects of using SPSS.
- *Case Studies* gives you real live examples of how SPSS can be applied.
- *Working with R* provides you with information about how to work with the open source statistics package named R.
- *Statistics Coach* walks you through procedures step-by-step.
- *Command Syntax Reference* helps you to learn and use SPSS's programming language.
- *SPSS Community* gives you access to other SPSS users and information.
- *About . . .* gives you some technical information about SPSS, including the version you are using.
- *Algorithms* focuses on the calculations that are used to produce the results you see in SPSS.
- *IBM SPSS Products Home* takes you to the home page for SPSS on the Internet.
- *Programmability* provides you with information about creating add-ons and other program enhancements for SPSS.
- *Diagnose* helps you to diagnose why SPSS may not be running properly.

| Figure A.4 | The Various Help Options |

A BRIEF TOUR OF SPSS

Now, sit back and enjoy a brief tour of what SPSS can do. Nothing fancy here. Just some simple descriptions of data, a test of significance, and a graph or two. What we are trying to show you is how easy it is to use SPSS.

Opening a File

You can enter your own data to create a new SPSS data file, use an existing file, or even import data from such applications as Microsoft Excel into SPSS. Any way you do it, you need to have data to work with. In Figure A.5, the data contained in Appendix C (named *Sample Data Set.sav*) are shown. This file is also available on the book's companion website or directly from the author.

A Simple Table and Graph

Now it is time to get to the reason why we are using SPSS in the first place—the various analytical tools that are available.

First, let's say we want to know the general distribution of males and females. That's all—just a count of how many males and how many females are in our total sample. We also want to create a simple bar graph of the distribution.

In Figure A.6, you see the output that provides exactly the information we asked for, which was the frequency of the number of males and

---

Figure A.5 An Open SPSS File

*Sample Data Set.sav [DataSet1] - IBM SPSS Statistics Data Editor

File   Edit   View   Data   Transform   Analyze   Graphs   Custom   Utilities   Add-ons

15 :

	ID	Gender	Treatment	Test1	Test2
1	1	Male	Control	98	32
2	2	Female	Experimental	87	33
3	3	Female	Control	89	54
4	4	Female	Control	88	44
5	5	Male	Experimental	76	64
6	6	Male	Control	68	54
7	7	Female	Control	78	44
8	8	Female	Experimental	98	32
9	9	Female	Experimental	93	64
10	10	Male	Experimental	76	37
11	11	Female	Control	75	43
12	12	Female	Control	65	56
13	13	Male	Control	76	78
14	14	Female	Control	78	99
15	15	Female	Control	89	87
16	16	Female	Experimental	81	56
17	17	Male	Control	78	78
18	18	Female	Control	83	56
19	19	Male	Control	88	67
20	20	Female	Control	90	88

---

females. We used the "Frequencies" option on the Descriptive Statistics menu (under the main menu Analyze) to compute these values, and then we created a simple bar graph.

Figure A.6 The Results of a Simple Descriptive Analysis

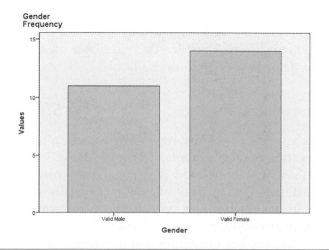

Gender

		Frequency	Percent	Valid Percent	Cumulative Percent
Valid	Male	11	44.0	44.0	44.0
	Female	14	56.0	56.0	100.0
	Total	25	100.0	100.0	

Gender
Frequency

A Simple Analysis

Let's see if males and females differ in their average Test1 scores. This is a simple analysis requiring a *t*-test for independent samples. The procedure is a comparison between males and females for the mean of Test1 for each group.

In Figure A.7, you can see a summary of the results of the *t*-test. Notice that the listing in the left pane (the outline view) of the SPSS Viewer now shows the Frequencies and *t*-test procedures listed. To see any part of the output, all we need do is click on that element. Almost always, when SPSS produces output in the Viewer, you will have to scroll to see the entire output.

| Figure A.7 | The Results of an Independent-Samples *t*-Test |

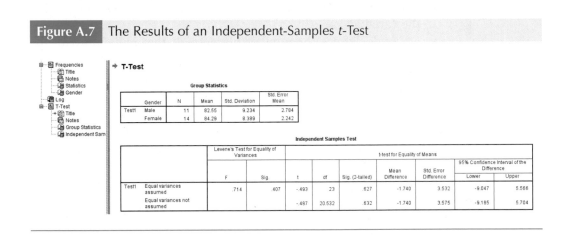

CREATING AND EDITING A DATA FILE

As a hands-on exercise, let's create the beginning of the sample data file you see in Appendix C. The first step is to define the variables in your data set, and the second step is to enter the data. You should have a new Data Editor window open (Click File → New → Data).

Defining Variables

SPSS cannot work unless variables are defined. You can have SPSS define the variables for you, or you can do the defining yourself, thereby having much more control over the way things look and work. SPSS will automatically name the first variable VAR00001. If you defined a variable in row 1, column 5, then SPSS would name the variable VAR00005 and number the other columns sequentially.

Custom Defining Variables: Using the Variable View Window

To define a variable yourself, you must first go to the Variable View window by clicking the Variable View tab at the bottom of the SPSS screen. Once that is done, you will see the Variable View window, as shown in Figure A.8, and be able to define any one variable as you see fit.

Once in the Variable View window, you can define variables along the following parameters:

- *Name* allows you to give a variable a name of up to eight characters.
- *Type* defines the type of variable, such as text, numeric, string, scientific notation, and so on.
- *Width* defines the number of characters of the column housing the variable.
- *Decimals* defines the number of decimals that will appear in the Data View.
- *Label* defines a label of up to 256 characters for the variable.
- *Values* defines the labels that correspond to certain numerical values (such as 1 for male and 2 for female).
- *Missing* indicates how any missing data will be dealt with.
- *Columns* defines the number of spaces allocated for the variable in the Data View window.
- *Align* defines how the data are to appear in the cell (right, left, or center aligned).
- *Measure* defines the scale of measurement that best characterizes the variable (nominal, ordinal, or interval).
- *Role* defines the part that the variable plays in the overall analysis (input, target, etc.).

If you place the cursor in the first cell under the "Name" column, enter any name, and press the Enter key, then SPSS will automatically provide you with the default values for all the variable characteristics. Even if you are not in the Data View screen (click the tab on the bottom of the window), SPSS will automatically name the variables var0001, var0002, and so on.

Figure A.8   The Variable View Window

In the Variable View, enter the names of the variables as you see in Figure A.9.

Now, if you wanted, you could switch to the Data View (see Figure A.10) and just enter the data as you see in Figure A.5. But first, let's look at just one of the cool SPSS bells and whistles.

Figure A.9 Defining Variables in the Variable View Window

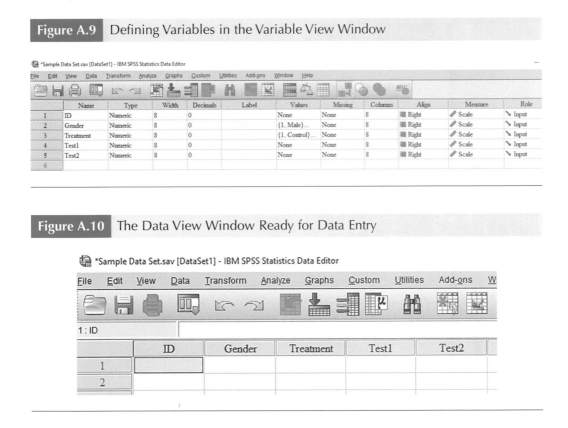

Figure A.10 The Data View Window Ready for Data Entry

Defining Value Labels

You can leave your data as numerical values in the SPSS Data Editor, or you can have labels represent the numerical values (as you saw in Figure A.5).

Why would you want to change the label of a variable? You probably already know that, in general, it makes more sense to work with numbers (like 1 or 2) than with string or alphanumeric variables (such as male or female). An often made error is entering data as a string (such as male or female) rather than as the number representing the variable named gender. When it comes time for an analysis, it is very difficult to work with nonnumerical entries (such as "male").

But it sure is a lot easier to look at a data file and see words rather than numbers. Just think about the difference between data files with numbers representing various levels (such as 1 and 2) of a variable and with the actual values (such as male and female). The Values option in the Variable View screen allows you to enter *values* in the cell, but what you will see are value *labels*.

If you click the ellipsis button in the Values column (see Figure A.11), you will see the Value Labels dialog box, as shown in Figure A.12.

Figure A.11   The Values Column in the Variable View Screen

Figure A.12   The Value Labels Dialog Box

Changing Value Labels

To assign or change a variable label, follow these steps. Here, we will label males as 1 and females as 2.

1. For the variable gender, click on the ellipsis (see Figure A.11) to open the Value Labels dialog box.

2. Enter a value for the variable—in this case, 1 for males.

3. Enter the value label for the value, which is male.

4. Click Add.

5. Do the same for female and value 2. When you finish your business in the Define Labels dialog box (see Figure A.13), click OK, and the new labels will take effect.

When you select View → Variable Labels from the main menu (in the Data View), you will see the labels in the Data Editor. Notice that the value of the entry in Figure A.14 is actually 2, even though the label in the cell reads Female.

Figure A.13 The Completed Value Labels Dialog Box

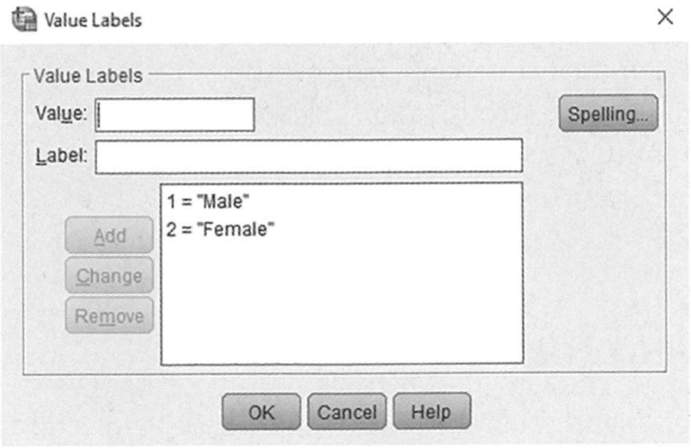

Figure A.14 Seeing Variable Labels

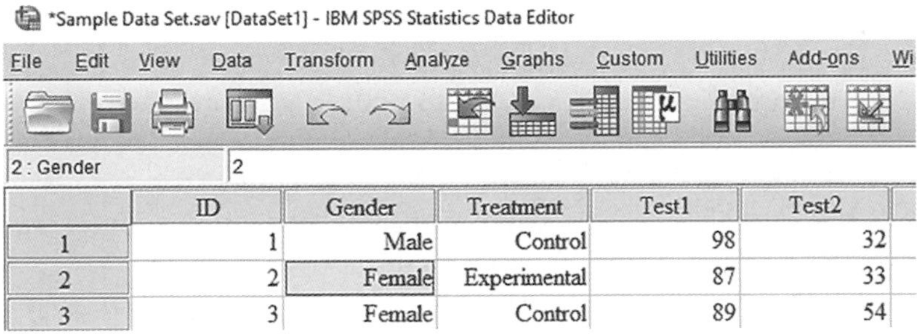

Opening a Data File

Once a file is saved, you have to open or retrieve it when you want to use it again. The steps are simple.

1. Click File → Open. You will see the Open Data File dialog box.

2. Find the data file you want to open and highlight it.

3. Click OK.

A quick way to find and open an SPSS file is to click on the Recently Used Data option on the File menu. SPSS lists the most recently used files there.

PRINTING WITH SPSS

Here comes information on the last thing you will do once a data file is created. Once you have created the data file you want or completed any type of analysis or chart, you probably will want to print out a hard copy for safekeeping or for inclusion in a report or paper. Then, when your SPSS document is printed and you want to stop working, it will be time to exit SPSS.

Printing is almost as important a process as editing and saving data files. If you cannot print, you have nothing to take away from your work session. You can export data from an SPSS file to another application, but getting a hard copy directly from SPSS is often more timely and more important.

Printing an SPSS Data File

It is simple to print either an entire data file or a selection from one.

1. Be sure that the data file you want to print is the active window.
2. Click File → Print. When you do this, you will see the Print dialog box.
3. Click OK, and whatever is active will print.

As you can see, you can choose to print the entire document or a specific selection (which you will have already made in the Data Editor window), and you can increase the number of copies from 1 to 99 (the maximum number of copies you can print). You can also configure the Print dialog box so that a PDF file is produced.

Printing a Selection From an SPSS Data File

Printing a selection from a data file follows exactly the steps that we listed above for printing a data file, except that in the Data Editor window, you select what you want to print and click on the Selection option in the Print dialog box. The steps go like this:

1. Be sure that the data you want to print are selected.
2. Click File → Print.
3. Click Selection in the Print dialog box.
4. Click OK, and whatever you selected will be printed.

CREATING AN SPSS CHART

A picture is worth a thousand words, and SPSS offers you just the features to create charts that bring the results of your analyses to life. We'll go through the steps to create several different types of charts and provide examples of different charts. Then, we will show you how to modify a chart, including adding a chart title; adding labels to axes; modifying scales; and working with patterns, fonts, and more. For whatever reason, SPSS uses the words *graphs* and *charts* interchangeably.

Creating a Simple Chart

The one thing that all charts have in common is that they are based on data. Although you may import data to create a chart, in this example, we will use the data from Appendix C to create a bar chart (like the one you saw in Figure A.6) of the number of males and females in each group.

Creating a Bar Chart

The steps for creating any chart are basically the same. You first enter the data you want to use in the chart, select the type of chart you want from the Graphs menu, define how the chart should appear, and then click OK. Here are the steps we followed to create the chart you see in Figure A.6.

1. Enter the data you want to use to create the chart.

2. Click Graphs → Legacy Dialogs → Bar. When you do this, you will see the Bar Charts dialog box you see in Figure A.15.

3. Click Simple.

4. Click Summaries for groups of cases.

5. Click Define. When you do this, you will see the Define Simple Bar: Summaries for Groups of Cases dialog box.

6. Click Cum n of cases.

7. Click gender. Then move the variable to the Category Axis area by dragging it.

8. Click OK, and you see the results of the chart in Figure A.16.

That's just the beginning of making the chart. To make any changes, you have to use the chart editor tools.

Figure A.15 The Bar Charts Dialog Box

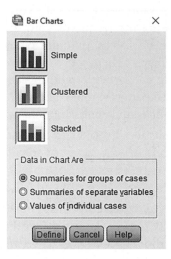

Figure A.16 A Simple Bar Chart

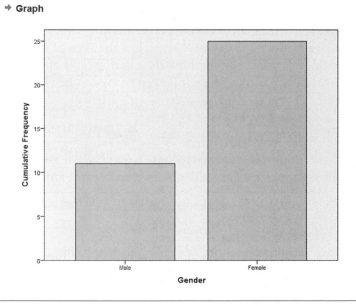

Saving a Chart

A chart is only one component of the Viewer window. A chart is part of the output generated when you perform some type of analysis. The chart is not a separate entity that stands by itself, and it cannot be saved as such. To save a chart, you need to save the contents of the entire Viewer. Follow these steps to do that:

1. Click File → Save.

2. Provide a name for the Viewer window.

3. Click OK. The output is saved under the name that you provide with an *.spo* extension.

ENHANCING SPSS CHARTS

Once you create a chart as we showed you in the previous section, you can edit the chart to reflect exactly what you want to say. Colors, shapes, scales, fonts, and more can be changed. We will be working with the bar chart that was first shown to you in Figure A.16.

Editing a Chart

The first step in editing a chart is to double-click on the chart and then click the maximize button. You will see the entire chart in the Chart Editor window, as shown in Figure A.17.

Figure A.17 The Chart Editor Window

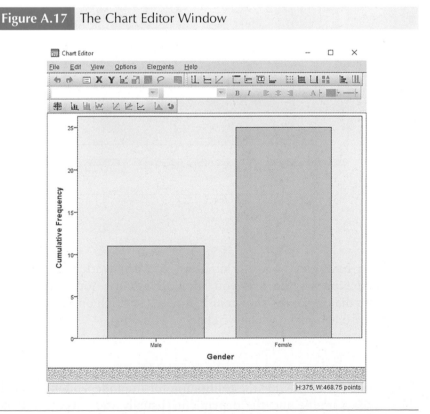

Working With Titles and Subtitles

Our first task is to enter a title and subtitle on the chart you saw in Figure A.16.

1. Click the Insert a Title icon on the Toolbar. When you do this, as you see in Figure A.18, you can edit the element Title right on the screen and enter what you wish.

2. To insert a subtitle (or, in fact, as many titles as you like), just keep clicking the Insert a Title icon.

Figure A.18   Inserting a Title

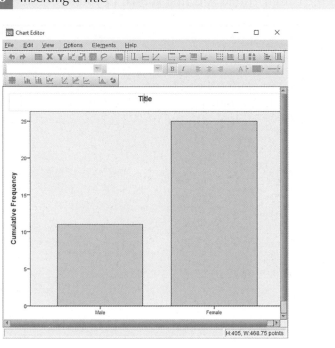

Working With Fonts

Once you have created a title or titles, you can work with fonts by double-clicking on the text you want to modify. You will see the Properties dialog box, as shown in Figure A.19. Click on the Text Style tab, and you can make whatever changes you wish.

Working With Axes

The x- and y-axes provide the calibration for the independent (usually the x-axis) variable and the dependent (usually the y-axis) variable.

 **Figure A.19** Working With Fonts

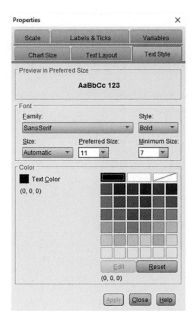

SPSS names the *y*-axis the Scale axis and the *x*-axis the Category axis. Each of these axes can be modified in a variety of ways. To modify either axis, double-click on the title of the axis.

How to Modify the Scale (y) Axis

To modify the *y*-axis, follow these steps:

1. Still in the chart editor? We hope so. Double-click on the axis (not the axis label).

2. Click the Scale tab in the Properties dialog box. When you do this, you will see the Scale Axis dialog box, as shown in Figure A.20.

3. Select the options you want from the Scale Axis dialog box.

How to Modify the Category (x) Axis

Working with the *x*-axis is no more difficult than working with the *y*-axis. Here is how the *x*-axis was modified.

1. Double-click on the *x*-axis. The Category Axis dialog box opens. It is very similar to the Scale Axis dialog box that you see in Figure A.20.

2. Select the options you want from the Category Axis dialog box.

Figure A.20 The Scale Axis Dialog Box

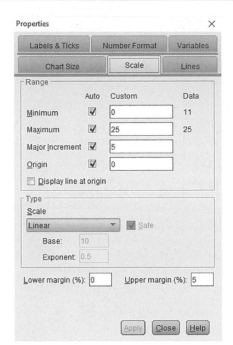

When you are done, close the Chart Editor by double-clicking on the window icon or selecting File → Close.

DESCRIBING DATA

Now you have some idea about how data files are created in SPSS. Let's move on to some examples of simple analysis.

Frequencies and Crosstab Tables

Frequencies simply compute the number of times that a particular value occurs. Crosstabs allow you to compute the number of times that a value occurs when categorized by one or more dimensions, such as gender and age. Both frequencies and crosstabs are often reported first in research reports because they give the reader an overview of what the data look like. To compute frequencies, follow these steps. You should be in the Data Editor window.

1. Click Analyze → Descriptive Statistics → Frequencies. When you do this, you will see the Frequencies dialog box shown in Figure A.21.

2. Double-click the variables for which you want frequencies computed. In this case, they are Test1 and Test2.

3. Click Statistics. You will see the Frequencies: Statistics dialog box shown in Figure A.22.

4. Click the Statistics button.

5. In the Dispersion area, click Std. deviation.

6. Under the Central Tendency area, click Mean.

7. Click Continue.

8. Click OK.

The output consists of a listing of the frequency of each value for Test1 and Test2, plus summary statistics (mean and standard deviation) for each, as you see in Figure A.23.

Figure A.21 The Frequencies Dialog Box

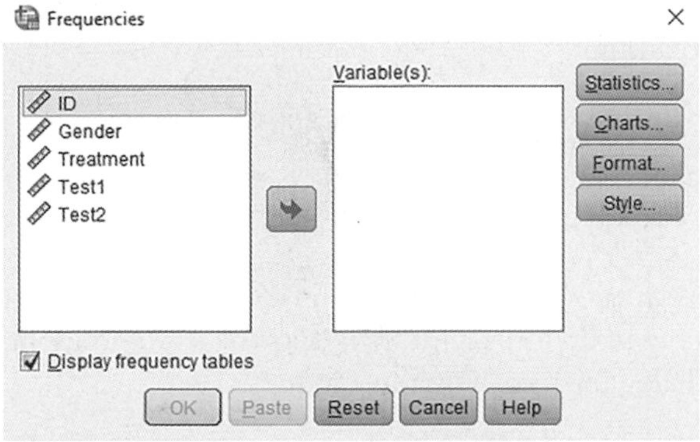

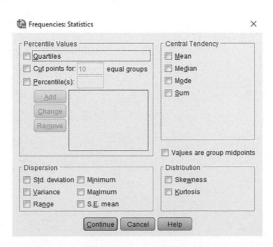

Figure A.22 The Frequencies: Statistics Dialog Box

Figure A.23 Summary Statistics for Test1 and Test2

➡ **Frequencies**

Statistics

		Test1	Test2
N	Valid	25	25
	Missing	0	0
Mean		83.52	64.24
Std. Deviation		8.627	21.642

Applying the Independent-Samples t-Test

Independent-samples *t*-tests are used to analyze data from a number of types of studies, including experimental, quasi-experimental, and field studies such as those shown in the following example, where we test the hypothesis that there are differences between males and females in reading.

How to Conduct an Independent-Samples t-Test

To conduct an independent-samples *t*-test, follow these steps.

1. Click Analyze → Compare Means → Independent-Samples T Test. When you do this, you will see the Independent-Samples T Test dialog box, as shown in Figure A.24.

Figure A.24 The Independent-Samples T Test Frequencies Dialog Box

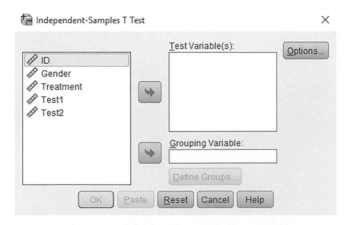

The Independent-Samples T Test Dialog Box

On the left-hand side of the dialog box, you see a listing of all the variables that can be used in the analysis. What you now need to do is define the test and the grouping variable.

2. Click Test1 and drag it to the Test Variable(s) area.

3. Click Gender and drag it to the Grouping Variable area.

4. Click on Gender under the Grouping Variable box.

5. Click Define Groups.

6. In the Group 1 box, type 1.

7. In the Group 2 box, type 2.

8. Click Continue.

9. Click OK.

The output contains the means and standard deviations for each variable, plus the results of the *t*-test, as shown in Figure A.25.

Figure A.25	Results of the Simple *t*-Test

T-Test

Group Statistics

	Gender	N	Mean	Std. Deviation	Std. Error Mean
Test1	Male	11	82.55	9.234	2.784
	Female	14	84.29	8.389	2.242

Independent Samples Test

		Levene's Test for Equality of Variances		t-test for Equality of Means						95% Confidence Interval of the Difference	
		F	Sig.	t	df	Sig. (2-tailed)	Mean Difference	Std. Error Difference		Lower	Upper
Test1	Equal variances assumed	.714	.407	-.493	23	.627	-1.740	3.532		-9.047	5.566
	Equal variances not assumed			-.487	20.532	.632	-1.740	3.575		-9.185	5.704

EXITING SPSS

To exit SPSS, click File → Exit. SPSS will be sure that you get the chance to save any unsaved or edited windows and will then close.

We have just given you the briefest of introductions to SPSS, and certainly none of these skills means anything if you don't know the value or meaningfulness of the data you originally entered. So, don't be impressed by your or others' skills at using programs like SPSS. Be impressed when those other people can tell you what the output means and how it reflects on your original question. And be really impressed if you can do it!

Appendix B

Tables

TABLE B.1: AREAS BENEATH THE NORMAL CURVE

How to use this table:

1. Compute the z score based on the raw score and the mean of the sample.

2. Read to the right of the z score to determine the percentage of area underneath the normal curve or the area between the mean and computed z score.

Table B.1 Areas Beneath the Normal Curve

z-Score	Area Between the Mean and the z-Score	z-Score	Area Between the Mean and the z-Score	z-Score	Area Between the Mean and the z-Score	z-Score	Area Between the Mean and the z-Score	z-Score	Area Between the Mean and the z-Score	z-Score	Area Between the Mean and the z-Score	z-Score	Area Between the Mean and the z-Score	z-Score	Area Between the Mean and the z-Score
0.00	0.00														
0.01	0.40	0.51	19.50	1.01	34.38	1.51	43.45	2.01	47.78	2.51	49.40	3.01	49.87	3.51	49.98
0.02	0.80	0.52	19.85	1.02	34.61	1.52	43.57	2.02	47.83	2.52	49.41	3.02	49.87	3.52	49.98
0.03	1.20	0.53	20.19	1.03	34.85	1.53	43.70	2.03	47.88	2.53	49.43	3.03	49.88	3.53	49.98
0.04	1.60	0.54	20.54	1.04	35.08	1.54	43.82	2.04	47.93	2.54	49.45	3.04	49.88	3.54	49.98
0.05	1.99	0.55	20.88	1.05	35.31	1.55	43.94	2.05	47.98	2.55	49.46	3.05	49.89	3.55	49.98
0.06	2.39	0.56	21.23	1.06	35.54	1.56	44.06	2.06	48.03	2.56	49.48	3.06	49.89	3.56	49.98
0.07	2.79	0.57	21.57	1.07	35.77	1.57	44.18	2.07	48.08	2.57	49.49	3.07	49.89	3.57	49.98
0.08	3.19	0.58	21.90	1.08	35.99	1.58	44.29	2.08	48.12	2.58	49.51	3.08	49.90	3.58	49.98
0.10	3.98	0.60	22.57	1.09	36.21	1.59	44.41	2.09	48.17	2.59	49.52	3.09	49.90	3.59	49.98
0.11	4.38	0.61	22.91	1.10	36.43	1.60	44.52	2.10	48.21	2.60	49.53	3.10	49.90	3.60	49.98
0.12	4.78	0.62	23.24	1.11	36.65	1.61	44.63	2.11	48.26	2.61	49.55	3.11	49.91	3.61	49.98
0.13	5.17	0.63	23.57	1.12	36.86	1.62	44.74	2.12	48.30	2.62	49.56	3.12	49.91	3.62	49.98
0.14	5.57	0.64	23.89	1.13	37.08	1.63	44.84	2.13	48.34	2.63	49.57	3.13	49.91	3.63	49.98
0.15	5.96	0.65	24.22	1.14	37.29	1.64	44.95	2.14	48.38	2.64	49.59	3.14	49.92	3.64	49.98
0.16	6.36	0.66	24.54	1.15	37.49	1.65	45.05	2.15	48.42	2.65	49.60	3.15	49.92	3.65	49.98
0.17	6.75	0.67	24.86	1.16	37.70	1.66	45.15	2.16	48.46	2.66	49.61	3.16	49.92	3.66	49.98
0.18	7.14	0.68	25.17	1.17	37.90	1.67	45.25	2.17	48.50	2.67	49.62	3.17	49.92	3.67	49.98
0.19	7.53	0.69	25.49	1.18	38.10	1.68	45.35	2.18	48.54	2.68	49.63	3.18	49.93	3.68	49.98
0.20	7.93	0.70	25.80	1.19	38.30	1.69	45.45	2.19	48.57	2.69	49.64	3.19	49.93	3.69	49.98
0.21	8.32	0.71	26.11	1.20	38.49	1.70	45.54	2.20	48.61	2.70	49.65	3.20	49.93	3.70	49.98
0.22	8.71	0.72	26.42	1.21	38.69	1.71	45.64	2.21	48.64	2.71	49.66	3.21	49.93	3.71	49.99
0.23	9.10	0.73	26.73	1.23	39.07	1.73	45.82	2.23	48.71	2.73	49.68	3.23	49.94	3.73	49.99
0.24	9.48	0.74	27.04	1.24	39.25	1.74	45.91	2.24	48.75	2.74	49.69	3.24	49.94	3.74	49.99

z-Score	Area Between the Mean and the z-Score	z-Score	Area Between the Mean and the z-Score	z-Score	Area Between the Mean and the z-Score	z-Score	Area Between the Mean and the z-Score	z-Score	Area Between the Mean and the z-Score	z-Score	Area Between the Mean and the z-Score	z-Score	Area Between the Mean and the z-Score	z-Score	Area Between the Mean and the z-Score
0.25	9.97	0.75	27.34	1.25	39.44	1.75	45.99	2.25	48.78	2.75	49.70	3.25	49.94	3.75	49.99
0.26	10.26	0.76	27.64	1.26	39.62	1.76	46.08	2.26	48.81	2.76	49.71	3.26	49.94	3.76	49.99
0.27	10.64	0.77	27.94	1.27	39.80	1.77	46.16	2.27	48.84	2.77	49.72	3.27	49.94	3.77	49.99
0.28	11.03	0.78	28.23	1.28	39.97	1.78	46.25	2.28	48.87	2.78	49.73	3.28	49.94	3.78	49.99
0.29	11.41	0.79	28.52	1.29	40.15	1.79	46.33	2.29	48.90	2.79	49.74	3.29	49.94	3.79	49.99
0.30	11.79	0.80	28.81	1.30	40.32	1.80	46.41	2.30	48.93	2.80	49.74	3.30	49.95	3.80	49.99
0.31	12.17	0.81	29.10	1.31	40.49	1.81	46.49	2.31	48.96	2.81	49.75	3.31	49.95	3.81	49.99
0.32	12.55	0.82	29.39	1.32	40.66	1.82	46.56	2.32	48.98	2.82	49.76	3.32	49.95	3.82	49.99
0.33	12.93	0.83	29.67	1.33	40.82	1.83	46.64	2.33	49.01	2.83	49.77	3.33	49.95	3.83	49.99
0.34	13.31	0.84	29.95	1.34	40.99	1.84	46.71	2.34	49.04	2.84	49.77	3.34	49.95	3.84	49.99
0.35	13.68	0.85	30.23	1.35	41.15	1.85	46.78	2.35	49.06	2.85	49.78	3.35	49.96	3.85	49.99
0.36	14.06	0.86	30.51	1.36	41.31	1.86	46.86	2.36	49.09	2.86	49.79	3.36	49.96	3.86	49.99
0.37	14.43	0.87	30.78	1.37	41.47	1.87	46.93	2.37	49.11	2.87	49.79	3.37	49.96	3.87	49.99
0.38	14.80	0.88	31.06	1.38	41.62	1.88	46.99	2.38	49.13	2.88	49.80	3.38	49.96	3.88	49.99
0.39	15.17	0.89	31.33	1.39	41.77	1.89	47.06	2.39	49.16	2.89	49.81	3.39	49.96	3.89	49.99
0.40	15.54	0.90	31.59	1.40	41.92	1.90	47.13	2.40	49.18	2.90	49.81	3.40	49.97	3.90	49.99
0.41	15.91	0.91	31.86	1.41	42.07	1.91	47.19	2.41	49.20	2.91	49.82	3.41	49.97	3.91	49.99
0.42	16.28	0.92	32.12	1.42	42.22	1.92	47.26	2.42	49.22	2.92	49.82	3.42	49.97	3.92	49.99
0.43	16.64	0.93	32.38	1.43	42.36	1.93	47.32	2.43	49.25	2.93	49.83	3.43	49.97	3.93	49.99
0.44	17.00	0.94	32.64	1.44	42.51	1.94	47.38	2.44	49.27	2.94	49.84	3.44	49.97	3.94	49.99
0.45	17.36	0.95	32.89	1.45	42.65	1.95	47.44	2.45	49.29	2.95	49.84	3.45	49.98	3.95	49.99
0.46	17.72	0.96	33.15	1.46	42.79	1.96	47.50	2.46	49.31	2.96	49.85	3.46	49.98	3.96	49.99
0.47	18.08	0.97	33.40	1.47	42.92	1.97	47.56	2.47	49.32	2.97	49.85	3.47	49.98	3.97	49.99
0.48	18.44	0.98	33.65	1.48	43.06	1.98	47.61	2.48	49.34	2.98	49.86	3.48	49.98	3.98	49.99
0.49	18.79	0.99	33.89	1.49	43.19	1.99	47.67	2.49	49.36	2.99	49.86	3.49	49.98	3.99	49.99

TABLE B.2: *T* VALUES NEEDED FOR REJECTION OF THE NULL HYPOTHESIS

How to use this table:

1. Compute the *t* value test statistic.

2. Compare the obtained *t* value with the critical value listed in this table. Be sure you have calculated the number of degrees of freedom correctly and you have selected an appropriate level of significance.

3. If the obtained value is greater than the critical or tabled value, the null hypothesis (that the means are equal) is not the most attractive explanation for any observed differences.

4. If the obtained value is less than the critical or table value, the null hypothesis is the most attractive explanation for any observed differences.

Table B.2 t Values Needed for Rejection of the Null Hypothesis

df	One-Tailed Test			df	Two-Tailed Test		
	0.10	0.05	0.01		0.10	0.05	0.01
1	3.078	6.314	31.821	1	6.314	12.706	63.657
2	1.886	2.92	6.965	2	2.92	4.303	9.925
3	1.638	2.353	4.541	3	2.353	3.182	5.841
4	1.533	2.132	3.747	4	2.132	2.776	4.604
5	1.476	2.015	3.365	5	2.015	2.571	4.032
6	1.44	1.943	3.143	6	1.943	2.447	3.708
7	1.415	1.895	2.998	7	1.895	2.365	3.5
8	1.397	1.86	2.897	8	1.86	2.306	3.356
9	1.383	1.833	2.822	9	1.833	2.262	3.25
10	1.372	1.813	2.764	10	1.813	2.228	3.17
11	1.364	1.796	2.718	11	1.796	2.201	3.106
12	1.356	1.783	2.681	12	1.783	2.179	3.055
13	1.35	1.771	2.651	13	1.771	2.161	3.013
14	1.345	1.762	2.625	14	1.762	2.145	2.977
15	1.341	1.753	2.603	15	1.753	2.132	2.947
16	1.337	1.746	2.584	16	1.746	2.12	2.921
17	1.334	1.74	2.567	17	1.74	2.11	2.898
18	1.331	1.734	2.553	18	1.734	2.101	2.879
19	1.328	1.729	2.54	19	1.729	2.093	2.861
20	1.326	1.725	2.528	20	1.725	2.086	2.846
21	1.323	1.721	2.518	21	1.721	2.08	2.832
22	1.321	1.717	2.509	22	1.717	2.074	2.819

(Continued)

df	One-Tailed Test			df	Two-Tailed Test		
	0.10	0.05	0.01		0.10	0.05	0.01
23	1.32	1.714	2.5	23	1.714	2.069	2.808
24	1.318	1.711	2.492	24	1.711	2.064	2.797
25	1.317	1.708	2.485	25	1.708	2.06	2.788
26	1.315	1.706	2.479	26	1.706	2.056	2.779
27	1.314	1.704	2.473	27	1.704	2.052	2.771
28	1.313	1.701	2.467	28	1.701	2.049	2.764
29	1.312	1.699	2.462	29	1.699	2.045	2.757
30	1.311	1.698	2.458	30	1.698	2.043	2.75
35	1.306	1.69	2.438	35	1.69	2.03	2.724
40	1.303	1.684	2.424	40	1.684	2.021	2.705
45	1.301	1.68	2.412	45	1.68	2.014	2.69
50	1.299	1.676	2.404	50	1.676	2.009	2.678
55	1.297	1.673	2.396	55	1.673	2.004	2.668
60	1.296	1.671	2.39	60	1.671	2.001	2.661
65	1.295	1.669	2.385	65	1.669	1.997	2.654
70	1.294	1.667	2.381	70	1.667	1.995	2.648
75	1.293	1.666	2.377	75	1.666	1.992	2.643
80	1.292	1.664	2.374	80	1.664	1.99	2.639
85	1.292	1.663	2.371	85	1.663	1.989	2.635
90	1.291	1.662	2.369	90	1.662	1.987	2.632
95	1.291	1.661	2.366	95	1.661	1.986	2.629
100	1.29	1.66	2.364	100	1.66	1.984	2.626
Infinity	1.282	1.645	2.327	Infinity	1.645	1.96	2.576

TABLE B.3: CRITICAL VALUES FOR ANALYSIS OF VARIANCE OR *F*-TEST

How to use this table:

1. Compute the *F* value.

2. Determine the number of degrees of freedom for the numerator $(k-1)$ and the number of degrees of freedom for the denominator $(n-k)$.

3. Locate the critical value by reading across to locate the degrees of freedom in the numerator and down to locate the degrees of freedom in the denominator. The critical value is at the intersection of this column and row.

4. If the obtained value is greater than the critical or tabled value, the null hypothesis (that the means are equal to one another) is not the most attractive explanation for any observed differences.

5. If the obtained value is less than the critical or tabled value, the null hypothesis is the most attractive explanation for any observed differences.

| Table B.3 | Critical Values for Analysis of Variance or *F*-Test |

df for the Denominator	Type I Error Rate	df for the Numerator					
		1	2	3	4	5	6
1	.01	4052.00	4999.00	5403.00	5625.00	5764.00	5859.00
	.05	162.00	200.00	216.00	225.00	230.00	234.00
	.10	39.90	49.50	53.60	55.80	57.20	58.20
2	.01	98.50	99.00	99.17	99.25	99.30	99.33
	.05	18.51	19.00	19.17	19.25	19.30	19.33
	.10	8.53	9.00	9.16	9.24	9.29	9.33
3	.01	34.12	30.82	29.46	28.71	28.24	27.91
	.05	10.13	9.55	9.28	9.12	9.01	8.94
	.10	5.54	5.46	5.39	5.34	5.31	5.28
4	.01	21.20	18.00	16.70	15.98	15.52	15.21
	.05	7.71	6.95	6.59	6.39	6.26	6.16
	.10	.55	4.33	4.19	4.11	4.05	4.01
5	.01	16.26	13.27	12.06	11.39	10.97	10.67
	.05	6.61	5.79	5.41	5.19	5.05	4.95
	.10	4.06	3.78	3.62	3.52	3.45	3.41
6	.01	13.75	10.93	9.78	9.15	8.75	8.47
	.05	5.99	5.14	4.76	4.53	4.39	4.28
	.10	3.78	3.46	3.29	3.18	3.11	3.06
7	.01	12.25	9.55	8.45	7.85	7.46	7.19
	.05	5.59	4.74	4.35	4.12	3.97	3.87
	.10	3.59	3.26	3.08	2.96	2.88	2.83
8	.01	11.26	8.65	7.59	7.01	6.63	6.37
	.05	5.32	4.46	4.07	3.84	3.69	3.58
	.10	3.46	3.11	2.92	2.81	2.73	2.67
9	.01	10.56	8.02	6.99	6.42	6.06	5.80
	.05	5.12	4.26	3.86	3.63	3.48	3.37
	.10	3.36	3.01	2.81	2.69	2.61	2.55
10	.01	10.05	7.56	6.55	6.00	5.64	5.39
	.05	4.97	4.10	3.71	3.48	3.33	3.22
	.10	3.29	2.93	2.73	2.61	2.52	2.46
11	.01	9.65	7.21	6.22	5.67	5.32	5.07
	.05	4.85	3.98	3.59	3.36	3.20	3.10
	.10	3.23	2.86	2.66	2.54	2.45	2.39

_df_ for the Denominator	Type I Error Rate	_df_ for the Numerator					
		1	2	3	4	5	6
12	.01	9.33	6.93	5.95	5.41	5.07	4.82
	.05	4.75	3.89	3.49	3.26	3.11	3.00
	.10	3.18	2.81	2.61	2.48	2.40	2.33
13	.01	9.07	6.70	5.74	5.21	4.86	4.62
	.05	4.67	3.81	3.41	3.18	3.03	2.92
	.10	3.14	2.76	2.56	2.43	2.35	2.28
14	.01	8.86	6.52	5.56	5.04	4.70	4.46
	.05	4.60	3.74	3.34	3.11	2.96	2.85
	.10	3.10	2.73	2.52	2.40	2.31	2.24
15	.01	8.68	6.36	5.42	4.89	4.56	4.32
	.05	4.54	3.68	3.29	3.06	2.90	2.79
	.10	3.07	2.70	2.49	2.36	2.27	2.21
16	.01	8.53	6.23	5.29	4.77	4.44	4.20
	.05	4.49	3.63	3.24	3.01	2.85	2.74
	.10	3.05	2.67	2.46	2.33	2.24	2.18
17	.01	8.40	6.11	5.19	4.67	4.34	4.10
	.05	4.45	3.59	3.20	2.97	2.81	2.70
	.10	3.03	2.65	2.44	2.31	2.22	2.15
18	.01	8.29	6.01	5.09	4.58	4.25	4.02
	.05	4.41	3.56	3.16	2.93	2.77	2.66
	.10	3.01	2.62	2.42	2.29	2.20	2.13
19	.01	8.19	5.93	5.01	4.50	4.17	3.94
	.05	4.38	3.52	3.13	2.90	2.74	2.63
	.10	2.99	2.61	2.40	2.27	2.18	2.11
20	.01	8.10	5.85	4.94	4.43	4.10	3.87
	.05	4.35	3.49	3.10	2.87	2.71	2.60
	.10	2.98	2.59	2.38	2.25	2.16	2.09
21	.01	8.02	5.78	4.88	4.37	4.04	3.81
	.05	4.33	3.47	3.07	2.84	2.69	2.57
	.10	2.96	2.58	2.37	2.23	2.14	2.08
22	.01	7.95	5.72	4.82	4.31	3.99	3.76
	.05	4.30	3.44	3.05	2.82	2.66	2.55
	.10	2.95	2.56	2.35	2.22	2.13	2.06

(Continued)

Table B.3 (Continued)

df for the Denominator	Type I Error Rate	df for the Numerator					
		1	2	3	4	5	6
23	.01	7.88	5.66	4.77	4.26	3.94	3.71
	.05	4.28	3.42	3.03	2.80	2.64	2.53
	.10	2.94	2.55	2.34	2.21	2.12	2.05
24	.01	7.82	5.61	4.72	4.22	3.90	3.67
	.05	4.26	3.40	3.01	2.78	2.62	2.51
	.10	2.93	2.54	2.33	2.20	2.10	2.04
25	.01	7.77	5.57	4.68	4.18	3.86	3.63
	.05	4.24	3.39	2.99	2.76	2.60	2.49
	.10	2.92	2.53	2.32	2.19	2.09	2.03
26	.01	7.72	5.53	4.64	4.14	3.82	3.59
	.05	4.23	3.37	2.98	2.74	2.59	2.48
	.10	2.91	2.52	2.31	2.18	2.08	2.01
27	.01	7.68	5.49	4.60	4.11	3.79	3.56
	.05	4.21	3.36	2.96	2.73	2.57	2.46
	.10	2.90	2.51	2.30	2.17	2.07	2.01
28	.01	7.64	5.45	4.57	4.08	3.75	3.53
	.05	4.20	3.34	2.95	2.72	2.56	2.45
	.10	2.89	2.50	2.29	2.16	2.07	2.00
29	.01	7.60	5.42	4.54	4.05	3.73	3.50
	.05	4.18	3.33	2.94	2.70	2.55	2.43
	.10	2.89	2.50	2.28	2.15	2.06	1.99
30	.01	7.56	5.39	4.51	4.02	3.70	3.47
	.05	4.17	3.32	2.92	2.69	2.53	2.42
	.10	2.88	2.49	2.28	2.14	2.05	1.98
35	.01	7.42	5.27	4.40	3.91	3.59	3.37
	.05	4.12	3.27	2.88	2.64	2.49	2.37
	.10	2.86	2.46	2.25	2.14	2.02	1.95
40	.01	7.32	5.18	4.31	3.91	3.51	3.29
	.05	4.09	3.23	2.84	2.64	2.45	2.34
	.10	2.84	2.44	2.23	2.11	2.00	1.93
45	.01	7.23	5.11	4.25	3.83	3.46	3.23
	.05	4.06	3.21	2.81	2.61	2.42	2.31
	.10	2.82	2.43	2.21	2.09	1.98	1.91
50	.01	7.17	5.06	4.20	3.77	3.41	3.19

df for the Denominator	Type I Error Rate	*df* for the Numerator					
		1	2	3	4	5	6
	.05	4.04	3.18	2.79	2.58	2.40	2.29
	.10	2.81	2.41	2.20	2.08	1.97	1.90
55	.01	7.12	5.01	4.16	3.72	3.37	3.15
	.05	4.02	3.17	2.77	2.56	2.38	2.27
	.10	2.80	2.40	2.19	2.06	1.96	1.89
60	.01	7.08	4.98	4.13	3.68	3.34	3.12
	.05	4.00	3.15	2.76	2.54	2.37	2.26
	.10	2.79	2.39	2.18	2.05	1.95	1.88
65	.01	7.04	4.95	4.10	3.65	3.31	3.09
	.05	3.99	3.14	2.75	2.53	2.36	2.24
	.10	2.79	2.39	2.17	2.04	1.94	1.87
70	.01	7.01	4.92	4.08	3.62	3.29	3.07
	.05	3.98	3.13	2.74	2.51	2.35	2.23
	.10	2.78	2.38	2.16	2.03	1.93	1.86
75	.01	6.99	4.90	4.06	3.60	3.27	3.05
	.05	3.97	3.12	2.73	2.50	2.34	2.22
	.10	2.77	2.38	2.16	2.03	1.93	1.86
80	.01	3.96	4.88	4.04	3.56	3.26	3.04
	.05	6.96	3.11	2.72	2.49	2.33	2.22
	.10	2.77	2.37	2.15	2.02	1.92	1.85
85	.01	6.94	4.86	4.02	3.55	3.24	3.02
	.05	3.95	3.10	2.71	2.48	2.32	2.21
	.10	2.77	2.37	2.15	2.01	1.92	1.85
90	.01	6.93	4.85	4.02	3.54	3.23	3.01
	.05	3.95	3.10	2.71	2.47	2.32	2.20
	.10	2.76	2.36	2.15	2.01	1.91	1.84
95	.01	6.91	4.84	4.00	3.52	3.22	3.00
	.05	3.94	3.09	2.70	2.47	2.31	2.20
	.10	2.76	2.36	2.14	2.01	1.91	1.84
100	.01	6.90	4.82	3.98	3.51	3.21	2.99
	.05	3.94	3.09	2.70	2.46	2.31	2.19
	.10	2.76	2.36	2.14	2.00	1.91	1.83
Infinity	.01	6.64	4.61	3.78	3.32	3.02	2.80
	.05	3.84	3.00	2.61	2.37	2.22	2.10
	.10	2.71	2.30	2.08	1.95	1.85	1.78

TABLE B.4: VALUES OF THE CORRELATION COEFFICIENT NEEDED FOR REJECTION OF THE NULL HYPOTHESIS

How to use this table:

1. Compute the value of the correlation coefficient.

2. Compare the value of the correlation coefficient with the critical value listed in this table.

3. If the obtained value is greater than the critical or tabled value, the null hypothesis (that the correlation coefficient is equal to zero) is not the most attractive explanation for any observed differences.

4. If the obtained value is less than the critical or tabled value, the null hypothesis is the most attractive explanation for any observed differences.

Table B.4	Values of the Correlation Coefficient Needed for Rejection of the Null Hypothesis				

One-Tailed Test			Two-Tailed Test		
df	.05	.01	*df*	.05	.01
1	.9877	.9995	1	.9969	.9999
2	.9000	.9800	2	.9500	.9900
3	.8054	.9343	3	.8783	.9587
4	.7293	.8822	4	.8114	.9172
5	.6694	.832	5	.7545	.8745
6	.6215	.7887	6	.7067	.8343
7	.5822	.7498	7	.6664	.7977
8	.5494	.7155	8	.6319	.7646
9	.5214	.6851	9	.6021	.7348
10	.4973	.6581	10	.5760	.7079
11	.4762	.6339	11	.5529	.6835
12	.4575	.6120	12	.5324	.6614
13	.4409	.5923	13	.5139	.6411
14	.4259	.5742	14	.4973	.6226
15	.412	.5577	15	.4821	.6055
16	.4000	.5425	16	.4683	.5897

One-Tailed Test			Two-Tailed Test		
df	.05	.01	*df*	.05	.01
17	.3887	.5285	17	.4555	.5751
18	.3783	.5155	18	.4438	.5614
19	.3687	.5034	19	.4329	.5487
20	.3598	.4921	20	.4227	.5368
25	.3233	.4451	25	.3809	.4869
30	.2960	.4093	30	.3494	.4487
35	.2746	.3810	35	.3246	.4182
40	.2573	.3578	40	.3044	.3932
45	.2428	.3384	45	.2875	.3721
50	.2306	.3218	50	.2732	.3541
60	.2108	.2948	60	.2500	.3248
70	.1954	.2737	70	.2319	.3017
80	.1829	.2565	80	.2172	.2830
90	.1726	.2422	90	.2050	.2673
100	.1638	.2301	100	.1946	.2540

TABLE B.5: CRITICAL VALUES FOR THE CHI-SQUARE TEST

How to use this table:

1. Compute the χ^2 value.

2. Determine the number of degrees of freedom for the rows $(R - 1)$ and the number of degrees of freedom for the columns $(C - 1)$. If your table is one-dimensional, then you have only columns.

3. Locate the critical value by locating the degrees of freedom in the titled (*df*) column. Then read across to the appropriate column for level of significance.

4. If the obtained value is greater than the critical or tabled value, the null hypothesis (that the frequencies are equal to one another) is not the most attractive explanation for any observed differences.

5. If the obtained value is less than the critical or tabled value, the null hypothesis is the most attractive explanation for any observed differences.

| Table B.5 | Critical Values for the Chi-Square Test | | |

	Level of Significance		
df	.10	.05	.01
1	2.71	3.84	6.64
2	4.00	5.99	9.21
3	6.25	7.82	11.34
4	7.78	9.49	13.28
5	9.24	11.07	15.09
6	10.64	12.59	16.81
7	12.02	14.07	18.48
8	13.36	15.51	20.09
9	14.68	16.92	21.67
10	16.99	18.31	23.21
11	17.28	19.68	24.72
12	18.65	21.03	26.22
13	19.81	22.36	27.69
14	21.06	23.68	29.14
15	22.31	25.00	30.58
16	23.54	26.30	32.00
17	24.77	27.60	33.41
18	25.99	28.87	34.80
19	27.20	30.14	36.19
20	28.41	31.41	37.57
21	29.62	32.67	38.93
22	30.81	33.92	40.29
23	32.01	35.17	41.64
24	33.20	36.42	42.98
25	34.38	37.65	44.81
26	35.56	38.88	45.64
27	36.74	40.11	46.96
28	37.92	41.34	48.28
29	39.09	42.56	49.59
30	40.26	43.77	50.89

Appendix C

Data Sets

These data files are referred to throughout *Statistics for People Who (Think They) Hate Statistics*. They are available here to be entered manually, or you can download them from one of two places.

- The website hosted by SAGE at **edge.sagepub.com/salkind6e**.
- The author's website at http://onlinefilefolder.com. The username is *ancillaries*, and the password is *files*.

Note that values (such as 1 and 2) are included, while labels (such as male and female) for those values are not. For example, for Chapter 11 Data Set 2, gender is represented by 1 (male) or 2 (female). If you use SPSS, you can use the labels feature to assign labels to these values.

Chapter 2 Data Set 1

Prejudice	Prejudice	Prejudice	Prejudice
87	87	76	81
99	77	55	82
87	89	64	99
87	99	81	93
67	96	94	94

Chapter 2 Data Set 2

Score 1	Score 2	Score 3
3	34	154
7	54	167
5	17	132
4	26	145
5	34	154
6	25	145
7	14	113
8	24	156
6	25	154
5	23	123

Chapter 2 Data Set 3

Number of Beds	Infection Rate	Number of Beds	Infection Rate
234	1.7	342	5.3
214	2.4	276	5.6
165	3.1	187	1.2
436	5.6	512	3.3
432	4.9	553	4.1

Chapter 2 Data Set 4

Group: 1 = little experience, 2 = moderate amount of experience, 3 = lots of experience

Group	Attitude	Group	Attitude
1	4	2	8
1	5	2	1
2	6	1	6
2	6	1	5
1	5	2	4
1	7	2	3
2	6	3	4
2	5	2	6
3	8	1	7
3	9	1	8

Chapter 3 Data Set 1

Reaction Time	Reaction Time	Reaction Time	Reaction Time	Reaction Time
0.4	0.3	1.1	0.5	0.5
0.7	1.9	1.3	2.6	0.7
0.4	1.2	0.2	0.5	1.1
0.9	2.8	0.6	2.1	0.9
0.8	0.8	0.8	2.3	0.6
0.7	0.9	0.7	0.2	0.2

Chapter 3 Data Set 2

Math Score	Reading Score	Math Score	Reading Score
78	24	72	77
67	35	98	89
89	54	88	76
97	56	74	56
67	78	58	78
56	87	98	99
67	65	97	83
77	69	86	69
75	98	89	89
68	78	69	73
78	85	79	60
98	69	87	96
92	93	89	59
82	100	99	89
78	98	87	87

Chapter 3 Data Set 3

Height	Weight	Height	Weight
53	156	57	154
46	131	68	166
54	123	65	153
44	142	66	140
56	156	54	143
76	171	66	156
87	143	51	173
65	135	58	143
45	138	49	161
44	114	48	131

Chapter 3 Data Set 4

Accuracy
12
15
11
5
3
8
19
16
23
19

Chapter 4 Data Set 1

Comprehension Score	Comprehension Score	Comprehension Score	Comprehension Score
12	36	49	54
15	34	45	56
11	33	45	57
16	38	47	59
21	42	43	54
25	44	31	56
21	47	12	43
8	54	14	44
6	55	15	41
2	51	16	42
22	56	22	7
26	53	29	
27	57	29	

Chapter 4 Data Set 2

Monday	Tuesday	Wednesday	Thursday	Friday
12	17	10	15	20
9	11	10	4	0
6	8	9	5	10
4	0	5	4	9
9	7	8	5	11
10	5	4	4	15
13	12	7	3	10
22	16	18	15	20
1	3	6	4	2
5	8	4	6	7
7	0	3	8	2
10	4	1	8	12
4	5	8	6	9
15	12	10	9	11
3	6	4	7	10

Chapter 4 Data Set 3

Pie preferences

Pie	
	Chocolate Cream
Cherry	Apple
Apple	Chocolate Cream
Chocolate Cream	Chocolate Cream
Cherry	Chocolate Cream
Chocolate Cream	Chocolate Cream
Chocolate Cream	Apple
Apple	Chocolate Cream

Chapter 5 Data Set 1

Income	Education	Income	Education
$36,577	11	$64,543	12
$54,365	12	$43,433	14
$33,542	10	$34,644	12
$65,654	12	$33,213	10
$45,765	11	$55,654	15
$24,354	7	$76,545	14
$43,233	12	$21,324	11
$44,321	13	$17,645	12
$23,216	9	$23,432	11
$43,454	12	$44,543	15

Chapter 5 Data Set 2

Number Correct	Attitude	Number Correct	Attitude
17	94	14	85
13	73	16	66
12	59	16	79
15	80	18	77
16	93	19	91

Chapter 5 Data Set 3

Speed	Strength	Speed	Strength
21.6	135	19.5	134
23.4	213	20.9	209
26.5	243	18.7	176
25.5	167	29.8	156
20.8	120	28.7	177

Chapter 5 Data Set 4

Ach Inc	Budget Inc
0.07	0.11
0.03	0.14
0.05	0.13
0.07	0.26
0.02	0.08
0.01	0.03
0.05	0.06
0.04	0.12
0.04	0.11

Chapter 5 Data Set 5

Exercise	GPA
25	3.6
30	4.0
20	3.8
60	3.0
45	3.7
90	3.9
60	3.5
0	2.8
15	3.0
10	2.5

Chapter 5 Data Set 6

Age	Level	Score	Age	Level	Score
25	1	78	24	5	84
16	2	66	25	5	87
8	2	78	36	4	69
23	3	89	45	4	87
31	4	87	16	4	88
19	4	90	23	1	92
15	4	98	31	2	97
31	5	76	53	2	69
21	1	56	11	3	79
26	1	72	33	2	69

Chapter 6 Data Set 1

Fall Results	Spring Results	Fall Results	Spring Results
21	7	3	30
38	13	16	26
15	35	34	43
34	45	50	20
5	19	14	22
32	47	14	25
24	34	3	50
3	1	4	17
17	12	42	32
32	41	28	46
33	3	40	10
15	20	40	48
21	39	12	11
8	46	5	23

Chapter 6 Data Set 2

ID	Form 1	Form 2	ID	Form 1	Form 2
1	89	78	51	73	93
2	98	75	52	91	87
3	83	70	53	81	78
4	78	97	54	97	84
5	70	91	55	97	85
6	86	82	56	91	79
7	83	97	57	71	99
8	73	88	58	82	97
9	86	81	59	95	97
10	83	80	60	70	76
11	83	95	61	70	88
12	94	75	62	96	96
13	90	96	63	70	77
14	81	87	64	71	70
15	82	93	65	87	89
16	98	82	66	97	71
17	99	84	67	81	75
18	83	78	68	89	75
19	72	77	69	71	73
20	86	94	70	71	82
21	80	85	71	75	81
22	80	86	72	72	97
23	93	92	73	88	78
24	100	98	74	86	77
25	84	98	75	70	92
26	89	99	76	79	88
27	87	83	77	96	81
28	82	95	78	82	88
29	95	90	79	97	74

ID	Form 1	Form 2	ID	Form 1	Form 2
30	99	92	80	93	72
31	82	78	81	70	82
32	94	89	82	76	84
33	97	100	83	74	88
34	71	81	84	81	81
35	91	96	85	88	86
36	83	85	86	70	90
37	95	75	87	91	73
38	72	88	88	96	94
39	98	74	89	81	99
40	89	88	90	95	86
41	83	80	91	72	100
42	100	81	92	93	90
43	72	100	93	76	78
44	97	82	94	91	90
45	71	81	95	100	78
46	74	93	96	76	92
47	79	82	97	78	87
48	91	70	98	74	88
49	81	90	99	80	92
50	87	85	100	93	96

Chapter 11 Data Set 1

Group: 1 = treatment, 2 = no treatment

Group	Memory Test	Group	Memory Test	Group	Memory Test
1	7	1	5	2	3
1	3	1	7	2	2
1	3	1	1	2	5
1	2	1	9	2	4
1	3	1	2	2	4
1	8	1	5	2	6
1	8	1	2	2	7
1	5	1	12	2	7
1	8	1	15	2	5
1	5	1	4	2	6
1	5	2	5	2	4
1	4	2	4	2	3
1	6	2	4	2	2
1	10	2	5	2	7
1	10	2	5	2	6
1	5	2	7	2	2
1	1	2	8	2	8
1	1	2	8	2	9
1	4	2	9	2	7
1	3	2	8	2	6
1	1	2	8	2	8
1	1	2	8	2	9
1	4	2	9	2	7
1	3	2	8	2	6

Chapter 11 Data Set 2

Gender: 1 = male, 2 = female

Gender	Hand Up	Gender	Hand Up
1	9	1	8
2	3	2	7

Gender	Hand Up	Gender	Hand Up
2	5	1	9
2	1	2	9
1	8	1	8
1	4	2	7
2	2	2	3
2	6	2	7
1	9	2	6
1	3	1	10
2	4	1	7
1	8	1	6
2	3	1	12
1	10	2	8
2	6	2	8

Chapter 11 Data Set 3

Group: 1 = urban, 2 = rural

Group	Attitude	Group	Attitude
1	6.50	1	4.23
2	7.90	1	6.95
2	4.30	2	6.74
2	6.80	1	5.96
1	9.90	2	5.25
1	6.80	2	2.36
1	4.80	1	9.25
2	6.50	1	6.36
2	3.30	1	8.99
1	4.00	1	5.58
2	13.17	2	4.25
1	5.26	1	6.60
2	9.25	2	1.00
1	8.00	1	5.00
2	1.25	2	3.50

Chapter 11 Data Set 4

Group: 1 = first group of fourth graders, 2 = second group of fourth graders

Group	Score	Group	Score
1	11	2	14
1	11	2	7
1	10	2	8
1	7	2	10
1	2	2	15
1	6	2	9
1	12	2	19
1	5	2	9
1	7	2	17
1	11	2	18
1	9	2	19
1	7	2	8
1	3	2	7
1	4	2	9
1	10	2	14

Chapter 11 Data Set 5

Group: 1 = treatment group 1, 2 = treatment group 2
    Used to compute the t score for the difference between two groups on the test variable

Group	Score
1	5
2	6
1	5
2	4
1	5
2	8
1	7
2	6
1	8
2	7

Chapter 11 Data Set 6

Used to compute the *t* score for the difference between two groups (Group 1 and Group 2) on the test variable

Group	Score
1	5
2	6
1	5
2	4
1	5
2	8
1	7
2	6
1	8
2	7
1	5
2	6
1	5
2	4
1	5
2	8
1	7
2	6
1	8
2	7

Chapter 12 Data Set 1

Pretest	Posttest	Pretest	Posttest	Pretest	Posttest
3	7	6	8	9	4
5	8	7	8	8	4
4	6	8	7	7	5
6	7	7	9	7	6
5	8	6	10	6	9
5	9	7	9	7	8
4	6	8	9	8	12
5	6	8	8		
3	7	9	8		

Chapter 12 Data Set 2

Tons of paper used before and after implementation of a recycling program

Before	After	Before	After
20	23	23	22
6	8	33	35
12	11	44	41
34	35	65	56
55	57	43	34
43	76	53	51
54	54	22	21
24	26	34	31
33	35	32	33
21	26	44	38
34	29	17	15
33	31	28	27
54	56		

Chapter 12 Data Set 3

Families' satisfaction levels with service centers before and after a social service intervention

Before	After	Before	After
1.30	6.50	9.00	8.40
2.50	8.70	7.60	6.40
2.30	9.80	4.50	7.20
8.10	10.20	1.10	5.80
5.00	7.90	5.60	6.90
7.00	6.50	6.20	5.90
7.50	8.70	7.00	7.60
5.20	7.90	6.90	7.80
4.40	8.70	5.60	7.30
7.60	9.10	5.20	4.60

Chapter 12 Data Set 4

Workers' shifts and feelings of stress

First Shift	Second Shift	First Shift	Second Shift
4	8	3	7
6	5	6	6
9	6	6	7
3	4	9	6
6	7	6	4
7	7	5	4
5	6	4	4
9	8	4	8
5	8	3	9
4	9	3	0

Chapter 12 Data Set 5

Bone density in adults who participated in a weight-lifting class

Fall_Score	Spring_Score
2	7
7	6
6	9
5	8
8	7
7	6
8	7
9	8
8	9
9	9
4	9

(Continued)

(Continued)

Fall_Score	Spring_Score
6	7
5	8
2	7
5	6
4	4
3	6
6	7
7	7
6	6
5	9
4	0
3	9
5	8
4	7

Chapter 13 Data Set 1

Group: 1 = 5 hours of preschool participation per week, 2 = 10 hours of preschool participation per week, 3 = 20 hours of preschool participation per week

Group	Language Score	Group	Language Score
1	87	2	81
1	86	2	82
1	76	2	78
1	56	2	85
1	78	2	91
1	98	3	89
1	77	3	91
1	66	3	96

Group	Language Score	Group	Language Score
1	75	3	87
1	67	3	89
2	87	3	90
2	85	3	89
2	99	3	96
2	85	3	96
2	79	3	93

Chapter 13 Data Set 2

Practice: 1 = less than 15 hours, 2 = 15–25 hours, 3 = 25 or more hours

Practice	Time	Practice	Time
1	58.7	2	54.6
1	55.3	2	51.5
1	61.8	2	54.7
1	49.5	2	61.4
1	64.5	2	56.9
1	61.0	3	68.0
1	65.7	3	65.9
1	51.4	3	54.7
1	53.6	3	53.6
1	59.0	3	58.7
2	64.4	3	58.7
2	55.8	3	65.7
2	58.7	3	66.5
2	54.7	3	56.7
2	52.7	3	55.4
2	67.8	3	51.5
2	61.6	3	54.8
2	58.7	3	57.2

Chapter 13 Data Set 3

Shift	Stress	Shift	Stress
4 PM to Midnight	7	Midnight to 8 AM	5
4 PM to Midnight	7	Midnight to 8 AM	5
4 PM to Midnight	6	Midnight to 8 AM	6
4 PM to Midnight	9	Midnight to 8 AM	7
4 PM to Midnight	8	Midnight to 8 AM	6
4 PM to Midnight	7	8 AM to 4 PM	1
4 PM to Midnight	6	8 AM to 4 PM	3
4 PM to Midnight	7	8 AM to 4 PM	4
4 PM to Midnight	8	8 AM to 4 PM	3
4 PM to Midnight	9	8 AM to 4 PM	1
Midnight to 8 AM	5	8 AM to 4 PM	1
Midnight to 8 AM	6	8 AM to 4 PM	2
Midnight to 8 AM	3	8 AM to 4 PM	6
Midnight to 8 AM	5	8 AM to 4 PM	5
Midnight to 8 AM	4	8 AM to 4 PM	4
Midnight to 8 AM	6	8 AM to 4 PM	3
Midnight to 8 AM	5	8 AM to 4 PM	4
Midnight to 8 AM	4	8 AM to 4 PM	5

Chapter 13 Data Set 4

Consumers' preferences for noodle thickness

Thickness	Pleasantness	Thickness	Pleasantness
Thick	1	Medium	3
Thick	3	Medium	4
Thick	2	Medium	5
Thick	3	Medium	4

Thickness	Pleasantness	Thickness	Pleasantness
Thick	4	Medium	3
Thick	4	Medium	4
Thick	5	Medium	3
Thick	4	Medium	3
Thick	3	Medium	2
Thick	4	Medium	3
Thick	3	Thin	1
Thick	2	Thin	3
Thick	2	Thin	2
Thick	3	Thin	1
Thick	3	Thin	1
Thick	4	Thin	1
Thick	5	Thin	2
Thick	4	Thin	2
Thick	3	Thin	3
Thick	2	Thin	2
Medium	3	Thin	1
Medium	4	Thin	1
Medium	3	Thin	2
Medium	2	Thin	2
Medium	3	Thin	3
Medium	4	Thin	3
Medium	4	Thin	2
Medium	4	Thin	1
Medium	3	Thin	2
Medium	3	Thin	1

Chapter 14 Data Set 1

Treatment: 1 = high impact, 2 = low impact; Gender: 1 = male, 2 = female

Treatment	Gender	Loss	Treatment	Gender	Loss
1	1	76	2	1	88
1	1	78	2	1	76
1	1	76	2	1	76
1	1	76	2	1	76
1	1	76	2	1	56
1	1	74	2	1	76
1	1	74	2	1	76
1	1	76	2	1	98
1	1	76	2	1	88
1	1	55	2	1	78
1	2	65	2	2	65
1	2	90	2	2	67
1	2	65	2	2	67
1	2	90	2	2	87
1	2	65	2	2	78
1	2	90	2	2	56
1	2	90	2	2	54
1	2	79	2	2	56
1	2	70	2	2	54
1	2	90	2	2	56

Chapter 14 Data Set 2

Severity: 1 = low severity, 2 = high severity

Severity	Treatment	Pain Score	Severity	Treatment	Pain Score
1	Drug #1	6	2	Drug #2	7
1	Drug #1	6	2	Drug #2	5
1	Drug #1	7	2	Drug #2	4
1	Drug #1	7	2	Drug #2	3
1	Drug #1	7	2	Drug #2	4
1	Drug #1	6	2	Drug #2	5
1	Drug #1	5	2	Drug #2	4
1	Drug #1	6	2	Drug #2	4
1	Drug #1	7	2	Drug #2	3
1	Drug #1	8	2	Drug #2	3
1	Drug #1	7	2	Drug #2	4
1	Drug #1	6	2	Drug #2	5
1	Drug #1	5	2	Drug #2	6
1	Drug #1	6	2	Drug #2	7
1	Drug #1	7	2	Drug #2	7
1	Drug #1	8	2	Drug #2	6
1	Drug #1	9	2	Drug #2	5
1	Drug #1	8	2	Drug #2	4
1	Drug #1	7	2	Drug #2	4
1	Drug #1	7	2	Drug #2	5
2	Drug #1	7	1	Placebo	2
2	Drug #1	8	1	Placebo	1
2	Drug #1	8	1	Placebo	3
2	Drug #1	9	1	Placebo	4
2	Drug #1	8	1	Placebo	5
2	Drug #1	7	1	Placebo	4
2	Drug #1	6	1	Placebo	3
2	Drug #1	6	1	Placebo	3
2	Drug #1	6	1	Placebo	3

(Continued)

(Continued)

Severity	Treatment	Pain Score	Severity	Treatment	Pain Score
2	Drug #1	7	1	Placebo	4
2	Drug #1	7	1	Placebo	5
2	Drug #1	6	1	Placebo	3
2	Drug #1	7	1	Placebo	1
2	Drug #1	8	1	Placebo	2
2	Drug #1	8	1	Placebo	4
2	Drug #1	8	1	Placebo	3
2	Drug #1	9	1	Placebo	5
2	Drug #1	0	1	Placebo	4
2	Drug #1	9	1	Placebo	2
2	Drug #1	8	1	Placebo	3
1	Drug #2	6	2	Placebo	4
1	Drug #2	5	2	Placebo	5
1	Drug #2	4	2	Placebo	6
1	Drug #2	5	2	Placebo	5
1	Drug #2	4	2	Placebo	4
1	Drug #2	3	2	Placebo	4
1	Drug #2	3	2	Placebo	6
1	Drug #2	3	2	Placebo	5
1	Drug #2	4	2	Placebo	4
1	Drug #2	5	2	Placebo	2
1	Drug #2	5	2	Placebo	1
1	Drug #2	5	2	Placebo	3
1	Drug #2	6	2	Placebo	2
1	Drug #2	6	2	Placebo	2
1	Drug #2	7	2	Placebo	3
1	Drug #2	6	2	Placebo	4
1	Drug #2	5	2	Placebo	3
1	Drug #2	7	2	Placebo	2
1	Drug #2	6	2	Placebo	2
1	Drug #2	8	2	Placebo	1

Gender: 1 = male, 2 = female

Gender	Caff_Consumption	Stress
1	5	1
1	6	3
2	7	3
1	7	2
1	5	3
1	6	3
1	8	2
1	8	2
2	9	1
2	8	1
2	9	1
2	7	2
2	4	1
2	3	1
1	0	1
2	4	2
1	5	1
2	6	2
1	2	2
1	4	3
1	5	3
2	5	3
1	4	2
1	3	2
1	7	3
2	8	2
1	9	2
1	11	1
1	2	2
1	3	1

Chapter 14 Data Set 4

Gender	Training	Skill
Male	Strength Emphasis	10
Male	Speed Emphasis	5
Male	Strength Emphasis	9
Male	Speed Emphasis	3
Male	Strength Emphasis	9
Male	Speed Emphasis	2
Male	Speed Emphasis	5
Male	Strength Emphasis	0
Male	Strength Emphasis	0
Male	Speed Emphasis	4
Male	Speed Emphasis	3
Male	Speed Emphasis	2
Male	Speed Emphasis	2
Male	Speed Emphasis	3
Male	Strength Emphasis	10
Male	Speed Emphasis	4
Male	Strength Emphasis	3
Male	Speed Emphasis	3
Male	Strength Emphasis	0
Female	Speed Emphasis	9
Female	Strength Emphasis	3
Female	Strength Emphasis	2
Female	Strength Emphasis	3
Female	Speed Emphasis	9
Female	Strength Emphasis	1
Female	Strength Emphasis	1

Gender	Training	Skill
Female	Speed Emphasis	9
Female	Speed Emphasis	8
Female	Speed Emphasis	7
Female	Speed Emphasis	9

Chapter 15 Data Set 1

Quality of Marriage: 1 = low, 2 = moderate, 3 = high

Quality of Marriage	Quality Parent–Child	Quality of Marriage	Quality Parent–Child
1	58.7	2	54.6
1	55.3	2	51.5
1	61.8	2	54.7
1	49.5	2	61.4
1	64.5	2	56.9
1	61.0	3	68.0
1	65.7	3	65.9
1	51.4	3	54.7
1	53.6	3	53.6
1	59.0	3	58.7
2	64.4	3	58.7
2	55.8	3	65.7
2	58.7	3	66.5
2	54.7	3	56.7
2	52.7	3	55.4
2	67.8	3	51.5
2	61.6	3	54.8
2	58.7	3	57.2

Chapter 15 Data Set 2

Motivation	GPA	Motivation	GPA
1	3.4	6	2.6
6	3.4	7	2.5
2	2.5	7	2.8
7	3.1	2	1.8
5	2.8	9	3.7
4	2.6	8	3.1
3	2.1	8	2.5
1	1.6	7	2.4
8	3.1	6	2.1
6	2.6	9	4.0
5	3.2	7	3.9
6	3.1	8	3.1
5	3.2	7	3.3
5	2.7	8	3.0
6	2.8	9	2.0

Chapter 15 Data Set 3

Level of Education: 1 = low, 2 = medium, 3 = high

Income	Level of Education	Income	Level of Education
$45,675	1	$74,776	3
$34,214	2	$89,689	3
$67,765	3	$96,768	2
$67,654	3	$97,356	3
$56,543	2	$38,564	2
$67,865	1	$67,375	3
$78,656	3	$78,854	3
$45,786	2	$78,854	3
$87,598	3	$42,757	1
$88,656	3	$78,854	3

Chapter 15 Data Set 4

Hours of Study	Grade
0	80
5	93
8	97
6	100
5	75
3	83
4	98
8	100
6	90
2	78

Chapter 15 Data Set 5

Age	Shoe Size	Intelligence	Level of Education
15	Small	110	7
22	Medium	109	12
56	Large	98	15
7	Small	105	4
25	Medium	110	15
57	Large	125	8
12	Small	110	11
45	Medium	98	15
76	Large	97	12
14	Small	107	10
34	Medium	125	12
56	Large	106	12
9	Small	110	5
44	Medium	123	12
56	Large	109	18

Chapter 16 Data Set 1

Training	Injuries	Training	Injuries
12	8	11	5
3	7	16	7
22	2	14	8
12	5	15	3
11	4	16	7
31	1	22	3
27	5	24	8
31	1	26	8
8	2	31	2
16	2	12	2
14	7	24	3
26	2	33	3
36	2	21	5
26	2	12	7
15	6	36	3

Chapter 16 Data Set 2

Time	Correct	Time	Correct
14.5	5	13.9	3
13.4	7	17.3	12
12.7	6	12.5	5
16.4	2	16.7	4
21	4	22.7	3

Chapter 16 Data Set 3

Number of Homes Sold	Years in Business	Level of Education
8	10	11
6	7	12
12	15	11
3	3	12
17	18	11

Number of Homes Sold	Years in Business	Level of Education
4	6	12
13	5	11
16	16	12
4	3	11
8	7	12
6	6	11
3	6	12
14	13	11
15	15	12
4	6	11
6	3	12
4	6	11
11	12	12
12	14	11
15	21	12

Chapter 17 Data Set 1

Voucher	Voucher	Voucher	Voucher	Voucher
1	1	2	3	3
1	1	2	3	3
1	1	2	3	3
1	1	2	3	3
1	1	3	3	3
1	2	3	3	3
1	2	3	3	3
1	2	3	3	3
1	2	3	3	3
1	2	3	3	3
1	2	3	3	3
1	2	3	3	3
1	2	3	3	3
1	2	3	3	3
1	2	3	3	3
1	2	3	3	3
1	2	3	3	3
1	2	3	3	3

Chapter 17 Data Set 2

Gender: 1 = male, 2 = female; Vote: 1 = yes, 2 = no

Gender	Vote	Gender	Vote
1	1	2	1
1	1	2	1
1	1	2	1
1	1	2	1
1	1	2	1
1	1	2	1
1	1	2	1
1	1	2	1
1	1	2	1
1	1	1	2
1	1	1	2
1	1	1	2
1	1	1	2
1	1	1	2
1	1	1	2
1	1	1	2
1	1	1	2
1	1	1	2
1	1	1	2
1	1	1	2
1	1	1	2
1	1	1	2
1	1	1	2
1	1	1	2
1	1	1	2
1	1	1	2
1	1	1	2

Gender	Vote	Gender	Vote
1	1	2	2
1	1	2	2
1	1	2	2
1	1	2	2
1	1	2	2
1	1	2	2
1	1	2	2
1	1	2	2
1	1	2	2
2	1	2	2
2	1	2	2
2	1	2	2
2	1	2	2
2	1	2	2
2	1	2	2
2	1	2	2
2	1	2	2
2	1	2	2
2	1	2	2
2	1	2	2
2	1	2	2
2	1	2	2
2	1	2	2
2	1	2	2
2	1	2	2
2	1	2	2
2	1	2	2
2	1	2	2
2	1	2	2
2	1	2	2
2	1	2	2
2	1		

Chapter 17 Data Set 3

Gender: 1 = male, 2 = female

Gender	Gender	Gender	Gender	Gender
1	1	1	2	2
1	1	1	2	2
1	1	1	2	2
1	1	1	2	2
1	1	1	2	2
1	1	2	2	2
1	1	2	2	2
1	1	2	2	2
1	1	2	2	2
1	1	2	2	2
1	1	2	2	2
1	1	2	2	2
1	1	2	2	2
1	1	2	2	2
1	1	2	2	2
1	1	2	2	2
1	1	2	2	2
1	1	2	2	2
1	1	2	2	2

Chapter 17 Data Set 4

Preference	Preference
Nuts & Grits	Dimples
Nuts & Grits	Dimples
Nuts & Grits	Froggy
Nuts & Grits	Froggy
Nuts & Grits	Froggy

Preference	Preference
Nuts & Grits	Froggy
Nuts & Grits	Froggy
Nuts & Grits	Froggy
Nuts & Grits	Froggy
Bacon Surprise	Froggy
Bacon Surprise	Froggy
Bacon Surprise	Froggy
Bacon Surprise	Froggy
Bacon Surprise	Froggy
Bacon Surprise	Froggy
Bacon Surprise	Froggy
Bacon Surprise	Froggy
Bacon Surprise	Froggy
Bacon Surprise	Froggy
Bacon Surprise	Chocolate Delight
Bacon Surprise	Chocolate Delight
Bacon Surprise	Chocolate Delight
Bacon Surprise	Chocolate Delight
Bacon Surprise	Chocolate Delight
Bacon Surprise	Chocolate Delight
Bacon Surprise	Chocolate Delight
Bacon Surprise	Chocolate Delight
Bacon Surprise	Chocolate Delight
Bacon Surprise	Chocolate Delight
Bacon Surprise	Chocolate Delight
Bacon Surprise	Chocolate Delight
Bacon Surprise	Chocolate Delight
Bacon Surprise	Chocolate Delight
Bacon Surprise	Chocolate Delight
Bacon Surprise	Chocolate Delight

(Continued)

(Continued)

Preference	Preference
Bacon Surprise	Chocolate Delight
Dimples	Chocolate Delight
Dimples	Chocolate Delight
Dimples	Chocolate Delight
Dimples	Chocolate Delight
Dimples	Chocolate Delight
Dimples	Chocolate Delight
Dimples	Chocolate Delight
Dimples	Chocolate Delight
Dimples	Chocolate Delight
Dimples	Chocolate Delight
Dimples	Chocolate Delight
Dimples	Chocolate Delight
Dimples	Chocolate Delight
Dimples	Chocolate Delight

Chapter 17 Data Set 5

Strength	Age
Strong	Young
Strong	Young
Strong	Young
Strong	Young
Strong	Young
Strong	Young
Strong	Young
Strong	Young
Strong	Young
Strong	Young
Strong	Young
Strong	Young

Strength	Age
Strong	Young
Strong	Young
Strong	Young
Strong	Young
Strong	Young
Strong	Young
Strong	Young
Strong	Young
Strong	Young
Strong	Young
Strong	Middle
Strong	Middle
Strong	Middle
Strong	Middle
Strong	Middle
Strong	Middle
Strong	Middle
Strong	Middle
Strong	Middle
Strong	Middle
Strong	Middle
Strong	Middle
Strong	Middle
Strong	Middle
Strong	Middle
Strong	Middle
Strong	Middle
Strong	Middle
Strong	Old
Strong	Old

(Continued)

(Continued)

Strength	Age
Strong	Old
Strong	Old
Strong	Old
Moderate	Young
Moderate	Young
Moderate	Young
Moderate	Young
Moderate	Young
Moderate	Young
Moderate	Young
Moderate	Young
Moderate	Young
Moderate	Young
Moderate	Young
Moderate	Young
Moderate	Young
Moderate	Young
Moderate	Young
Moderate	Young
Moderate	Young
Moderate	Young
Moderate	Middle
Moderate	Middle
Moderate	Middle
Moderate	Middle
Moderate	Middle
Moderate	Middle
Moderate	Middle
Moderate	Middle
Moderate	Middle
Moderate	Middle
Moderate	Middle

Strength	Age
Moderate	Middle
Moderate	Middle
Moderate	Middle
Moderate	Middle
Moderate	Middle
Moderate	Middle
Moderate	Middle
Moderate	Middle
Moderate	Middle
Moderate	Middle
Moderate	Middle
Moderate	Old
Moderate	Old
Moderate	Old
Moderate	Old
Moderate	Old
Moderate	Old
Moderate	Old
Moderate	Old
Moderate	Old
Moderate	Old
Weak	Young
Weak	Young
Weak	Young
Weak	Young
Weak	Young
Weak	Young
Weak	Young
Weak	Young
Weak	Young
Weak	Young
Weak	Young

(Continued)

(Continued)

Strength	Age
Weak	Young
Weak	Middle
Weak	Middle
Weak	Middle
Weak	Middle
Weak	Middle
Weak	Middle
Weak	Middle
Weak	Middle
Weak	Middle
Weak	Middle
Weak	Middle
Weak	Middle
Weak	Middle
Weak	Middle
Weak	Middle
Weak	Middle
Weak	Middle
Weak	Middle
Weak	Middle
Weak	Middle
Weak	Old
Weak	Old
Weak	Old
Weak	Old
Weak	Old
Weak	Old
Weak	Old
Weak	Old
Weak	Old

Chapter 19 Data Set 1 and Chapter 19 Data Set 2

These are pretty large data sets, so it's best for you to get them from the website at **edge.sagepub.com/salkind6e**.

Sample Data Set

Gender: 1 = male, 2 = female; Treatment: 1 = low, 2 = high

Gender	Treatment	Test1	Test2
1	1	98	32
2	2	87	33
2	1	89	54
2	1	88	44
1	2	76	64
1	1	68	54
2	1	78	44
2	2	98	32
2	2	93	64
1	2	76	37
2	1	75	43
2	1	65	56
1	1	76	78
2	1	78	99
2	1	89	87
2	2	81	56
1	1	78	78
2	1	83	56
1	1	88	67
2	1	90	88
1	1	93	81
1	2	89	93
2	2	86	87
1	1	77	80
1	1	89	99

Appendix D

Answers to Practice Questions

All of these questions are exploratory with no necessarily right answer. Their purpose is to get you involved in thinking about statistics as a field of study and a useful tool.

1. By hand (easy test!)

   Mean = 87.55

   Median = 88

   Mode = 94

2.

	Score 1	Score 2	Score 3
Mean	5.6	27.6	144.3
Median	5.5	25.0	149.5
Mode	5	25, 34	154

3. Here's the SPSS output:

Statistics

		Hospital Size	Infection Rate
N	Valid	10	10
	Missing	0	0
Mean		335.10	3.7200

4. Here's what your one paragraph might look like.

   As usual, the Chicken Littles (the mode) led the way in sales. The total value of food sold was $303, for an average of $2.55 for each special.

5. There's nothing too dramatic here in terms of values that are really big or small or look weird (all reasons to use the median), so we'll just use the mean. You can see the mean for the three stores in the last column. Seems like these might be the numbers you want the new store to approximate for it to be around the average of all the stores you manage.

Average	Store 1	Store 2	Store 3	Mean
Sales (in thousands of dollars)	$323	$234	$308	$288
Number of items purchased	3,454	5,645	4,565	4,555
Number of visitors	4,534	6,765	6,654	5,984

6. Well, it seems like beer chicken is the winner and the fruit cup comes in last. We computed the mean since this is a rating of values across a scale that is interval (at least in its intent). Here are the data . . .

Snack Food	North Fans	East Fans	South Fans	West Fans	Mean Rating
Loaded Nachos	4	4	5	4	4.25
Fruit Cup	2	1	2	1	1.5
Spicy Wings	4	3	3	3	3.25
Gargantuan Overstuffed Pizza	3	4	4	5	4
Beer Chicken	5	5	5	4	4.75

7. You use the median when you have extreme scores that would disproportionately bias the mean. One situation in which the median is preferable to the mean is when income is reported. Because it varies so much, you want a measure of central tendency that is insensitive to extreme scores. As another example, say you are studying the speed with which a group of adolescents can run 100 yards when one or two individuals are exceptionally fast.

8. You would use the median because it is insensitive to extreme scores.

9. The median is the best measure of central tendency, and it is the one score that best represents the entire set of scores. Why? Because it is relatively unaffected by the (somewhat) extreme data point $199,000. As you can see in the table below, the mean is affected (it is more than $83,000 when the highest score is included).

Before Removal of Highest Score	Mean	$83,111
	Median	$77,153
After Removal of Highest Score	Mean	$75,318
	Median	$76,564

10. The averages are as follows: Group 1 = 5.88, Group 2 = 5.00, and Group 3 = 7.00.

11. This should be an easy one. Anytime values are expressed as categories, the only type of average that makes sense is the mode. So, who likes pie? Well, the first week, the winner is apple, and week 2 finds apple once again the leader. For week 3, it's Douglas County Pi, and the month is finished up in week 5 with lots of helpings of chocolate silk.

CHAPTER 3

1. The range is the most convenient measure of dispersion, because it requires only that you subtract one number (the lowest value) from another number (the highest value). It's imprecise because it does not take into account the values that fall between the highest and the lowest values in a distribution. Use the range when you want a very gross (not very precise) estimate of the variability in a distribution.

2.

High Score	Low Score	Inclusive Range	Exclusive Range
7	6	10.1	9.1
89	45	42	41
34	17	1	0
15	2	2.5	1.5
1	1	2	24

3. For the most part, first-year students have stopped growing by that time, and the enormous variability that one sees in early childhood and adolescence has evened out. On a personality measure, however, those individual differences seem to be constant and are expressed similarly at any age.

4. As individuals score more similarly, they are closer to the mean, and the deviation about the mean is smaller. Hence, the standard deviation is smaller as well. There's strength in numbers, and the larger the data set, the more inclusive it is. That is, it includes more, rather than fewer, values that are similar to one another—hence, less variability.

5. The exclusive range is 39. The unbiased sample standard deviation equals 13.10. The biased estimate equals 12.42. The difference is due to dividing by a sample size of 9 (for the unbiased estimate) as compared with a sample size of 10 (for the biased estimate). The unbiased estimate of the variance is 171.66, and the biased estimate is 154.49.

6. The unbiased estimate is always larger than the biased estimate because the unbiased estimate actually overestimates the value of the statistic intentionally to be more conservative. And the numerator $(n - 1)$ for the unbiased statistic is always less than for the biased estimate (which is n), resulting in a larger value.

7.

Test 1		Test 2		Test 3	
Mean	49.00	Mean	50.10	Mean	49.30
Median	49.00	Median	49.50	Median	48.00
Mode	49	Mode	49	Mode	45
Range	5	Range	9	Range	10
Stan dev	1.41	Stan dev	2.69	Stan dev	3.94
Variance	2	Variance	7.21	Variance	15.57

Test 2 has the highest average score, and Test 1 has the smallest amount of variability. Also, there are multiple modes, and SPSS computes and reports the smallest value.

8. The standard deviation is 12.39, and the variance is 153.53.

9. The standard deviation is the square root of the variance (which is 36), making the standard deviation 6. You can't possibly

know what the range is by knowing only the standard deviation or the variance. You can't even tell whether the range is large or small, because you don't know what's being measured and you don't know the scale of the measurement (itsy-bitsy bugs or output from steam engines).

10.

    a. Range = 6, standard deviation = 2.58, variance = 6.67.

    b. Range = 0.6, standard deviation = 0.25, variance = 0.06.

    c. Range = 3.5, standard deviation = 1.58, variance = 2.49.

    d. Range = 123, standard deviation = 55.7, variance = 3,102.

11. Here's a chart that summarizes the results. Look familiar? It should—it's exactly what the SPSS output looks like.

Statistics

		Height	Weight
N	Valid	20	20
	Missing	0	0
Std. Deviation		11.44	15.66
Variance		130.78	245.00
Range		43	59

12. Okay, here's how you answer it. Compute the standard deviation for any set of 10 or so numbers by hand using the formula for the standard deviation that has $n - 1$ in the denominator (the unbiased estimate). Then compare it, using the same numbers, with the SPSS output. As you will see, they are the same, indicating that SPSS produces an unbiased estimate. If you got this one correct, you know what you're doing—go to the head of the class. Another way would be to write to the SPSS folks and ask them. No kidding—see what happens.

13. The unbiased estimate of the standard deviation is 6.49, and the biased estimate is 6.15. The unbiased estimate of the variance is 42.1, and the biased estimate is 37.89. The biased estimates are (always) smaller because they are based on a larger n and are a more conservative estimate.

14. A standard deviation of 0.94 means that on average, each score in the set of scores is a distance of 0.94 correct words from the average of all scores.

CHAPTER 4

1.

   a. Here's the frequency distribution.

Class Interval	Frequency
55–59	7
50–54	5
45–49	5
40–44	7
35–39	2
30–34	3
25–29	5
20–24	4
15–19	4
10–14	4
5–9	3
0–4	1

Figure D.1 shows what the histogram (created using SPSS) should look like.

Figure D.1   Histogram of Data in Chapter 4 Data Set 1

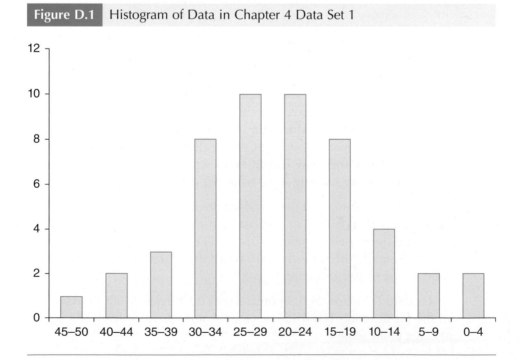

You may note that the *x*-axis in the histogram that you created is different from the one you see here. We double-clicked on it and, entering the Chart Editor, changed the number of ticks, range, and starting point for this axis. Nothing too different—it just looks a bit nicer.

b. We settled on a class interval of 5 because it allowed us to have close to 10 class intervals; it fit the criteria that we discussed in this chapter for deciding on a class interval.

c. The distribution is negatively skewed because the mean is less than the median.

2. In Figure D.2 you can see what the histogram could look like, but remember that how it appears depends upon your choice of metrics for the *x*- and *y*-axes.

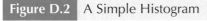

Figure D.2   A Simple Histogram

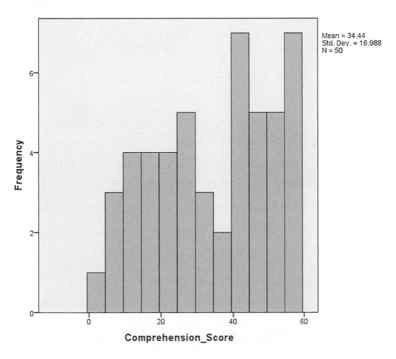

3. Your bar chart will look different depending upon the length of your *x*- and *y*-axes and other variables. The first thing you need to compute is the sum for each day (use the Analyze → Descriptive Statistics → Frequencies options). Then use those values to create the bar chart as shown in Figure D.3.

Figure D.3 A Simple Bar Chart

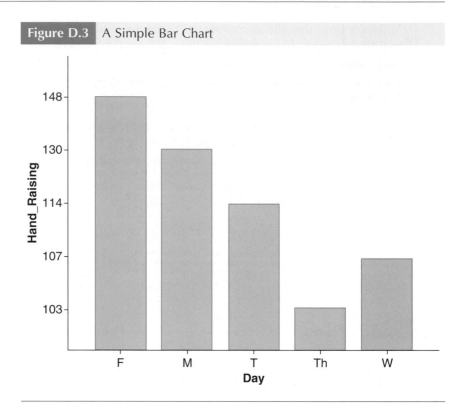

4.

  a. This is negatively skewed because the majority of athletes scored in the upper range.

  b. This is not skewed at all. In fact, the distribution is like a rectangle because everyone scored exactly the same.

  c. This distribution is positively skewed, because most of the spellers scored very low.

5. We used SPSS to create a simple pie chart and then changed the fill in each section of the pie, as shown in Figure D.4. What kind of pie do you like?

6.

  a. Pie because you are interested in looking at proportions.

  b. Line because you are looking at a trend (over time).

  c. Bar because you are looking at a number of discrete categories.

  d. Pie because you are looking at categories defined as proportions.

  e. Bar because you are looking at a number of discrete categories.

Figure D.4	A Simple Pie Chart

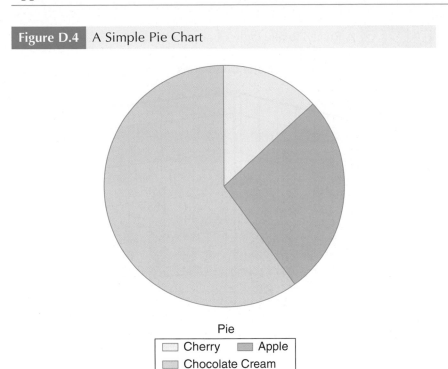

Pie
☐ Cherry   ■ Apple
☐ Chocolate Cream

7. You will come up with examples of your own, but here are some of ours. Draw your own examples.

   a. The number of words that a child knows as a function of ages from 12 months to 36 months

   b. The percentage of senior citizens who belong to AARP as a function of gender and ethnicity

   c. The plotting of paired scores such as height and weight for each individual. We haven't discussed this type of chart much, but take a few glances at the Graph menu on SPSS and you'll get it.

8. On your own!

9. We created Figure D.5 using SPSS and the chart editor, and it's as uninformative as it is ugly. Major chart junk attack.

10. A picture (a chart or graph) is worth more than 1,000 words. In other words, there are many possible answers to such a question, but in general, the purpose of a chart or a graph is to visually illustrate information in a way that is as simple as possible while communicating a central message that is clear and direct.

Figure D.5	A Really, Really Ugly Chart

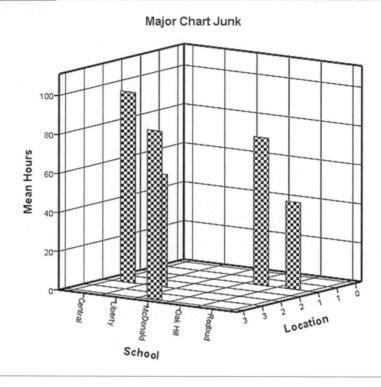

CHAPTER 5

1.

    a. $r = .596$.

    b. From the answer to 1a, you already know that the correlation is direct. But from the scatterplot shown in Figure D.6 (we used SPSS, but you should do it by hand), you can predict it to be such (without actually knowing the sign of the coefficient) because the data points group themselves from the lower left corner of the graph to the upper right corner and assume a positive slope.

2.

    a. $r = .609$.

    b. According to Table 5.3 on page 92, the general strength of the correlation of this magnitude is strong. The coefficient of determination is $.609^2$, so .371 or 37.1% of the variance is accounted for. The subjective analysis (strong) and the objective one (37.1% of the variance accounted for) are consistent with one another.

| Figure D.6 | Scatterplot of Data From Chapter 5 Data Set 2 |

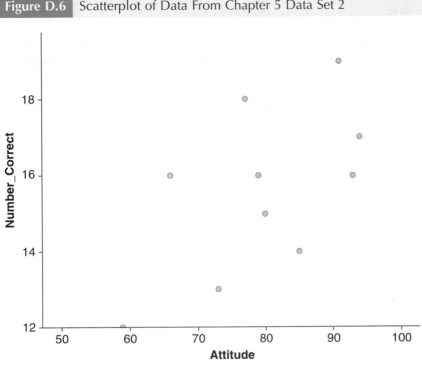

3. Note that the .71 and the .47 have no sign, and in that case, we always assume that (like any other number) the values are positive.

+.36

−.45

+.47

−.62

+.71

4. The correlation is .64, meaning that increases in budget and increases in classroom achievement are positively related to one another (and note that we have to test for significance). From a descriptive perspective, a bit more than 40% of the variance is shared between the two variables.

5. The correlation between number of minutes of exercise per day and GPA is .39, showing that as exercise increases, so does GPA. And of course, as exercise decreases, so does GPA. And want to be buff and get good grades? Work out, study, read broadly (and shower often). Keep in mind that GPA has nothing to do with

how good you look at the gym. That would be causality and we're only dealing with associations here.

6. The correlation is −.022 and is so low because both the hours of studying and the set of GPA scores has very little variability. When there is so little variability, there is nothing to share, and the two sets of scores have little in common—hence, the low correlation.

7.

   a. .9

   b. Very strong

   c. 1.00 − .81, or .19 (19%)

8. Here's the matrix . . .

	Age at Injury	Level of Treatment	12-Month Treatment Score
Age at Injury	1		
Level of Treatment	0.0557	1	
12-Month Treatment Score	−0.154	0.389	1

9. To examine the relationship between sex (defined as male or female) and political affiliation, you would use the phi coefficient because both variables are nominal in nature. To examine the relationship between family configuration and high school GPA, you would use the point biserial correlation because one variable is nominal (family configuration) and the other is interval (GPA).

10. Just because two things are related does not mean that one causes the other. There are plenty of runners with average strength who can run fast, and plenty of very strong people who run slowly. Strength can help people run faster—but technique is more important (and, by the way, accounts for more of the variance).

11. You will come up with your own explanations, but here are two examples.

   a. We bet you expected this one—consumption of ice cream and number of crimes.

   b. Amount of money spent on political ads and the number of people who vote

     c. The introduction of abstinence-only programs in middle school and the number of sexually active teenagers

     d. Income and job satisfaction

     e. Level of education and drug use

12. An excellent question. Correlations are a reflection of how much two variables have in common, but what they have in common may have nothing to do with what causes one, or the other, variable to increase or decrease. You saw that at work regarding ice cream consumption and rate of crime (both are related to outdoor temperature, not to each other). The same could be said about the relationship between school achievement scores and parents' level of education, where many variables can be said to intervene, such as class size, family structure, and family income. So, what is shared is what's important, not what appears to be conveniently related.

13. Partial correlations are used when you want to examine the correlation between two variables with the influence of a third removed.

CHAPTER 6

1. Do this one on your own.

2. Test–retest reliability should be established when you are interested in the consistency of an assessment over time, such as pre- and posttest studies or longitudinal studies. Parallel forms reliability is important to establish to make sure those different forms of the same test are similar to one another.

3. Test–retest reliability is established through the calculation of a simple correlation coefficient over the two testings. In this case, the correlation between fall and spring scores is .139, which has a probability of occurring by chance of .483. The correlation of .138 is not even close to what one might need (at least .85) to conclude that the test is reliable over time.

4. In general, a test that is reliable but not valid does what it does over and over, but it does not do what it is supposed to. And, oops! A test cannot be valid without being reliable, because if it does not do anything consistently, then it certainly cannot do one thing consistently.

5. The reliability coefficient for these parallel forms of the same test is −.09. You did a great job as far as calculating the coefficient, but as for the reliability of the test? Not so great.

6. It's simple. A test must first be able to do what it does over and over again (reliability) before one can conclude that it does what it should (validity). If a test is inconsistent (or unreliable), then it can't possibly be valid. For example, although a test consisting of items like this one . . .

$$15 \times 3 = ?$$

is surely reliable, if the 15 items on the test were labeled "Spelling Test," the test surely would not be valid.

7. You need to use both a reliable and a valid test because if you get a null result, you will never be sure whether the instrument is not measuring what it is supposed to or the hypothesis is actually faulty.

8. Content validity looks at a test to see whether, "on the face of it," it samples from an entire universe of possible items. For example, does a high school history test on the American Revolution contain items that reflect that subject area of American history?

Predictive validity examines how well a test predicts a particular outcome. For example, how well does a test of spatial skills predict success as a mechanical engineer?

Construct validity is present when a testing instrument assesses an underlying construct. For example, how well does an observation tool assess one dimension of bipolar disorder in young adolescents?

9. This is a tough one. Let's assume (as you would need to) that you have a strong theoretical basis for "out-of-the-box thinking" that can form the basis of test items. You would locate an independent group of "out-of-the-box experts," have them "take" the test, and have scorers observe and score the behaviors on the items you created. In theory, if your construct holds up and if the test works, then those people who have been independently verified as being out-of-the-box thinkers would score differently than those verified independently as not thinking in unique ways. Those items that differentiate between the two groups become the set of test items. This can be approached in many ways, and what we just presented is only one.

CHAPTER 7

1. On your own!

2. On your own again!

3.

   a. *Null:* Children with short attention spans, as measured by the Attention Span Observation Scale, have the same frequency of out-of-seat behavior as those with long attention spans.

   Directional: Children with short attention spans, as measured by the Attention Span Observation Scale, have a higher frequency of out-of-seat behavior than those with long attention spans.

   Nondirectional: Children with short attention spans, as measured by the Attention Span Observation Scale, differ in their frequency of out-of-seat behavior from those with long attention spans.

   b. *Null:* There is no relationship between the overall quality of a marriage and spouses' relationships with their siblings.

   Directional: There is a positive relationship between the overall quality of a marriage and spouses' relationships with their siblings.

   Nondirectional: There is a relationship between the overall quality of a marriage and spouses' relationships with their siblings.

   c. *Null:* Pharmacological treatment combined with traditional psychotherapy has the same effect in treating anorexia nervosa as does traditional psychotherapy alone.

   Directional: Pharmacological treatment combined with traditional psychotherapy is more effective in treating anorexia nervosa than is traditional psychotherapy alone.

   Nondirectional: Pharmacological treatment combined with traditional psychotherapy has a different effect than does treating anorexia nervosa with traditional psychotherapy alone.

4. On your own!

5. The most significant problem is that the test of the research hypothesis will probably be inconclusive. Regardless of the outcome, poor language leads to misleading conclusions. In addition, the study may be impossible to replicate and the findings to be generalized—all in all, a questionable effort.

6. The null (which literally means "void") represents the lack of any observed difference between groups of outcomes. It is *the* starting point in any research endeavor and differs from the research hypothesis is many ways, but primarily it is a statement of equality while the research hypothesis is a statement of inequality.

7. If you're at the beginning of exploring a question (which then becomes a hypothesis) and you have little knowledge about the outcome (which is why you are asking the question and performing the test), then the null is the perfect starting point. It is a statement of equality that basically says, "Given no other information about the relationships that we are studying, I should start at the beginning, where I know very little." The null is the perfect, unbiased, and objective starting point because it is the place where everything is thought to be equal unless proven otherwise.

8. As you already know, the null hypothesis states that there is no relationship between variables. Why? Simply, that's the best place to start given no other information. For example, if you are investigating the role of early events in the development of linguistic skills, then it is best to assume that they have no role whatsoever unless you prove otherwise. That's why we set out to *test* null hypotheses and not *prove* them. We want to be as unbiased as possible.

CHAPTER 8

1. In a normal curve, the mean, median, and mode are equal to one another; the curve is symmetrical about the mean; and the tails are asymptotic. Height and weight are examples, as are intelligence and problem-solving skills.

2. The raw score, the mean, and the standard deviation

3. Because they all use the same metric—the standard deviation—and we can compare scores in units of standard deviations.

4. A z score is a standard score (comparable to others of the same type of score) because it is based on the degree of variability within its respective distribution of scores. A z score is always a measure of the distance between the mean and some point on the x-axis (regardless of the mean and standard deviation differences from one distribution to the next). Because the same units are used (units of standard deviation), the z scores can be compared with one another. That's the magic—comparability.

5. The z score for a raw score of 55, where the mean is 50 and the standard deviation is 2.5, is 2. So, the amount of variability in the set of scores is half as much, and the corresponding z score is twice as large (it went from 1 to 2). This indicates that as variability increases, and all other things are held constant, the same raw score becomes more extreme. The less the difference between scores (and the lower the variability), the less extreme the same score will be.

6.

Raw Score	z Score
68.0	−0.61
58.16	−1.60
82.0	0.79
92.09	1.80
69.0	−0.51
69.14	−0.50
85.0	1.09
91.10	−0.50
72.0	−0.21

7.

18	−1.01
19	−0.88
15	−1.41
20	−0.75
25	−0.10
31	0.69
17	−1.14
35	1.21
27	0.17
22	−0.49
34	1.08
29	0.43
40	1.87
33	0.95
21	−0.62

8.

a. The probability of a score falling between a raw score of 70 and a raw score of 80 is .5646. A z score for a raw score of 70 is −0.78, and a z score for a raw score of 80 is 0.78. The area between the mean and a z score of 0.78 is 28.23%. The area between the two scores is 28.23 times 2, or 56.46%.

b. The probability of a score falling above a raw score of 80 is .2177. A z score for a raw score of 80 is 0.78. The area between the mean and a z score of 0.78 is 28.23%. The area below a z score of 0.78 is .50 + .2823, or .7823. The difference between 1 (the total area under the curve) and .7823 is .2177 or 21.77%.

c. The probability of a score falling between a raw score of 81 and a raw score of 83 is .068. A z score for a raw score of 81 is 0.94, and a z score for a raw score of 83 is 1.25. The area between the mean and a z score of 0.94 is 32.64%. The area between the mean and a z score of 1.25 is 39.44%. The difference between the two is .3944 − .3264 = .068, or 6.8%.

d. The probability of a score falling below a raw score of 63 is .03. A z score for a raw score of 63 is −1.88. The area between the mean and a z score of −1.88 is 46.99%. The area below a z score of 1.88 is 1 − (.50 + .4699) = .03, or 3%.

9. A little magic lets us solve for the raw score using the same formulas for computing the z score that you've seen throughout this chapter. Here's the transformed formula . . .

$$X = (s \times z) + \overline{X}$$

And taking this one step further, all we really need to know is the z score of 90% (or 40% in Table B.1), which is 1.29.

So we have the following formula:

$$X = (s \times z) + \overline{X}$$

or

$$X = 78 + (5.5 \times 1.29) = 85.095$$

Jake is home free if he gets that score and, along with it, his certificate.

10. It doesn't make sense because raw scores are not comparable to one another when they belong to different distributions. A raw score of 80 on the math test, where the class mean was 40, is just not comparable to an 80 on the essay-writing skills test, where everyone got the one answer correct. Distributions, like people, are not always comparable to one another. Not everything (or everyone) is comparable to something else.

11. Here's the info with the unknown values in bold.

Math		
Class Mean	81	
Class Standard Deviation	2	
Reading		
Class Mean	87	
Class Standard Deviation	10	
Raw Scores		
	Math Score	Reading Score	Average
Noah	85	88	86.5
Talya	87	81	84.0
z Scores		
	Math Score	Reading Score	Average
Noah	**2**	**0.1**	**1.05**
Talya	**3**	**−0.6**	**1.2**

Noah has the higher average raw score (86.5 vs. 84 for Talya), but Talya has the higher average z score (1.2 vs. 1.05 for Noah). Remember that we asked who was the better student relative to the rest, which requires the use of a standard score (we used z scores). But why is Talya the better student relative to Noah? It's because on the tests with the lowest variability (Math with an $SD = 2$), Talya really stands out with a z score of 3. That put her ahead to stay.

12. The fact that the tails do not touch the x-axis indicates that there is always a chance—even though it may be very, very small—that extreme scores (all the way far out in either direction on the x-axis) are possible. If the tails did touch the x-axis, it would mean there is a limit to how improbable an outcome can be. In other words, no matter what the outcome, there's always a chance it will occur.

a. Null hypothesis: There is no difference between the amount of money spent on food by undergraduate students and the amount spent by undergraduate student–athletes. H_0: $\overline{X}_{us} = \overline{X}_{sa}$	a. Research hypothesis: Undergraduate student–athletes spend more money on food than do undergraduate students. H_1: $\overline{X}_{us} < \overline{X}_{sa}$

(Continued)

(Continued)

b. Null hypothesis: There is no difference between white and brown rats in the average amount of time taken to get out of a maze. $H_0: \overline{X}_W = \overline{X}_B$	b. Research hypothesis: There is a difference between white and brown rats in the average amount of time taken to get out of a maze. $H_1: \overline{X}_W \neq \overline{X}_B$
c. Null hypothesis: The effects of Drug A on a disease are not different from the effects of Drug B. $H_0: \overline{X}_A = \overline{X}_B$	c. Research hypothesis: Drug A has stronger effects on a disease than Drug B. $H_1: \overline{X}_A > \overline{X}_B$
d. Null hypothesis: There is no difference between Method 1 and Method 2 in terms of the time needed to complete a task. $H_0: \overline{X}_1 = \overline{X}_2$	d. Research hypothesis: There is a difference between Method 1 and Method 2 in terms of the time needed to complete a task. $H_1: \overline{X}_1 \neq \overline{X}_2$

CHAPTER 9

1. The concept of significance is crucial to the study and use of inferential statistics, because significance (reflected in the idea of a significance level) sets the criterion for being confident that the outcomes we observe are "truthful." It further determines to what extent these outcomes can be generalized to the larger population from which the sample was selected.

2. Statistical significance is the idea that certain outcomes are due not to chance, but to factors that have been identified and tested by the researcher. These outcomes can be assigned a value that represents the probability that they are due to chance or some other factor or set of factors. The statistical significance of these outcomes is the value of that probability.

3. The critical value represents the minimum value at which the null hypothesis is no longer the accepted explanation for any differences that are observed. It is the cut point: Obtained values that are more extreme indicate that there is no equality but instead a difference (and the nature of that difference depends on the questions being asked). Do remember that this cut point is set by the researcher (even if .01 and .05 are conventional and often used cut points).

4.

    a. Reject the null hypothesis. Because the level of significance is less than 5%, there is a relationship between a person's choice of music and his or her crime rate.

    b. Fail to reject the null hypothesis. The probability is greater than .05, meaning there is no relationship between the amount of coffee consumed and GPA.

    c. Nope—no relationship with a probability that high.

5.

    a. *Level of significance* refers only to a single, independent test of the null hypothesis and not to multiple tests.

    b. It is impossible to set the error rate to zero because it is not possible that we might not reject a null hypothesis when it is actually true. There's always a chance that we would.

    c. The level of risk that you are willing to take to reject the null hypothesis when it is true has nothing to do with the meaningfulness of the outcomes of your research. You can have a highly significant outcome that is meaningless, or you can have a relatively high Type I error rate (.10) and a very meaningful finding.

6. At the .01 level, less room is left for errors or mistakes because the test is more rigorous. In other words, it is "harder" to find an outcome that is sufficiently removed from what you would expect by chance (the null hypothesis) when the probability associated with that income is smaller (such as .01) rather than larger (such as .05).

7. Here's a fine point but one that people get pretty excited about. We can't "reject" the null because we never directly test it. Remember, nulls reflect population characteristics, and the whole point is that we cannot directly test populations, only samples. If we can't test it, how can we reject it?

8. *Significance* is a statistical term that simply defines an area at which the probability associated with the null hypothesis is too small for us to conclude anything other than "The research hypothesis is the most attractive explanation." Meaningfulness has to do with the application of the findings and whether, within the broad context of the question being asked, the findings have relevance or importance.

9. You can come up with the substance of these yourself, but how about the following examples?

    a. A significant difference between two groups of readers was found, with the group that received intensive comprehension training outperforming the group that got no training (but the same amount of attention) on a test of reading comprehension.

    b. Examining a huge sample (which is why the finding was significant), a researcher found a very strong positive correlation between shoe size and calories eaten on a daily basis. Silly, but true . . .

10. Chance is reflected in the degree of risk (Type I error) that we are willing to take in the possible rejection of a true null hypothesis. It is, first and foremost, always a plausible explanation for any difference that we might observe, and it is the most attractive explanation given no other information.

11.

    a. The striped area represents values that are extreme enough that none of them reflects a finding that supports the null hypothesis.

    b. A larger number of values reflects a higher likelihood of a Type I error.

CHAPTER 10

1. The one-sample Z-test is used when you want to compare a sample mean with a population parameter. Actually, think of it as a test to see whether one number (which is a mean) belongs to a huge set of numbers.

2. The (BIG) Z is similar to the small z for one very good reason: It is a standard score. The z score has the sample standard deviation as the denominator, whereas the Z-test value has the standard error of the mean (or a measure of the variability of all the means from the population) as the denominator. In other words, they both use a standard measure that allows us to use the normal curve table (in Appendix B) to understand how far the values are from what we would expect by chance alone.

3.

    a. The weight loss of Bob's group on the chocolate-only diet is not representative of weight loss in a large population of middle-aged men who are on a chocolate-only diet.

    b. The rate of flu infections per 1,000 citizens for this past flu season is not comparable to the rate of infection over the past 50 seasons.

    c. Blair's costs for this month are not different from his average monthly costs over the last 20 years.

4. The Z-test results in a value of −1.48. This is not nearly extreme enough (we would need a value of ±1.96) to conclude that the average number of cases (15) is any different from that of the state (which is 16).

5. The Z-test results in a value of 0.31, which is not as extreme as would be needed to say that the group of stockers in the sample specialty stores (at 500 products stocked) does better (or even differently) than the larger group of stockers (at 496 products) who work in the chain.

6. What's missing here is the significance level at which the hypothesis is being tested. If the level is .01, then the critical z value for rejecting the null and concluding that the sample is different from the population is 1.96. If the level at which the hypothesis is being tested is .05, then the critical z value is 1.65. This presents the most interesting of questions regarding the trade-off between making a Type I error (1% or 5% of true nulls being rejected) versus stating statistical significance. You should be able to justify your answer as to which level of significance you chose.

7. If you plug in the data according to Formula 10.1, you get a Z-test score of 12.22, which is way, way off the charts (at the extended end of the curve). But, remember, the lower the golf score, the better, so Millman's group is in no way similar in membership to the pros. Too bad for Millman.

8. Take a look at the SPSS results shown in Figure D.7. As you can see, the value of 31,456 lies well within ($t = −1.681, p = .121$) the general range of values of last year's months of unit sales. Nope, not different.

Figure D.7 The Results of a One-Sample t-Test

One-Sample Statistics

	N	Mean	Std. Deviation	Std. Error Mean
Score	12	30162.8333	2665.47587	769.45661

One-Sample Test

	Test Value = 31456					
					95% Confidence Interval of the Difference	
	t	df	Sig. (2-tailed)	Mean Difference	Lower	Upper
Score	-1.681	11	.121	-1293.16667	-2986.7292	400.3959

CHAPTER 11

1. The mean for boys equals 7.93, and the mean for girls equals 5.31. The obtained t value is 3.006, and the critical t value at the .05 level for rejection of the null hypothesis for a one-tailed test (boys *more* than girls, remember?) is 1.701. Conclusion? Boys raise their hands significantly more!

2. Now this is very interesting. We have the same exact data, of course, but a different hypothesis. Here, the hypothesis is that the number of times is *different* (not just more or less), necessitating a two-tailed test. So, using Table B.2 and at the .01 level for a two-tailed test, the critical value is 2.764. The obtained value of 3.006 (same results as when you did the analysis for Question 1 in this chapter) does exceed what we would expect by chance, and given this hypothesis, there is a difference. So in comparison with one another, a one-tailed finding (see Question 1) need not be as extreme as a two-tailed finding, given the same data, to support the same conclusion (that the research hypothesis is supported).

3.
   a. $t(18) = 1.58$.
   b. $t(46) = 0.88$.
   c. $t(32) = 2.43$.

4.
   a. $t_{crit}(18) = 2.101$. Because the observed t value exceeds the critical t value, we do not reject the null hypothesis.
   b. $t_{crit}(46) = 2.009$ (using the value for 50 degrees of freedom). Because the observed t value does not exceed the critical t value, we fail to reject the null hypothesis.
   c. $t_{crit}(32) = 2.03$ (using the value for 35 degrees of freedom). Because the observed t value exceeds the critical t value, we reject the null hypothesis.

5. First, here's the output for the t-test between independent samples that was conducted using SPSS.

Group Statistics

	Group	N	Mean	Std. Deviation	Std. Error Mean
Score	In Home	20	4.1500	2.20705	.49351
	Out of Home	20	5.5000	1.73205	.38730

	Levene's Test for Equality of Variances		t-Test for Equality of Means							
	F	Sig.	t	df	Sig (2-tailed)	Mean Difference	Std. Error Difference	95% Confidence Interval of the Difference	
								Lower	Upper
Score Equal variances assumed	.938	.339	−2.152	38	.038	−1.35000	.62734	−2.61998	−.08002
Equal variances not assumed			−2.152	35.968	.038	−1.35000	.62734	−2.62234	−.07766

And here's one summary paragraph . . .

There was a significant difference between the mean performance for the group that received treatment in the home versus the group that received treatment out of the home. The mean for the In Home group was 4.15, and the mean for the Out of Home group was 5.5. The probability associated with this difference was .038, meaning that there is a less than 4% chance that the observed difference is due to chance. It is much more likely that the Out of Home program was more effective.

6. You can see the SPSS output in Figure D.8, where there is no significant difference ($p = .253$) between urban and rural dwellers in their attitude toward gun control.

7. This certainly is a good one to think about, as it has many different "correct" answers and raises a bunch of issues. If all you are concerned about is the level of Type I error, then we suppose that

Figure D.8 SPSS Output for a *t*-Test of Independent Means

Group Statistics

	Group	N	Mean	Std. Deviation	Std. Error Mean
Attitude	Urban	16	6.5112	1.77221	.44305
	Rural	14	5.3979	3.31442	.88582

Independent Samples Test

		Levene's Test for Equality of Variances		t-test for Equality of Means						
		F	Sig.	t	df	Sig. (2-tailed)	Mean Difference	Std. Error Difference	95% Confidence Interval of the Difference	
									Lower	Upper
Attitude	Equal variances assumed	4.463	.044	1.168	28	.253	1.11339	.95311	-.83897	3.06576
	Equal variances not assumed			1.124	19.273	.275	1.11339	.99044	-.95763	3.18442

Dr. L's findings are more trustworthy because these results imply that one is less likely to make a Type I error. However, both of these findings are significant, even if one of them is marginally so. Thus, if your personal system of evaluating these kinds of outcomes says, "I believe that significant is significant—that's what's important," then both should be considered equally valid and equally trustworthy. However, do remember to keep in mind that the meaningfulness of outcomes is critical as well (and you should get extra points if you bring this into the discussion). It seems to us that regardless of the level of Type I error, the results from both studies are highly meaningful because the program will probably result in safer children.

8. Here are the data plus the answers:

Experiment	Effect Size
1	2.6
2	1.3
3	0.65

As you can see, as the standard deviation doubles, the effect size is half as much. Why? If you remember, effect size gives you another indication of how meaningful the differences between groups are. If there is very little variability, there is not much difference between individuals, and any mean differences become more interesting (and probably more meaningful). When the standard deviation is 2 in the first experiment, then the effect size is 2.6. But when the variability increases to 8 (the third experiment), interestingly, the effect size is reduced to 0.65. It's tough to talk about how meaningful differences between groups might be when there is less and less similarity among members of those groups.

9. And the answer? As you can see, Group 2 scored a higher mean of 12.20 (compared with a mean for Group 1 of 7.67), and based on the results you see in Figure D.9, the difference is significant at the .004 level. Our conclusion is that the second group of fourth graders spells better.

10. For Chapter 11 Data Set 5, the t value is $-.218$. For Chapter 11 Data Set 6, the t value is $-.327$. And in both cases, the means are 6.00 (for Group 1) and 6.2 (for Group 2). Also, the variability of the larger set is slightly less. The reason why the second set of data has a larger (or more extreme) t score is that it is

Figure D.9 *t*-Test of the Mean Spelling Scores of Two Groups of Fourth Graders

Group Statistics

	Group	N	Mean	Std. Deviation	Std. Error Mean
Score	Group 1	15	7.67	3.200	.826
	Group 2	15	12.20	4.539	1.172

Independent Samples Test

		Levene's Test for Equality of Variances		t-test for Equality of Means						95% Confidence Interval of the Difference	
		F	Sig.	t	df	Sig. (2-tailed)	Mean Difference	Std. Error Difference		Lower	Upper
Score	Equal variances assumed	5.482	.027	-3.162	28	.004	-4.533	1.434		-7.470	-1.596
	Equal variances not assumed			-3.162	25.159	.004	-4.533	1.434		-7.485	-1.581

based on a sample size that is twice as large (20 versus 10). The *t*-test takes this into account, producing a more extreme score (and one headed in the direction of "significance"), because the large sample more closely approximates the population size. One might make the argument that with the same mean and less variability, the score would not only be more extreme but also be related to the effect of sample size.

11. This question actually has an easier answer than you might initially think. Specifically, the difference between two groups is somewhat (if not entirely) independent of the effect size. In this example, the statistically significant difference is probably due to the very large sample size, yet the effect is not very pronounced. This result is entirely possible and feasible and adds a new dimension to understanding the dual importance of significance and effect size.

CHAPTER 12

1. A *t*-test for independent means tests two distinct groups of participants, and each group is tested once. A *t*-test for dependent means tests one group of participants, and each participant is tested twice.

2.

   a. Independent

   b. Independent

   c. Dependent

   d. Independent

   e. Dependent

3. The mean before the recycling program was 34.44, and the mean afterward was 34.84. There is an increase in recycling. Is the difference across the 25 districts significant? The obtained t value is 0.262, and with 24 degrees of freedom, the difference is not significant at the .01 level—the level at which the research hypothesis is being tested. Conclusion: The recycling program does not result in an increase in paper recycled.

4. Here's the SPSS output for this t-test between dependent or paired means.

Paired Samples Statistics

	Mean	N	Std. Deviation	Std. Error Mean
Pair 1 Before_Treatment	32.8500	20	9.05117	2.02390
After_Treatment	36.9500	20	7.41602	1.65827

Paired Samples Test

	Paired Differences					t	df	Sig (2-tailed)
	Mean	Std. Deviation	Std. Error Mean	95% Confidence Interval of the Difference				
				Lower	Upper			
Pair 1 Before_ Treatment After_ Treatment	−4.10000	10.59245	2.36854	−9.05742	.85742	−1.731	19	.100

The conclusion? The before-treatment mean was lower (showing less tolerance) than the after-treatment mean. However, the difference between the two means is not significant, so the direction of the difference is irrelevant.

5. There was an increase in level of satisfaction, from 5.480 to 7.595, resulting in a t value of −3.893. This difference has an associated probability level of .001. It's very likely that the social service intervention worked.

6. The average score for Nibbles preference was 5.1, and the average score for Wribbles preference was 6.5. With 19 degrees of freedom, the t value for this test of dependent means was −1.965, with a critical value for rejecting the null of 2.093. Because the obtained t value of −1.965 does not exceed the critical value, the

marketing consultant's conclusion is that both crackers are preferred about the same.

7. Nope. The means are not that different (5.35 for the first shift and 6.15 for the second), and the t value of -1.303 (so there is a bit less stress) is not significant. It doesn't matter what shift one works: The level of stress is the same.

8. Absolutely. The scores increased from 5.52 to 7.04, and the difference between the two sets is significant at the .005 level. And, as a bonus ☺, the effect size is .75—quite large.

CHAPTER 13

1. Although both of these techniques look at differences between means, ANOVA is appropriate when more than two means are being compared. It can be used for a simple test between means, but it assumes that the groups are independent of one another.

2. A one-way ANOVA only examines differences across different levels of one variable, while a factorial ANOVA looks at differences across two or more levels when more than one variable is involved.

3.

Design	Grouping Variable(s)	Test Variable
Simple ANOVA	Four levels of hours of training—2, 4, 6, and 8 hours	Typing accuracy
	Three age-groups—20-, 25-, and 30-year-olds	Strength
	Six levels of job types	Job performance
Two-factor ANOVA	Two levels of training and gender (2 × 2 design)	Typing accuracy
	Three levels of age (5, 10, and 15 years) and number of siblings	Social skills
Three-factor ANOVA	Curriculum type (Type 1 or Type 2), GPA (above or below 3.0), and activity participation (participates or not)	ACT scores

4. The means for the three groups are 58.05 seconds, 57.96 seconds, and 59.03 seconds, and the probability of this F value ($F_{(2,33)} = 0.160$) occurring by chance is .853, far above what we would expect due to the treatment. Our conclusion? The number of hours of practice makes no difference in how fast you swim!

5. The F value is 28.773, which is significant at the .000 (!) level, indicating that the amount of stress differs across the three groups. As you can see, the least stressed is the 8:00 AM to 4:00 PM group.

6. For this problem, post hoc analysis (and we use the Bonferroni procedure) is required. As you can see by the output shown in Figure D.10, thin noodles have the lowest average score (1.80—which you already know about, right?), with significant differences between pairings of all three types of noodles.

7. Remember that the F-test in ANOVA is a robust or overall test of the difference between means and does not specify where that difference lies. So, if the overall F ratio is not significant, comparing individual groups through a post hoc procedure does not make sense because there is no difference to investigate!

Figure D.10 One-Way ANOVA Output With Post Hoc Comparisons

Tests of Between-Subjects Effects

Dependent Variable: Pleasantness

Source	Type III Sum of Squares	df	Mean Square	F	Sig.
Corrected Model	29.233[a]	2	14.617	19.398	.000
Intercept	464.817	1	464.817	616.870	.000
Thickness	29.233	2	14.617	19.398	.000
Error	42.950	57	.754		
Total	537.000	60			
Corrected Total	72.183	59			

a. R Squared = .405 (Adjusted R Squared = .384)

Post Hoc Tests

Thickness

Multiple Comparisons

Dependent Variable: Pleasantness

Bonferroni

(I) Thickness	(J) Thickness	Mean Difference (I-J)	Std. Error	Sig.	95% Confidence Interval	
					Lower Bound	Upper Bound
Thick	Medium	-.15	.275	1.000	-.83	.53
	Thin	1.40*	.275	.000	.72	2.08
Medium	Thick	.15	.275	1.000	-.53	.83
	Thin	1.55*	.275	.000	.87	2.23
Thin	Thick	-1.40*	.275	.000	-2.08	-.72
	Medium	-1.55*	.275	.000	-2.23	-.87

1. Easy. Factorial ANOVA is used only when you have more than one factor or independent variable! And actually, this is not so easy an answer to get (but if you get it, you really understand the material) when you hypothesize an interaction.

2. Here's one of many different possible examples. There are three levels of one treatment (or factor) and two levels of severity of illness.

		Treatment		
		Drug #1	Drug #2	Placebo
Severity of Illness	Severe			
	Mild			

3. And the source table looks like this . . .

Tests of Between-Subjects Effects

Dependent Variable: PAIN_SCO

Source	Type III Sum of Squares	df	Mean Square	F	Sig.
Corrected Model	266.742	5	53.348	26.231	.000
Intercept	3070.408	1	3070.408	1509.711	.000
SEVERITY	.075	1	.075	.037	.848
TREATMEN	263.517	2	131.758	64.785	.000
SEVERITY * TREATMEN	3.150	2	1.575	.774	.463
Error	231.850	114	2.034		
Total	3569.000	120			
Corrected Total	498.592	119			

As far as our interpretation, in this data set, there is no main effect for severity, there is a main effect for treatment, and there is no interaction between the two main factors.

4.

    a. No. The F value for the ANOVA test is insignificant ($F = 0.004$, $p = .996$).

    b. No. The F value for the ANOVA test is insignificant ($F = 0.899$, $p = .352$).

    c. There are no significant interactions.

5. Indeed, once you use SPSS to do the analysis, you can see that there is no main effect for gender, but there is one for training and also a significant interaction. But then, when you plot the means and actually create a visual image of the interaction (as shown in Figure D.11), you can see that the skill levels of males are higher under strength-training conditions and those levels for females are higher under speed-training conditions. When evaluating interactions, it is very important to look carefully at these visual plots.

Figure D.11 Interaction Effect Between Gender and Training Conditions

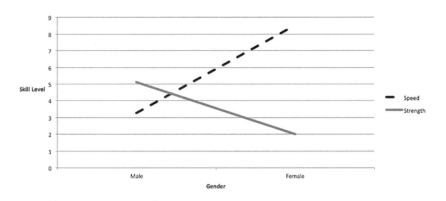

CHAPTER 15

1.

    a. With 18 degrees of freedom ($df = n - 2$) at the .01 level, the critical value for rejection of the null hypothesis is 0.516. There is a significant correlation between speed and strength, and the correlation accounts for 32.15% of the variance.

    b. With 78 degrees of freedom at the .05 level, the critical value for rejection of the null hypothesis is 0.183 for a one-tailed test. There is a significant correlation between number correct and time. A one-tailed test was used because the research hypothesis was that the relationship was indirect or negative, and approximately 20% of the variance is accounted for.

c. With 48 degrees of freedom at the .05 level, the critical value for rejection of the null hypothesis is 0.273 for a two-tailed test. There is a significant correlation between number of friends a child might have and GPA, and the correlation accounts for 13.69% of the variance.

2.

Correlations

		Motivation	GPA
Motivation	Pearson Correlation	1	.434*
	Sig. (2-tailed)		.017
	N	30	30
GPA	Pearson Correlation	.434*	1
	Sig. (2-tailed)	.017	
	N	30	30

*Correlation is significant at the.05 level (two-tailed).

a. and b. We used SPSS to compute the correlation as .434, which is significant at the .017 level using a two-tailed test.

c. True. The more motivated you are, the more you will study; and the more you study, the more you are motivated. But (and this is a big "but") studying more does not cause you to be more highly motivated, nor does being more highly motivated cause you to study more.

3.

		Income	Level_Education
Income	Pearson Correlation	1	.629(*)
	Sig. (2-tailed)		.003
	N	20	20
Level_Education	Pearson Correlation	.629(*)	1
	Sig. (2-tailed)	.003	
	N	20	20

*Correlation is significant at the .05 level (2-tailed).

    a. The correlation between income and education is .629, as you can see by examining the SPSS output.

    b. The correlation is significant at the .003 level.

    c. The only argument you can make is that these two variables share something in common (and the more they share, the higher the correlation) and that neither one can cause the other.

4.

    a. $r_{(8)} = .69$

    b. $t_{crit}(8) = 0.6319$. The observed correlation of .69 exceeds the critical value using 8 degrees of freedom. Therefore, our observed correlation coefficient is statistically significant at the .05 level.

    c. The shared variance equals 47% ($r^2_{hours.grade} = .47$).

    d. There is a strong positive relationship between the number of hours spent studying and the grade earned on a test, and this relationship is statistically significant. The more hours a student studies, the higher his or her test grade; or the fewer hours spent studying, the lower the test grade!

5. The correlation is significant at the .01 level, and what's wrong with the statement is that an association between two variables does not imply that one causes the other. These two are correlated, which may be for many other reasons, but regardless of how much coffee one drinks, we cannot say it results in a change in stress levels as a function of cause and effect.

6.

    a. The correlation is .671.

    b. With 8 degrees of freedom, and at the .05 level, the critical value for rejection of the null hypothesis that the correlation equals zero is 0.5494 (see Table B.4). The obtained value of .671 is greater than the critical value (or what you would expect by chance), and our conclusion is that the correlation is significant and the two variables are related.

    c. If you recall, the best way to interpret any Pearson product-moment correlation is by squaring it, which gives us a coefficient of determination of .58, meaning that 58% of the variance in age is accounted for by the variance in number of words known. That's not huge, but as correlations between variables in human behavior go, it's pretty substantial.

7. The example here is the number of hours you study and your performance on your first test in statistics. These variables are not causally related. For example, you will have classmates who studied

for hours and did poorly because they never understood the material and classmates who did very well without any studying at all because they had some of the same material in another class. Just imagine if we forced someone to stay at his or her desk and study for 10 hours each of four nights before the exam. Would that ensure that he or she got a good grade? Of course not. Just because the variables are related does not mean that one causes the other.

8. It's really up to you and the way that you interpret these correlations how you decide which significant correlations are meaningful. For example, shoe size and age are of course significantly correlated ($r = .938$), because as one gets older, the size of his or her feet increases. But meaningful? We think not. On the other hand, for example, intelligence is unrelated to all three other variables. And you may expect such, given that the test is standardized for age so that at all ages, 100 is average. In other words, for this test, you don't get smarter as you get older (despite what parents tell children).

CHAPTER 16

1. The major difference is that linear regression is used to explore whether one variable predicts another. Analysis of variance examines differences between group means but does not have predictive capabilities.

2.

a. The regression equation is $Y' = -0.214$ (number correct) + 17.202.

b. $Y' = -0.214(8) + 17.202 = 15.49$.

c.

Time (Y)	# Correct (X)	Y'	Y – Y'
14.5	5	16.13	−1.6
13.4	7	15.70	−2.3
12.7	6	15.92	−3.2
16.4	2	16.77	−0.4
21.0	4	16.35	4.7
13.9	3	16.56	−2.7
17.3	12	14.63	2.7
12.5	5	16.13	−3.6
16.7	4	16.35	0.4
22.7	3	16.56	6.1

3.

    a.   The other predictor variables should not be related to any other predictor variable. Only when they are independent of one another can they each contribute unique information to predicting the outcome or dependent variable.

    b.   Examples are living arrangements (single or in a group) and access to health care (high, medium, or low).

    c.   Presence of Alzheimer's = (level of education)X_{IV1} + (general physical health)X_{IV2} + (living arrangements)X_{IV3} + (access to health care)X_{IV4} + a.

4. This one you do on your own, being sure to focus on your major area or some question that you are really interested in.

5.

    a.   You could compute the correlation between the two variables, which is .204. According to the information in Chapter 5, the magnitude of such a correlation is quite low. You could reach the conclusion that the number of wins is not a very good predictor of whether a team ever won a Super Bowl.

    b.   Many variables are categorical by nature (gender, race, social class, and political party) and cannot be easily measured on a scale from 1 to 100, for example. Using categorical variables allows us more flexibility.

    c.   Some other variables might be number of All-American players, win–loss record of coaches, and home game attendance.

6.

    a.   Caffeine consumption

    b.   Stress group

    c.   The correlation coefficient calculated from Chapter 15 is .373, which you would square to get .139, or R^2. Pretty nifty!

7.

    a.   The best predictor of the three is years of experience, but because none is significant (.102), one could say that the three are equally good (or bad)!

    b.   And here's the regression equation:

$$Y' = 0.959(X_1) - 5.786(X_2) - 1.839(X_3) + 96.377$$

If we substitute the values for X_1, X_2, and X_3, we get the following:

$$Y' = 0.959(12) - 5.786(2) - 1.839(5) + 96.377$$

And the chef's predicted score is 64.062.

	Unstandardized Coefficients	Standardized Coefficients	Sig.		
	B	Std. Error	Beta	t	
(Constant)	97.237	7.293		13.334	.000
Years_Ex	1.662	0.566	1.163	2.937	.102
Level_Ed	−7.017	2.521	−0.722	−2.783	.162
Num_Pos	−2.555	1.263	−0.679	−2.022	.212

Note also that years of experience and level of education were significant predictors.

8. The reason why level of education is such a poor predictor is that it has low variability, which means it has a very small correlation with number of homes sold (−.074, contributing nothing) and, for that matter, little correlation with any other variable. Therefore, it has no predictive value. The best predictor? Years in business, which is significant way beyond the .001 level. When you do the multiple regression analysis, you find that, as a predictor, level of education (as you would expect given the simple correlation) contributes nothing as well to our understanding of why these real estate agents (and others, if we want to infer such) differ in the number of homes sold.

9. To maximize the value of the predictors, they should all be related to the predicted or outcome variable but not have anything in common with the other predictors (if at all possible).

CHAPTER 17

1. This occurs when the expected and the observed values are identical. One example would be when you expected an unequal number of first-year and second-year students to show up and they really did!

2. Here's the worksheet for computing the chi-square value:

Category	O (observed frequency)	E (expected frequency)	D (difference)	$(O - E)^2$	$(O - E)^2/E$
Republican	800	800	0	0	0.00
Democrat	700	800	100	10,000	12.50
Independent	900	800	100	10,000	12.50

With 2 degrees of freedom at the .05 level, the critical value needed for rejection of the null hypothesis is 5.99. The obtained value of 25.00 allows us to reject the null and conclude that there is a significant difference in the numbers of people who voted as a function of political party.

3. Here's the worksheet for computing the chi-square value:

Category	O (observed frequency)	E (expected frequency)	D (difference)	$(O - E)^2$	$(O - E)^2/E$
Boys	45	50	5	25	0.50
Girls	55	50	5	25	0.50

With 1 degree of freedom at the .01 level of significance, the critical value needed for rejection of the null hypothesis is 6.64. The obtained value of 1.00 means that the null cannot be rejected, and there is no difference between the number of boys and girls who play soccer.

4. Some fun facts first. The total number of students enrolled in all six grades is 2,217, and the expected frequency value of each cell is 2,217/6 or 369.50.

    The obtained chi-square value is 36.98. With 5 degrees of freedom at the .05 level, the value needed for rejection of the null hypothesis is 11.07. Because the obtained value of 36.98 exceeds the critical value, the conclusion is that the enrollment numbers are not those expected, and indeed, significantly different proportions are enrolled in each grade.

5. Okay, gang, here's the deal. The chi-square value is 15.8, significant at the .003 level, meaning that there certainly is a preference among candy bar tasters.

6. The chi-square value for this test of independence is 8.1, and with 2 degrees of freedom, it is more than large enough (actually $p = .0174$) to conclude that the null hypothesis of independence cannot be accepted and indeed preference for plain or peanut is dependent upon exercise level.

7. As you can see by the following SPSS output, the chi-square value of 3.991 with 4 degrees of freedom (2 × 2) is significant at the .41 level—not significant and not extreme enough for us to conclude that these two variables are dependent upon one another or related.

CHAPTER 19

1. You can see in Figure D.12 that the most industrious (category 10) and the least proficient (proficiency level 1) numbers 6.

| Figure D.12 | A Crosstabs Table Showing Industriousness by Proficiency Level for 500 Participants |

Case Processing Summary

	Cases					
	Valid		Missing		Total	
	N	Percent	N	Percent	N	Percent
Industriousness * Proficiency	500	100.0%	0	0.0%	500	100.0%

Industriousness * Proficiency Crosstabulation

Count

		Proficiency					Total
		1	2	3	4	5	
Industriousness	1	6	12	11	9	6	44
	2	9	8	10	6	8	41
	3	6	10	12	10	18	56
	4	9	6	10	9	12	46
	5	16	9	13	11	11	60
	6	10	13	11	9	9	52
	7	10	11	12	11	9	53
	8	13	12	8	10	9	52
	9	6	5	7	6	10	34
	10	15	13	12	10	12	62
Total		100	99	106	91	104	500

2. Figure D.13 shows you a table and a graph that you could easily create.

| Figure D.13 | A Bar Graph of Level by Gender for 1,000 Participants |

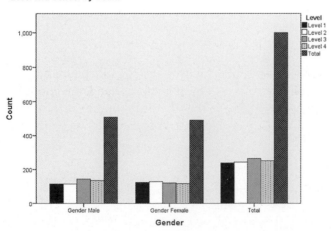

Level and Gender by Count

Appendix E

Math: Just the Basics

If you're reading this, then you know you may need a bit of help with your basic math skills. Lots of people need such help, especially after coming back to school after a break. There's nothing wrong with taking this little side trip before you continue working in *Statistics for People Who (Think They) Hate Statistics*.

Most of the skills you need to work the examples in this book and to complete the practice exercises at the end of the chapters you already know. For example, you can add, subtract, multiply, and divide. You also probably know how to use a calculator to compute the square root of a number.

The confusion begins when we start dealing with equations and the various operations that can take place within parentheses, like these → ( ), and brackets, like these → [ ].

That's where we will spend most of our time, and we'll show you examples so you better understand how to work through what appear to be complex operations. Rest assured, once they are reduced to their individual parts, completing them is a cinch.

THE BIG RULES: SAY HELLO TO BODMAS

Sounds like an alien life form or something out of the Borg, right?

Nope. It is simply an acronym that indicates the order in which operations take place in an expression or an equation.

BODMAS goes like this . . .

B is for brackets, such as [ ], or sometimes parens (short for parentheses), like this ( ), both of which are sometimes present in an expression or equation.

O is for order or power, as in raise 4 to the order of 2 or 4^2.

D is division, as in 6/3.

M is multiplication, as in 6×3.

A is addition, as in 2 + 3.

S is subtraction, as in 5 − 1.

You perform these operations in the above order. For example, the first thing you do is work within brackets (or parentheses), then take numbers to a power (such as squaring), then divide, then multiply, and so on. If there's nothing to square, skip the *O* step, and if there's nothing to add, skip the *A* step. If both brackets and parens are present, do the brackets first, and then the parentheses.

For example, take a look at this simple expression . . .

$$(3 + 2) \times 2 = ?$$

Using the BODMAS acronym, we know that

1. the first thing we do is any operation within brackets. So, 3 + 2 = 5.

2. Moving on, the next available step is multiplying 5 times 2 for a total of 10, and 10 is the answer.

Here's another . . .

$$(4/2 \times 5) + 7 = ?$$

And here's what we do . . .

1. First is the division of 4 by 2, which equals 2, and that 2 is multiplied by 5 for a value of 10.

2. Then, 10 is added to 7 for a final sum of 17.

Let's get a bit fancier and use some squaring of numbers.

$$(10^2 \times 3)/150 = ?$$

Here's what we do . . .

1. Look within the brackets (as always for the first step).

2. Square 10 to equal 100, and then multiply it by 3 for a value of 300.

3. Divide 300 by 150 for a value of 2.

Let's add another layer of complexity—sometimes we have more than one set of brackets or parens. All you need to remember is that we always begin with the innermost brackets first and work our way out as we work through the acronym.

Here we go . . .

$$[(15 \times 2) - (5 + 7)]/6 = ?$$

1. 15 times 2 is 30.

2. 5 plus 7 is 12.

3. 30 − 12 is 18.

4. 18 divided by 6 is 3.

THE LITTLE RULES

Just a few more. There are negative and positive numbers used throughout the book, and you need to know how they act when combined in different ways.

When multiplying negative and positive numbers, the following is true . . .

1. A negative number multiplied by a positive number (or a positive number multiplied by a negative number) always, always, always equals a negative number. For example . . .

$$-3 \times 2 = -6$$

or

$$4 \times -5 = -20$$

2. A negative number multiplied by another negative number results in a positive number. For example . . .

$$-4 \times -3 = 12$$

or

$$-2.5 \times -3 = 7.5$$

And a negative number divided by a positive number results in a negative number. For example . . .

$$-10/2 = -5$$

or

$$25/-5 = 5$$

And finally, a negative number divided by a negative number results in a positive number. For example . . .

$$-10/-5 = 2$$

Practice makes perfect, so here are 10 problems with the answers following. If you can't get the right answer, review the order of operations above and also ask someone in your study group to review where you might have gone wrong.

Here we go . . .

1. $75 + 10$

2. $104 - 50$

3. $50 - 104$

4. 42×-2

5. -50×-60

6. $25/5 - 6/3$

7. $6,392 - (-700)$

8. $(510 - 500)/-10$

9. $[(40^2 - 207) - (80^2 - 400)]/35 \times 24$

10. $([(502 - 300) - 25] - [(242 - 100) - 50])/20 \times 30$

Answers

1. 85

2. 54

3. −54

4. −84

5. 3,000

6. 3

7. 7,092

8. −1

9. −5.48

10. 0.141

Want some more help and more practice? Take a look at these sites . . .

http://www.webmath.com/index.html

http://www.math.com/homeworkhelp/BasicMath.html

There's nothing worse than starting a course and being so anxious that any meaningful learning just can't take place. Thousands of students less well prepared than you have succeeded, and you can as well. Reread the tips in Chapter 1 on how to approach the material in this course—and good luck!

Appendix F
The Reward
The Brownie Recipe

What the heck is a brownie recipe doing in an introductory statistics book? Good question.

In all seriousness, you have probably worked hard on this material, whether for a course, as a review, or just for your own edification. And, because of all your effort, you deserve a reward. Here it is. The recipe is based on several different recipes and some tweaking, and it's all your author's and he is happy to share it with you. There, the secret is out.

Right out of the pan, not even cooled, these brownies are terrific with ice cream. Once they've aged a bit, they get very nice and chewy and are great from the fridge. If you freeze them, note that it takes more calories to defrost them in your mouth than are contained in the brownies themselves, so there is a net loss. Eat as many frozen as you want. ☺

1 stick (8 tablespoons) butter

4 ounces unsweetened chocolate (or more) ½ tablespoon salt

2 eggs

1 cup flour

2 cups sugar

1 tablespoon vanilla

2 tablespoons mayonnaise (I know)

6 ounces chocolate chips (or more)

1 cup whole walnuts (optional)

How to do it . . .

1. Preheat oven to 325 °F.

2. Melt unsweetened chocolate and butter in a saucepan.

501

3. Mix flour and salt together in a bowl.

4. Add sugar, vanilla, nuts, mayonnaise, and eggs to melted chocolate-butter stuff and mix well.

5. Add all of #4 to flour mixture and mix well.

6. Add chocolate chips.

7. Pour into an 8″ × 8″ greased baking dish.

8. Bake for about 35 to 40 minutes or until tester comes out clean.

NOTES . . .

- I know about the mayonnaise thing. If you think it sounds weird, then don't put it in. These brownies are not delicious for nothing, though, so leave out this ingredient at your own risk.
- Use good chocolate—the higher the fat content, the better. And you can use up to 6 ounces of unsweetened chocolate and even more of the chocolate chips.

Glossary

Analysis of variance

A test for the difference between two or more means. A simple analysis of variance (or ANOVA) has only one independent variable, whereas a factorial analysis of variance tests the means of more than one independent variable. One-way analysis of variance looks for differences between the means of more than two groups.

Arithmetic mean

A measure of central tendency calculated by summing all the scores and dividing by the number of scores. *See* Mean.

Asymptotic

The quality of the normal curve such that the tails never touch the horizontal axis.

Average

The most representative score in a set of scores.

Bell-shaped curve

A distribution of scores that is symmetrical about the mean, median, and mode and has asymptotic tails. Often called the *normal curve*.

Class interval

A fixed range of values, used in the creation of a frequency distribution.

Coefficient of alienation

The amount of variance in one variable that is not accounted for by the variance in another variable.

Coefficient of determination

The amount of variance in one variable that is accounted for by the variance in another variable.

Coefficient of nondetermination

See Coefficient of alienation.

Concurrent criterion validity

How well a test outcome is consistent with a criterion that exists in the present.

Confidence interval

The best estimate of the range of a population value given the sample value.

Construct validity

How well a test reflects an underlying idea, such as intelligence or aggression.

Content validity

A type of validity that examines how well a test samples a universe of items.

Correlation coefficient

A numerical index that reflects the relationship between two variables, specifically how the value of one variable changes when the value of the other variable changes.

Correlation matrix

A table showing correlation coefficients among more than two variables.

Criterion

See Dependent variable.

Criterion validity

A type of validity that examines how well a test reflects some criterion that exists in either the present (*concurrent*) or the future (*predictive*).

Critical value

The value resulting from application of a statistical test that is necessary for rejection (or nonacceptance) of the null hypothesis.

Cross-tabulation table

A table that shows frequencies by two or more variables. The levels of one variable become column labels, and the levels of the other variable become row labels. Often called a *crosstab*.

Cumulative frequency distribution

A frequency distribution that shows frequencies for class intervals along with the cumulative frequency at each.

Data

A record of an observation or an event such as a test score, a grade in math class, or response time.

Data mining

Examining large data sets for patterns.

Data point

An observation.

Data set

A set of data points.

Degrees of freedom

A value, which is different for different statistical tests, that approximates the sample size of number of individual cells in an experimental design.

Dependent variable

The outcome variable or the predicted variable in a regression equation.

Descriptive statistics

Values that organize and describe the characteristics of a collection of data, sometimes called a *data set*.

Direct correlation

A positive correlation where the values of both variables change in the same direction.

Directional research hypothesis

A research hypothesis that posits a difference between groups in one direction. *See* Nondirectional research hypothesis.

Effect size

A measure of the magnitude of difference between two groups, usually calculated as Cohen's d.

Error in prediction

The difference between the observed score (Y) and the predicted score (Y').

Error score

The part of a test score that is random and contributes to the unreliability of a test.

Exabyte

1,152,921,504,606,846,976 bytes of data—lots and lots of data, and the amount of data in the world grew just as you read this. Wow.

Factorial analysis of variance

An analysis of variance with more than one factor or independent variable.

Factorial design

A research design used to explore more than one treatment variable.

Frequency distribution

A method for illustrating how often scores occur in groups called class intervals.

Frequency polygon

A graphical representation of a frequency distribution that uses a continuous line to show the number of values that fall within a *class interval*.

Goodness of fit test

A chi-square test on one dimension, which examines whether the distribution of frequencies is different than what one would expect by chance.

Histogram

A graphical representation of a frequency distribution that uses bars of different heights to show the number of values that fall within each *class interval*.

Hypothesis

An if–then statement of conjecture that relates variables to one another and is used to reflect the general problem statement or question that is the motivation for asking a research question.

Independent variable

The treatment variable that is manipulated or the predictor variable in a regression equation.

Indirect correlation

A negative correlation where the values of variables move in opposite directions.

Inferential statistics

Tools that are used to infer characteristics of a population based on data from a sample of that population.

Interaction effect

The outcome where the effect of one factor is differentiated across another factor.

Internal consistency reliability

A type of reliability that examines whether items on a test measure only one dimension, construct, or area of interest.

Interrater reliability

A type of reliability that examines whether observers are consistent with one another.

Interval level of measurement

A level of measurement that places a variable's values into categories that are equidistant from each other, as when points are evenly spaced along a scale.

Kurtosis

The quality of a distribution that defines how flat or peaked it is.

Leptokurtic

The quality of a normal curve that is relatively peaked compared with a normal distribution.

Line of best fit

The regression line that best fits the observed scores and minimizes the error in prediction.

Linear correlation

A correlation that is best expressed visually as a straight line.

Main effect

In analysis of variance, when a factor or an independent variable has a significant effect upon the outcome variable.

Mean

A type of average calculated by summing values and dividing that sum by the number of values.

Mean deviation

The average deviation for all scores from the mean of a distribution, calculated as the sum of the absolute value of the scores' deviations from the mean divided by the number of scores.

Measures of central tendency

The mean, the median, and the mode.

Median

The midpoint in a set of values, such that 50% of the cases in a distribution fall below the median and 50% fall above it.

Midpoint

The central point in a class interval.

Mode

The most frequently occurring score in a distribution.

Multiple regression

A statistical technique whereby several variables are used to predict one.

Negative correlation

See Indirect correlation.

Nominal level of measurement

The most gross level of measurement by which a variable's value can be placed in one and only one category.

Nondirectional research hypothesis

A research hypothesis that posits a difference between groups but not in either direction. *See* Directional research hypothesis.

Nonparametric statistics

Distribution-free statistics that do not require the same assumptions as do parametric statistics.

Normal curve

See Bell-shaped curve.

Null hypothesis

A statement of equality between sets of variables. *See* Research hypothesis.

Observed score

The score that is recorded or observed. *See* True score.

Obtained value

The value that results from the application of a statistical test.

Ogive

A visual representation of a cumulative frequency distribution.

One-sample Z-test

Used to compare a sample mean to a population mean.

One-tailed test

A directional test, reflecting a directional hypothesis.

One-way analysis of variance

See Analysis of variance.

Ordinal level of measurement

A level of measurement that places a variable's value into a category and assigns that category an order with respect to other categories.

Outliers

Those scores in a distribution that are noticeably much more extreme than the majority of scores. Whether a score is an outlier or not is usually an arbitrary decision made by the researcher.

Parallel forms reliability

A type of reliability that examines consistency across different forms of the same test.

Parametric statistics

Statistics used for the inference from a sample to a population that assume the variances of each group are similar and that the sample is large enough to represent the population. *See* Nonparametric statistics.

Partial correlation

A numerical index that reflects the relationship between two variables with the removal of the influence of a third variable (called a mediating or confounding variable).

Pearson product-moment correlation

See Correlation coefficient.

Percentile point

The percentage of cases equal to and below a particular score in a distribution or set of scores.

Pivot table

A tool in statistical software, such as SPSS or Excel, that allows the user to easily manipulate the rows, columns, and frequencies included in cross-tabulation tables.

Platykurtic

The quality of a normal curve that is relatively flat compared with a normal distribution.

Population

All the possible subjects or cases of interest. *See* Sample.

Positive correlation

See Direct correlation.

Post hoc

After the fact, referring to tests done to determine the true source of a difference among three or more groups.

Predictive validity

How consistent a test outcome is with a criterion that occurs in the future.

Predictor

See Independent variable.

Range

The positive difference between the highest and lowest score in a distribution. It is a gross measure of variability. Exclusive range is the highest score minus the lowest score. Inclusive range is the highest score minus the lowest score plus 1.

Ratio level of measurement

A level of measurement defined as having an absolute zero.

Regression equation

The equation that defines the points and the line that are closest to the observed scores.

Regression line

The line drawn based on values in a regression equation. Also known as a *trend line*.

Reliability

The consistency of a test.

Research hypothesis

A statement of inequality between two variables. *See* Null hypothesis.

Sample

A subset of a population. *See* Population.

Sampling error

The difference between sample and population values.

Scales of measurement

Different ways of categorizing measurement outcomes: nominal, ordinal, interval, and ratio.

Scattergram or scatterplot

A plot of paired data points on an x- and a y-axis, used to visually represent a correlation.

Significance level

The risk set by the researcher for rejecting a null hypothesis when it is true.

Simple analysis of variance

See Analysis of variance.

Skew or skewness

The quality of a distribution that defines the disproportionate frequency of certain scores. A longer right tail than left corresponds to a smaller number of occurrences at the high end of the distribution; this is a *positively skewed distribution*. A shorter right tail than left corresponds to a larger number of occurrences at the high end of the distribution; this is a *negatively skewed distribution*.

Source table

An analysis of variance summary table that lists sources of variance.

Standard deviation

The average amount of variability in a set of scores or the scores' average deviation from the mean.

Standard error of estimate

A measure of accuracy in prediction that reflects variability about the regression line. *See* Error in prediction.

Standard score

See z score.

Statistical significance

See Significance level.

Statistics

A set of tools and techniques used to describe, organize, and interpret information or data.

Test of independence

A chi-square test of two dimensions or more that examines whether the distribution of frequencies on a variable is independent of other variables.

Test–retest reliability

A type of reliability that examines a test's consistency over time.

Test statistic value

See Obtained value.

Trend line

See Regression line.

True score

The score that, if it could be observed, would reflect the actual ability or behavior being measured. *See* Observed score.

Two-tailed test

A nondirectional test, reflecting a nondirectional hypothesis.

Type I error

The probability of rejecting a null hypothesis when it is true.

Type II error

The probability of accepting a null hypothesis when it is false.

Unbiased estimate

A conservative estimate of a population parameter.

Validity

How well a test measures what it says it does.

Variability

How much scores differ from one another or, put another way, the amount of spread or dispersion in a set of scores.

Variance

The square of the standard deviation and another measure of a distribution's spread or dispersion.

Y' or Y prime

The predicted Y value in a regression equation.

z score

A raw score that is adjusted for the mean and standard deviation of the distribution from which the raw score comes.

Index